QUANTUM MECHANICS FOR ELECTRICAL ENGINEERS

IEEE Press
445 Hoes Lane
Piscataway, NJ 08854

QUANTUM MECHANICS FOR ELECTRICAL ENGINEERS

DENNIS M. SULLIVAN

IEEE PRESS

A JOHN WILEY & SONS, INC., PUBLICATION

Published by John Wiley & Sons, Inc., Hoboken, New Jersey. All rights reserved.
Published simultaneously in Canada

MATLAB and Simulink are registered trademarks of The MathWorks, Inc. See www. mathworks.com/trademarks for a list of additional trade marks. **The MathWorks Publisher Logo identifies books that contain MATLAB® content. Used with permission. The MathWorks does not warrant the accuracy of the text or exercises in this book or in the software downloadable from** http://www.wiley.com/WileyCDA/WileyTitle/productCd-047064477X.html **and** http://www.mathworks.com/matlabcentral/fileexchage/?term=authored%3A80973. **The book's or downloadable software's use or discussion of MATLAB® software or related products does not constitute endorsement or sponsorship by The MathWorks of a particular use of the MATLAB® software or related products.**

For MATLAB® and Simulink® product information, in information on other related products, please contact:

The MathWorks, Inc.
3 Apple Hill Drive
Natick, MA 01760-2098 USA
Tel: 508-647-7000
Fax: 508-647-7001
E-mail: info@mathworks.com
Web: www.mathworks.com

For general information on our other products and services or for technical support, please contact our Customer Care Department within the United States at (800) 762-2974, outside the United States at (317) 572-3993 or fax (317) 572-4002.

Wiley also publishes its books in a variety of electronic formats. Some content that appears in print may not be available in electronic formats. For more information about Wiley products, visit our web site at www.wiley.com.

Library of Congress Cataloging-in-Publication Data

ISBN: 978-0-470-87409-7

Printed in the United States of America

10 9 8 7 6 5 4 3 2 1

To
My Girl

CONTENTS

MATLAB Codes are downloadable from http://booksupport.wiley.com

PREFACE

A physics professor once told me that electrical engineers were avoiding learning quantum mechanics as long as possible. The day of reckoning has arrived. Any electrical engineer hoping to work in the field of modern semiconductors will have to understand some quantum mechanics.

Quantum mechanics is not normally part of the electrical engineering curriculum. An electrical engineering student taking quantum mechanics in the physics department may find it to be a discouraging experience. A quantum mechanics class often has subjects such as statistical mechanics, thermodynamics, or advanced mechanics as prerequisites. Furthermore, there is a greater cultural difference between engineers and physicists than one might imagine.

This book grew out of a one-semester class at the University of Idaho titled "Semiconductor Theory," which is actually a crash course in quantum mechanics for electrical engineers. In it there are brief discussions on statistical mechanics and the topics that are needed for quantum mechanics. Mostly, it centers on quantum mechanics as it applies to transport in semiconductors. It differs from most books in quantum mechanics in two other very important aspects: (1) It makes use of Fourier theory to explain several concepts, because Fourier theory is a central part of electrical engineering. (2) It uses a simulation method called the finite-difference time-domain (FDTD) method to simulate the Schrödinger equation and thereby provides a method of illustrating the behavior of an electron. The simulation method is also used in the exercises. At the same time, many topics that are normally covered in an introductory quantum mechanics text, such as angular momentum, are not covered in this book.

THE LAYOUT OF THE BOOK

Intended primarily for electrical engineers, this book focuses on a study of quantum mechanics that will enable a better understanding of semiconductors. Chapters 1 through 7 are primarily fundamental topics in quantum mechanics. Chapters 8 and 9 deal with the Green's function formulation for transport in semiconductors and are based on the pioneering work of Supriyo Datta and his colleagues at Purdue University. The Green's function is a method for calculating transport through a channel. Chapter 10 deals with approximation methods in quantum mechanics. Chapter 11 talks about the harmonic oscillator, which is used to introduce the idea of creation and annihilation operators that are not otherwise used in this book. Chapter 12 describes a simulation method to determine the eigenenergies and eigenstates in complex structures that do not lend themselves to easy analysis.

THE SIMULATION PROGRAMS

Many of the figures in this book have a title across the top. This title is the name of the MATLAB program that was used to generate that figure. These programs are available to the reader. Appendix D lists all the programs, but they can also be obtained from the following Internet site:

http://booksupport.wiley.com.

The reader will find it beneficial to use these programs to duplicate the figures and perhaps explore further. In some cases the programs must be used to complete the exercises at the end of the chapters. Many of the programs are time-domain simulations using the FDTD method, and they illustrate the behavior of an electron in time. Most readers find these programs to be extremely beneficial in acquiring some intuition for quantum mechanics. A request for the solutions manual needs to be emailed to pressbooks@ieee.org.

DENNIS M. SULLIVAN

Department of Electrical and Computer Engineering
University of Idaho

ACKNOWLEDGMENTS

I am deeply indebted to Prof. Supriyo Datta of Purdue University for his help, not only in preparing this book, but in developing the class that led to the book. I am very grateful to the following people for their expertise in editing this book: Prof. Richard Ziolkowski from the University of Arizona; Prof. Fred Barlow, Prof. F. Marty Ytreberg, and Paul Wilson from the University of Idaho; Prof. David Citrin from the Georgia Institute of Technology; Prof. Steven Hughes from Queens University; Prof. Enrique Navarro from the University of Valencia; and Dr. Alexey Maslov from Canon U.S.A. I am grateful for the support of my department chairman, Prof. Brian Johnson, while writing this book. Mr. Ray Anderson provided invaluable technical support. I am also very grateful to Ms. Judy LaLonde for her editorial assistance.

D.M.S.

ABOUT THE AUTHOR

Dennis M. Sullivan graduated from Marmion Military Academy in Aurora, Illinois in 1966. He spent the next 3 years in the army, including a year as an artillery forward observer with the 173rd Airborne Brigade in Vietnam. He graduated from the University of Illinois with a bachelor of science degree in electrical engineering in 1973, and received master's degrees in electrical engineering and computer science from the University of Utah in 1978 and 1980, respectively. He received his Ph.D. degree in electrical engineering from the University of Utah in 1987.

From 1987 to 1993, he was a research engineer with the Department of Radiation Oncology at Stanford University, where he developed a treatment planning system for hyperthermia cancer therapy. Since 1993, he has been on the faculty of electrical and computer engineering at the University of Idaho. His main interests are electromagnetic and quantum simulation. In 1997, his paper "Z Transform Theory and the FDTD Method," won the R. P. W. King Award from the IEEE Antennas and Propagation Society. In 2001, he received a master's degree in physics from Washington State University while on sabbatical leave. He is the author of the book *Electromagnetic Simulation Using the FDTD Method*, also from IEEE Press.

1

INTRODUCTION

This chapter serves as a foundation for the rest of the book. Section 1.1 provides a brief history of the physical observations that led to the development of the Schrödinger equation, which is at the heart of quantum mechanics. Section 1.2 describes a time-domain simulation method that will be used throughout the book as a means of understanding the Schrödinger equation. A few examples are given. Section 1.3 explains the concept of observables, the operators that are used in quantum mechanics to extract physical quantities from the Schrödinger equation. Section 1.4 describes the potential that is the means by which the Schrödinger equation models materials or external influences. Many of the concepts of this chapter are illustrated in Section 1.5, where the simulation method is used to model an electron interacting with a barrier.

1.1 WHY QUANTUM MECHANICS?

In the late nineteenth century and into the first part of the twentieth century, physicists observed behavior that could not be explained by classical mechanics [1]. Two experiments in particular stand out.

1.1.1 Photoelectric Effect

When monochromatic light—that is, light at just one wavelength—is used to illuminate some materials under certain conditions, electrons are emitted from

Quantum Mechanics for Electrical Engineers, First Edition. Dennis M. Sullivan.
© 2012 The Institute of Electrical and Electronics Engineers, Inc.
Published 2012 by John Wiley & Sons, Inc.

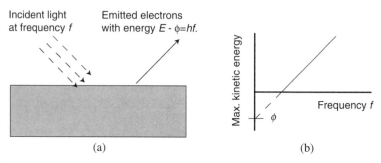

FIGURE 1.1   The photoelectric effect. (a) If certain materials are irradiated with light, electrons within the material can absorb energy and escape the material. (b) It was observed that the KE of the escaping electron depends on the frequency of the light.

the material. Classical physics dictates that the energy of the emitted particles is dependent on the intensity of the incident light. Instead, it was determined that at a constant intensity, the kinetic energy (KE) of emitted electrons varies linearly with the frequency of the incident light (Fig. 1.1) according to:

$$E - \phi = hf,$$

where, ϕ, the work function, is the minimum energy that the particle needs to leave the material.

Planck postulated that energy is contained in discrete packets called quanta, and this energy is related to frequency through what is now known as Planck's constant, where $h = 6.625 \times 10^{-34}$ J·s,

$$E = hf. \tag{1.1}$$

Einstein suggested that the energy of the light is contained in discrete wave packets called photons. This theory explains why the electrons absorbed specific levels of energy dictated by the frequency of the incoming light and became known as the photoelectric effect.

1.1.2 Wave–Particle Duality

Another famous experiment suggested that particles have wave properties. When a source of particles is accelerated toward a screen with a single opening, a detection board on the other side shows the particles centered on a position right behind the opening as expected (Fig. 1.2a). However, if the experiment is repeated with two openings, the pattern on the detection board suggests points of constructive and destructive interference, similar to an electromagnetic or acoustic wave (Fig. 1.2b).

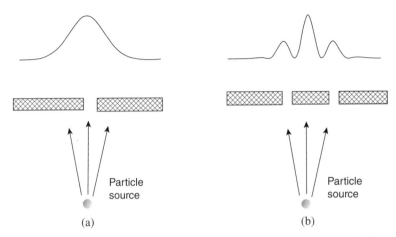

FIGURE 1.2   The wave nature of particles. (a) If a source of particles is directed at a screen with one opening, the distribution on the other side is centered at the opening, as expected. (b) If the screen contains two openings, points of constructive and destructive interference are observed, suggesting a wave.

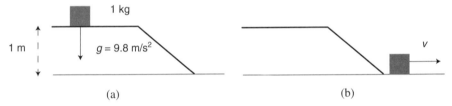

FIGURE 1.3   (a) A block with a mass of 1 kg has been raised 1 m. It has a PE of 9.8 J. (b) The block rolls down the frictionless incline. Its entire PE has been turned into KE.

Based on observations like these, Louis De Broglie postulated that matter has wave-like qualities. According to De Broglie, the momentum of a particle is given by:

$$p = \frac{h}{\lambda},$$

(1.2)

where λ is the wavelength. Observations like Equations (1.1) and (1.2) led to the development of quantum mechanics.

1.1.3   Energy Equations

Before actually delving into quantum mechanics, consider the formulation of a simple energy problem. Look at the situation illustrated in Figure 1.3 and think about the following problem: If the block is nudged onto the incline and rolls to the bottom, what is its velocity as it approaches the flat area, assuming that we can ignore friction? We can take a number of approaches to solve this problem. Since the incline is 45°, we could calculate the gravitational force exerted on the block while it is on the incline. However, physicists like to deal with energy. They would say that the block initially has a potential energy (PE) determined by the mass multiplied by the height multiplied by the acceleration of gravity:

$$PE = (1 \text{ kg})(1 \text{ m})\left(9.8\frac{\text{m}}{\text{s}^2}\right) = 9.8\frac{\text{kg} - \text{m}^2}{\text{s}^2} = 9.8 \text{ J}.$$

Once the block has reached the bottom of the incline, the PE has been all converted to KE:

$$KE = 9.8\left(\frac{\text{kg} \cdot \text{m}^2}{\text{s}^2}\right) = \frac{1}{2}(1 \text{ kg})v^2.$$

It is a simple matter to solve for the velocity:

$$v = \left(2 \times 9.8\frac{\text{m}^2}{\text{s}^2}\right)^{1/2} = 4.43\frac{\text{m}}{\text{s}}.$$

This is the fundamental approach taken in many physics problems. Very elaborate and elegant formulations, like Lagrangian and Hamiltonian mechanics, can solve complicated problems by formulating them in terms of energy. This is the approach taken in quantum mechanics.

Example 1.1

An electron, initially at rest, is accelerated through a 1 V potential. What is the resulting velocity of the electron? Assume that the electron then strikes a block of material, and all of its energy is converted to an emitted photon, that is, $\phi = 0$. What is the wavelength of the photon? (Fig. 1.4)

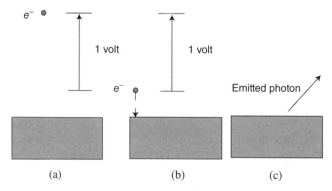

(a)   (b)   (c)

FIGURE 1.4   (a) An electron is initially at rest. (b) The electron is accelerated through a potential of 1 V. (c) The electron strikes a material, causing a photon to be emitted.

Solution. By definition, the electron has acquired energy of 1 electron volt (eV). To calculate the velocity, we first convert to joules. One electron volt is equal to 1.6×10^{-19} J. The velocity of the electron as it strikes the target is:

$$v = \sqrt{\frac{2 \cdot E}{m_e}} = \sqrt{\frac{2 \cdot 1.6 \times 10^{-19} \text{ J}}{9.11 \times 10^{-31} \text{ kg}}} = 0.593 \times 10^6 \text{m/s}.$$

The emitted photon also has 1 eV of energy. From Equation (1.1),

$$f = \frac{E}{h} = \frac{1 \text{ eV}}{4.135 \times 10^{-15} \text{ eV} \cdot \text{s}} = 2.418 \times 10^{14} \text{ s}^{-1}.$$

(Note that the Planck's constant is written in *electron volt-second* instead of *joule-second*.) The photon is an electromagnetic wave, so its wavelength is governed by:

$$c_0 = \lambda f,$$

where c_0 is the speed of light in a vacuum. Therefore:

$$\lambda = \frac{c_0}{f} = \frac{3 \times 10^8 \text{ m/s}}{2.418 \times 10^{14} \text{ s}^{-1}} = 1.24 \times 10^{-6} \text{ m}.$$

1.1.4 The Schrödinger Equation

Theoretical physicists struggled to include observations like the photoelectric effect and the wave–particle duality into their formulations. Erwin Schrödinger, an Austrian physicist, was using advanced mechanics to deal with these phenomena and developed the following equation [2]:

$$\frac{\partial^2}{\partial t^2} \psi = -\frac{1}{\hbar^2} \left(\frac{\hbar^2}{2m} \nabla^2 - V \right)^2 \psi, \tag{1.3}$$

where $\hbar$ is another version of Planck's constant, $\hbar = h/2\pi$, and m represents the mass. The parameter ψ in Equation (1.3) is called a state variable, because all meaningful parameters can be determined from it even though it has no direct physical meaning itself. Equation (1.3) is second order in time and fourth order in space. Schrödinger realized that so complicated an equation, requiring so many initial and boundary conditions, was completely intractable. Recall that computers did not exist in 1925. However, Schrödinger realized that if he considered ψ to be a complex function, $\psi = \psi_{\text{real}} + i\psi_{\text{imag}}$, he could solve the simpler equation:

$$i\hbar \frac{\partial}{\partial t} \psi = \left(-\frac{\hbar^2}{2m} \nabla^2 + V \right) \psi. \tag{1.4}$$

Putting $\psi = \psi_{real} + i\psi_{imag}$ into Equation (1.4) gives:

$$i\hbar\frac{\partial\psi_{real}}{\partial t} - \hbar\frac{\partial\psi_{imag}}{\partial t} = \left(-\frac{\hbar^2}{2m}\nabla^2 + V\right)\psi_{real} + i\left(-\frac{\hbar^2}{2m}\nabla^2 + V\right)\psi_{imag}.$$

Then setting real and imaginary parts equal to each other results in two coupled equations:

$$\frac{\partial\psi_{real}}{\partial t} = \frac{1}{\hbar}\left(-\frac{\hbar^2}{2m}\nabla^2 + V\right)\psi_{imag}, \tag{1.5a}$$

$$\frac{\partial\psi_{imag}}{\partial t} = \frac{-1}{\hbar}\left(-\frac{\hbar^2}{2m}\nabla^2 + V\right)\psi_{real}. \tag{1.5b}$$

If we take the time derivative of Equation (1.5a),

$$\hbar\frac{\partial^2\psi_{real}}{\partial t^2} = \left(-\frac{\hbar^2}{2m}\nabla^2 + V\right)\frac{\partial\psi_{imag}}{\partial t},$$

and use the time derivative of the imaginary part from Equation (1.5b), we get:

$$\frac{\partial^2\psi_{real}}{\partial t^2} = \frac{1}{\hbar}\left(-\frac{\hbar^2}{2m}\nabla^2 + V\right)\frac{-1}{\hbar}\left(-\frac{\hbar^2}{2m}\nabla^2 + V\right)\psi_{real}$$

$$= \frac{-1}{\hbar^2}\left(-\frac{\hbar^2}{2m}\nabla^2 + V\right)^2\psi_{real},$$

which is the same as Equation (1.3). We could have operated on the two equations in reverse order and gotten the same result for ψ_{imag}. Therefore, both the real and imaginary parts of ψ solve Equation (1.3). (An elegant and thorough explanation of the development of the Schrödinger equation is given in Borowitz [2].)

This probably seems a little strange, but consider the following problem. Suppose we are asked to solve the following equation where a is a real number:

$$x^2 + a^2 = 0.$$

Just to simplify, we will start with the specific example of $a = 2$:

$$x^2 + 2^2 = (x - i2)(x + i2) = 0.$$

We know one solution is $x = i2$ and another solution is $x^* = -i2$. Furthermore, for any a, we can solve the factored equation to get one solution, and the other will be its complex conjugate.

Equation (1.4) is the celebrated *time-dependent Schrödinger equation*. It is used to get a solution of the state variable ψ. However, we also need the complex conjugate ψ^* to determine any meaningful physical quantities. For instance,

$$|\psi(x,t)|^2 dx = \psi^*(x,t)\psi(x,t)dx$$

is the probability of finding the particle between x and $x + dx$ at time t. For this reason, one of the basic requirements in finding the solution to ψ is *normalization*:

$$\langle \psi(x)|\psi(x)\rangle = \int_{-\infty}^{\infty} |\psi(x)|^2 dx = 1. \tag{1.6}$$

In other words, the probability that the particle is somewhere is 1.

Equation (1.6) is an example of an *inner product*. More generally, if we have two functions, their inner product is defined as:

$$\langle \psi_1(x)|\psi_2(x)\rangle = \int_{-\infty}^{\infty} \psi_1^*(x)\psi_2(x)dx.$$

This is a very important quantity in quantum mechanics, as we will see.

The spatial operator on the right side of Equation (1.4) is called the *Hamiltonian*:

$$H = -\frac{\hbar^2}{2m_e}\nabla^2 + V(x).$$

Equation (1.4) can be written as:

$$i\hbar \frac{\partial}{\partial t}\psi = H\psi. \tag{1.7}$$

1.2 SIMULATION OF THE ONE-DIMENSIONAL, TIME-DEPENDENT SCHRÖDINGER EQUATION

We have seen that quantum mechanics is dictated by the time-dependent Schrödinger equation:

$$i\hbar \frac{\partial \psi(x,t)}{\partial t} = -\frac{\hbar^2}{2m_e}\frac{\partial^2 \psi(x,t)}{\partial x^2} + V(x)\psi(x,t). \tag{1.8}$$

The parameter $\psi(x,t)$ is a state variable. It has no direct physical meaning, but all relevant physical parameters can be determined from it. In general, $\psi(x,t)$

is a function of both space and time. $V(x)$ is the potential. It has the units of energy (usually electron volts for our applications.) $\hbar$ is Planck's constant. m_e is the mass of the particle being represented by the Schrödinger equation. In most instances in this book, we will be talking about the mass of an electron.

We will use computer simulation to illustrate the Schrödinger equation. In particular, we will use a very simple method called the finite-difference time-domain (FDTD) method. The FDTD method is one of the most widely used in electromagnetic simulation [3] and is now being used in quantum simulation [4].

1.2.1 Propagation of a Particle in Free Space

The advantage of the FDTD method is that it is a "real-time, real-space" method—one can observe the propagation of a particle in time as it moves in a specific area. The method will be described briefly.

We will start by rewriting the Schrödinger equation in one dimension as:

$$\frac{\partial \psi}{\partial t} = i \frac{\hbar}{2m_e} \frac{\partial^2 \psi(x,t)}{\partial x^2} - \frac{i}{\hbar} V(x) \psi(x,t). \tag{1.9}$$

To avoid using complex numbers, we will split $\psi(x,t)$ into two parts, separating the real and imaginary components:

$$\psi(x,t) = \psi_{\text{real}}(x,t) + i \cdot \psi_{\text{imag}}(x,t).$$

Inserting this into Equation (1.9) and separating into the real and imaginary parts leads to two coupled equations:

$$\frac{\partial \psi_{\text{real}}(x,t)}{\partial t} = -\frac{\hbar}{2m_e} \frac{\partial^2 \psi_{\text{imag}}(x,t)}{\partial x^2} + \frac{1}{\hbar} V(x) \psi_{\text{imag}}(x,t), \tag{1.10a}$$

$$\frac{\partial \psi_{\text{imag}}(x,t)}{\partial t} = \frac{\hbar}{2m_e} \frac{\partial^2 \psi_{\text{real}}(x,t)}{\partial x^2} - \frac{1}{\hbar} V(x) \psi_{\text{real}}(x,t). \tag{1.10b}$$

To put these equations in a computer, we will take the finite-difference approximations. The time derivative is approximated by:

$$\frac{\partial \psi_{\text{real}}(x,t)}{\partial t} \cong \frac{\psi_{\text{real}}(x,(m+1)\cdot\Delta t) - \psi_{\text{real}}(x,(m)\cdot\Delta t)}{\Delta t}, \tag{1.11a}$$

where Δt is a time step. The Laplacian is approximated by:

$$\frac{\partial^2 \psi_{\text{imag}}(x,t)}{\partial x^2} \cong \frac{1}{(\Delta x)^2} [\psi_{\text{imag}}(\Delta x \cdot (n+1), m \cdot \Delta t) \\ - 2\psi_{\text{imag}}(\Delta x \cdot n, m \cdot \Delta t) + \psi_{\text{imag}}(\Delta x \cdot (n-1), m \cdot \Delta t)] \tag{1.11b}$$

where Δx is the size of the cells being used for the simulation. For simplicity, we will use the following notation:

$$\psi(n \cdot \Delta x, m \cdot \Delta t) = \psi^m(n), \tag{1.12}$$

that is, the superscript ^{m} indicates the time in units of time steps ($t = m \cdot \Delta t$) and n indicates position in units of cells ($x = n \cdot \Delta x$).

Now Equation (1.10a) can be written as:

$$\frac{\psi_{real}^{m+1}(n) - \psi_{real}^m(n)}{\Delta t} = -\frac{\hbar}{2m} \frac{\psi_{imag}^{m+1/2}(n+1) - 2\psi_{imag}^{m+1/2}(n) + \psi_{imag}^{m+1/2}(n-1)}{(\Delta x)^2}$$

$$+ \frac{1}{\hbar} V(n) \psi_{imag}^{m+1/2}(n),$$

which we can rewrite as:

$$\psi_{real}^{m+1}(n) = \psi_{real}^m(n) - \frac{\hbar}{2m_e} \frac{\Delta t}{(\Delta x)^2} \left[\psi_{imag}^{m+1/2}(n+1) - 2\psi_{imag}^{m+1/2}(n) + \psi_{imag}^{m+1/2}(n-1) \right] \tag{1.13a}$$

$$+ \frac{\Delta t}{\hbar} V(n) \psi_{imag}^{m+1/2}(n).$$

A similar procedure converts Equation (1.10b) to the same form

$$\psi_{imag}^{m+3/2}(n) = \psi_{imag}^{m+1/2}(n) + \frac{\hbar}{2m_e} \frac{\Delta t}{(\Delta x)^2} \left[\psi_{real}^{m+1}(n+1) - 2\psi_{real}^{m+1}(n) + \psi_{real}^{m+1}(n-1) \right] \tag{1.13b}$$

$$- \frac{\Delta t}{\hbar} V(n) \psi_{real}^{m+1}(n).$$

Equation (1.13) tells us that we can get the value of ψ at time $(m + 1)\Delta t$ from the previous value and the surrounding values. Notice that the real values of ψ in Equation (1.13a) are calculated at integer values of m while the imaginary values of ψ are calculated at the half-integer values of m. This represents the leapfrogging technique between the real and imaginary terms that is at the heart of the FDTD method [3]. $\hbar$ is Planck's constant and m_e is the mass of a particle, which we will assume is that of an electron. However, Δx and Δt have to be chosen. For now, we will take $\Delta x = 0.1$ nm. We still have to choose Δt. Look at Equation (1.13). We will define a new parameter to combine all the terms in front of the brackets:

$$ra \equiv \frac{\hbar}{2m_e} \frac{\Delta t}{(\Delta x)^2}. \tag{1.14}$$

To maintain stability, this term must be small, no greater than about 0.15. All of the terms in Equation (1.14) have been specified except Δt. If $\Delta t = 0.02$

femtoseconds (fs), then $ra = 0.115$, which is acceptable. Actually, Δt must also be small enough so that the term $(\Delta t \cdot V(n)/h)$ is also less than 0.15, but we will start with a "free space" simulation where $V(n) = 0$. This leaves us with the equations:

$$\psi_{real}^{m+1}(n) = \psi_{real}^{m}(n) - ra \cdot \left[\psi_{imag}^{m+1/2}(n+1) - 2\psi_{imag}^{m+1/2}(n) + \psi_{imag}^{m+1/2}(n-1)\right], \qquad (1.15a)$$

$$\psi_{imag}^{m+3/2}(n) = \psi_{imag}^{m+1/2}(n) + ra \cdot \left[\psi_{real}^{m+1}(n+1) - 2\psi_{real}^{m+1}(n) + \psi_{real}^{m+1}(n-1)\right], \qquad (1.15b)$$

which can easily be implemented in a computer.

Figure 1.5 shows a simulation of an electron in free space traveling to the right in the positive x direction. It is initialized at time $T = 0$. (See program Se1_1.m in Appendix D.) After 1700 iterations, which represents a time of

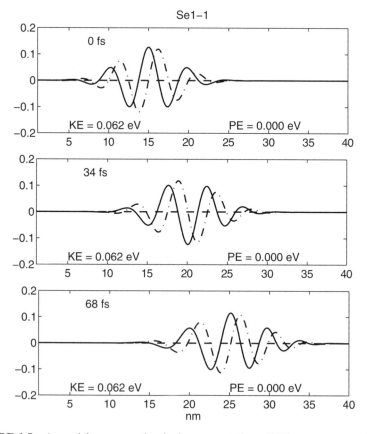

FIGURE 1.5   A particle propagating in free space. The solid line represents the real part of ψ and the dashed line represents the imaginary part.

$$T = 1700 \times \Delta T = 34 \text{ fs},$$

we see the electron has moved about 5 nm. After another 1700 iterations the electron has moved a total of about 10 nm. Notice that the waveform has real and imaginary parts and the imaginary part "leads" the real part. If it were propagating the other way, the imaginary part would be to the left of the real part.

Figure 1.5 indicates that the particle being simulated has 0.062 eV of KE. We will discuss how the program calculates this later. But for now, we can check and see if this is in general agreement with what we have learned. We know that in quantum mechanics, momentum is related to wavelength by Equation (1.2). So we can calculate KE by:

$$\text{KE} = \frac{1}{2} m_e v^2 = \frac{p^2}{2m_e} = \frac{1}{2m_e}\left(\frac{h}{\lambda}\right)^2.$$

In Figure 1.5, the wavelength appears to be 5 nm. The mass of an electron in free space is 9.1×10^{-31} kg.

$$\text{KE} = \frac{1}{2(9.1 \times 10^{-31} \text{ kg})}\left(\frac{6.625 \times 10^{-34} \text{ J} \cdot \text{s}}{5 \times 10^{-9} \text{ m}}\right)^2$$

$$= \frac{1}{2(9.1 \times 10^{-31} \text{ kg})}\left(1.325 \times 10^{-25} \frac{\text{J} \cdot \text{s}}{\text{m}}\right)^2 = \frac{1.756 \times 10^{-50} \text{ J}^2\text{s}^2}{18.2 \times 10^{-31} \text{ kg} \cdot \text{m}^2}$$

$$= 9.65 \times 10^{-21} \text{ J}\left(\frac{1 \text{ eV}}{1.6 \times 10^{-19} \text{ J}}\right) = 0.0603 \text{ eV}.$$

Let us see if simulation agrees with classical mechanics. The particle moved 10 nm in 68 fs, so its velocity is:

$$v = \frac{10 \times 10^{-9} \text{ m}}{68 \times 10^{-15} \text{ s}} = 0.147 \times 10^6 \text{ m/s},$$

$$\text{KE} = \frac{1}{2} m_e v^2 = \frac{1}{2}(9.1 \times 10^{-31} \text{ kg})(1.47 \times 10^5 \text{ m/s})^2$$

$$= 9.83 \times 10^{-21} \text{ J}\left(\frac{1 \text{ eV}}{1.6 \times 10^{-19}}\right) = 0.0614 \text{ eV}.$$

1.2.2 Propagation of a Particle Interacting with a Potential

Next we move to a simulation of a particle interacting with a potential. In Section 1.4 we will discuss what might cause this potential, but we will ignore that for right now. Figure 1.6 shows a particle initialized next to a barrier that is 0.1 eV high. The potential is specified by setting $V(n)$ of Equation (1.13) to 0.1 eV for those value of n corresponding to the region between 20 and 40 nm.

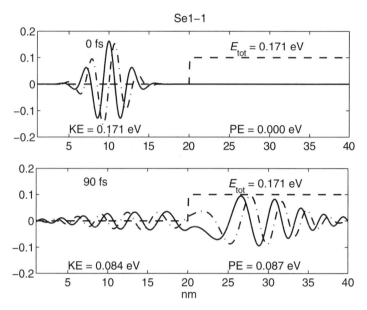

FIGURE 1.6 A particle is initialized in free space and strikes a barrier with a potential of 0.1 eV.

After 90 fs, part of the waveform has penetrated into the potential and is continuing to propagate in the same direction.

Notice that part of the waveform has been reflected. You might assume that the particle has split into two, but it has not. Instead, there is some probability that the particle will enter the potential and some probability that it will be reflected. These probabilities are determined by the following equations:

$$P_{\text{reflected}} = \int_0^{20\,\text{nm}} |\psi(x)|^2 \, dx, \tag{1.16a}$$

$$P_{\text{penetrated}} = \int_{20\,\text{nm}}^{40\,\text{nm}} |\psi(x)|^2 \, dx. \tag{1.16b}$$

Also notice that as the particle enters the barrier it exchanges some of its KE for PE. However, the total energy remains the same.

Now let us look at the situation where the particle is initialized at a potential of 0.1 eV, as shown in the top of Figure 1.7. This particle is also moving left to right. As it comes to the interface, most of the particle goes to the zero potential region, but some is actually reflected and goes back the other way. This is another purely quantum mechanical phenomena. According to classical

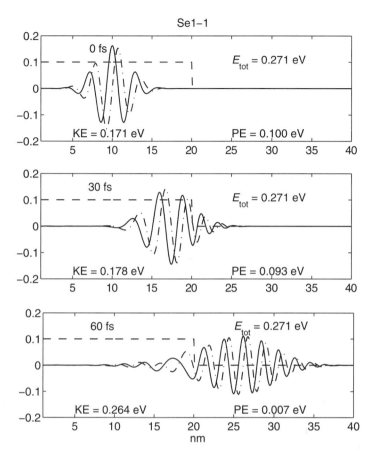

FIGURE 1.7   A particle moving left to right is initialized at a potential of 0.1 eV. Note that the particle initially has both KE and PE, but after most of the waveform moves to the zero potential region, it has mostly KE.

physics, a particle coming to the edge of a cliff would drop off with 100% certainty. Notice that by 60 fs, most of the PE has been converted to KE, although the total energy remains the same.

Example 1.2

The particle in the figure is an electron moving toward a potential of 0.1 eV. If the particle penetrates into the barrier, explain how you would estimate its total energy as it keeps propagating. You may write your answer in terms of known constants.

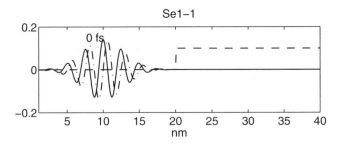

Solution. The particle starts with only KE, which can be estimated by:

$$KE = \frac{(h/\lambda)^2}{2m_e} = 3.84 \times 10^{-20} \text{ J} = 0.24 \text{ eV}.$$

In this case, $\lambda = 2.5$ nm and $m = 9.11 \times 10^{-31}$ kg the mass of an electron. If the particle penetrates into the barrier, 0.1 eV of this KE is converted to PE, but the total energy remains the same.

1.3 PHYSICAL PARAMETERS: THE OBSERVABLES

We said that the solution of the Schrödinger equation, the state variable ψ, contains all meaningful physical parameters even though its amplitude had no direct physical meaning itself. To find these physical parameters, we must do something to the waveform $\psi(x)$. In quantum mechanics, we say that we apply an *operator* to the function. It may seem strange that we have to do something to a function to obtain the information, but this is not as uncommon as you might first think. For example, if we wanted to find the total area under some waveform $F(x)$ we would apply the integration operator to find this quantity. That is what we do now. The operators that lead to specific physical quantities in quantum mechanics are called *oberservables*.

Let us see how we would go about extracting a physical property from $\psi(x)$. Suppose we have a waveform like the one shown in Figure 1.5, and that we can write this function as:

$$\psi(x) = A(x)e^{ikx}. \tag{1.17}$$

The e^{ikx} is the oscillating complex waveform and $A(x)$ describes the spatial variation, in this case a Gaussian envelope. Let us assume that we want to determine the momentum. We know from Equation (1.2) that in quantum mechanics, momentum is given by:

$$p = \frac{h}{\lambda} = \frac{2\pi h}{2\pi\lambda} = \hbar k.$$

So if we could get that k in the exponential of Equation (1.17), we could just multiply it by $\hbar$ to have momentum. We can get that k if we take the derivative with respect to x. Try this:

$$\frac{\partial}{\partial x}\psi(x) = \left(\frac{\partial}{\partial x}A(x)\right)e^{ikx} + ikA(x)e^{ikx}.$$

We know that the envelope function $A(x)$ is slowly varying compared to e^{ik}, so we will make the approximation

$$\frac{d}{dx}\psi(x) \cong ikA(x)e^{ikx}.$$

Similar to the way we multiplied a state variable with its complex conjugate and integrated to get the normalization factor in Equation (1.6), we will now multiply the above function with the complex conjugate of $\psi(x)$ and integrate:

$$\int_{-\infty}^{\infty} A^*(x)e^{-ikx}ikA(x)e^{ikx}dx = ik\int_{-\infty}^{\infty} A^*(x)A(x)e^{-ikx}e^{ikx}dx = ik.$$

We know that last part is true because $\psi(x)$ is a normalized function. If instead of just the derivative, we used the operator

$$p = \frac{\hbar}{i}\frac{d}{dx}, \tag{1.18}$$

when we take the inner product, we get $\hbar k$, the momentum. The p in Equation (1.18) is the momentum observable and the quantity we get after taking the inner product is the *expectation value* of the momentum, which has the symbol $\langle p \rangle$.

If you were to guess what the KE operator is, your first guess might be

$$KE = \frac{p^2}{2m_e} = \frac{1}{2m_e}\left(\frac{\hbar}{i}\frac{\partial}{\partial x}\right)^2 = -\frac{\hbar^2}{2m_e}\frac{\partial^2}{\partial x^2}. \tag{1.19}$$

You would be correct. The *expectation value* of the KE is actually the quantity that the program calculates for Figure 1.6.

We can calculate the KE in the FDTD program by taking the Laplacian, similar to Equation (1.11b),

$$Lap\_\psi(k) = \psi(k+1) - 2\psi(k) + \psi(k-1),$$

and then calculating:

$$\langle KE \rangle = -\frac{\hbar^2}{2m_e \cdot \Delta x^2} \sum_{k=1}^{NN} \psi^*(k) Lap\_\psi(k). \tag{1.20}$$

The number NN is the number of cells in the FDTD simulation.

What other physical quantities might we want, and what are the corresponding observables? The simplest of these is the *position operator*, which in one dimension is simply x. To get the expectation value of the operator x, we calculate

$$\langle x \rangle = \int_{-\infty}^{\infty} \psi^*(x) x \psi(x) dx. \tag{1.21}$$

To calculate it in the FDTD format, we use

$$\langle x \rangle = \sum_{n=1}^{NN} \left[\psi_{\text{real}}(n) - i\psi_{\text{imag}}(n) \right] (n \cdot \Delta x) \left[\psi_{\text{real}}(n) + i\psi_{\text{imag}}(n) \right]$$

$$= \sum_{n=1}^{NN} \left[\psi_{\text{real}}^2(n) + \psi_{\text{imag}}^2(n) \right] (n \cdot \Delta x),$$

where Δx is the cell size in the program and NN is the total number of cells. This can be added to the FDTD program very easily.

The expectation value of the PE is also easy to calculate:

$$\langle PE \rangle = \int_{-\infty}^{\infty} \psi^*(x) V(x) \psi(x) dx. \tag{1.22}$$

The potential $V(x)$ is a real function, so Equation (1.22) simplifies to

$$\langle PE \rangle = \int_{-\infty}^{\infty} |\psi(x)|^2 V(x) dx,$$

which is calculated in the FDTD program similar to $\langle x \rangle$ above,

$$\langle PE \rangle = \sum_{n=1}^{NN} \left[\psi_{\text{real}}^2(n) + \psi_{\text{imag}}^2(n) \right] V(n). \tag{1.23}$$

These expectation values of KE and PE are what appear in the simulation in Figure 1.6.

Probably the most important observable is the Hamiltonian itself, which is the sum of the KE and PE observable. Therefore, the expectation value of the Hamiltonian is the expectation of the total energy

$$\langle H \rangle = \langle KE \rangle + \langle PE \rangle.$$

1.4 THE POTENTIAL $V(x)$

Remember that we said that the Schrödinger equation is an energy equation, and that $V(x)$ represents the PE. In this section we will give two examples of how physical phenomena are represented through $V(x)$.

1.4.1 The Conduction Band of a Semiconductor

Suppose our problem is to simulate the propagation of an electron in an n-type semiconductor. The electrons travel in the conduction band [5]. A key reference point in a semiconductor is the Fermi level. The more the n-type semiconductor is doped, the closer the Fermi level is moved toward the conduction band. If two n-type semiconductors with different doping levels are placed next to each other, the Fermi levels will align, as shown in Figure 1.8. In this case, the semiconductor to the right of the junction is more lightly doped than the one on the left. This results in the step in the conduction band. An electron going from left to right will see this potential, and there will be some chance it will penetrate and some chance it will be reflected, similar to the simulation of Figure 1.6.

In actuality, one more thing must be changed to simulate Figure 1.8. If the simulation is in a semiconductor material, we can no longer use the free space mass of the electron, given by $m_e = 9.109 \times 10^{-31}$ kg. It must be altered by a quantity called the effective mass. We will discuss this in Chapter 6. For now, just understand that if the material we are simulating is silicon, which has an effective mass of 1.08, we must use a mass of $m_e = 1.08 \times (9.109 \times 10^{-31}$ kg) in determining the parameters for the simulation. Figure 1.9 is a simulation of a particle interacting with the junction of Figure 1.8.

1.4.2 A Particle in an Electric Field

Suppose we have the situation illustrated in Figure 1.10 on the following page. The voltage of U_0 volts results in an electric field through the material of

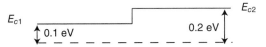

FIGURE 1.8 A junction formed by two n-type semiconductors with different doping levels. The material on the left has heavier doping because the Fermi level (dashed line) is closer to the conduction band.

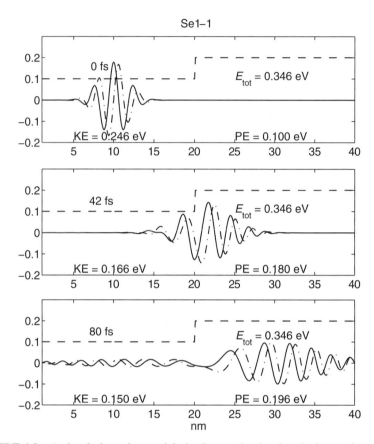

FIGURE 1.9   A simulation of a particle in the conduction band of a semiconductor, similar to the situation shown in Fig. 1.8. Note that the particle initially has a PE of 0.1 eV because it begins in a conduction band at 0.1 eV. After 80 fs, most of the waveform has penetrated to the conduction band at 0.2 eV, and much of the initial KE has been exchanged for PE.

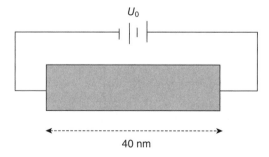

FIGURE 1.10   A semiconductor material with a voltage across it.

$$E_e = -\frac{U_0}{40 \text{ nm}}.$$

This puts the right side at a higher potential of U_0 volts.

To put this in the Schrödinger equation, we have to express this in terms of energy. For an electron to be at a potential of V_0 volts, it would have to have a PE of

$$V_e = -eU_0. \tag{1.24}$$

What are the units of the quantity V_e? Volts have the units of joules per coulomb, so V_e has the units of joules. As we have seen, it is more convenient to work in electron volts: to convert V_e to electron volts, we divide by $1/1.6 \times 10^{-19}$. That means the application of U_0 volts lowers the potential by V_e electron volts. That might seem like a coincidence, but it is not. We saw earlier that an electron volt is defined as the energy to move charge of one electron through a potential difference of 1 V. To quantify our discussion we will say that $U_0 = 0.2$ V. With the above reasoning, we say that the left side has a PE that is 0.2 eV higher than the right side. We write this as:

$$V_e(x) = 0.2 - \frac{0.2}{40} x \text{ eV}, \tag{1.25}$$

as shown by the dashed line in Figure 1.11 (x is in nanometers). This potential can now be incorporated into the Schrödinger equation:

$$i\hbar \frac{\partial}{\partial t} \psi = \left(-\frac{\hbar^2}{2m_e} \nabla^2 + V_e(x) \right) \psi. \tag{1.26}$$

Note that the potential induces an electric field given by [6]:

$$\mathbf{E} = -\nabla V_e(x) = -\frac{\partial V_e(x)}{dx}.$$

If we take $U_0 = 0.2$ V, then

$$\mathbf{E} = -\frac{0.02 \text{ V}}{40 \times 10^{-9} \text{m}} = -10^6 \text{ V/m}. \tag{1.27}$$

This seems like an extremely intense E field but it illustrates how intensive E fields can appear when we are dealing with very small structures.

Figure 1.11 is a simulation of a particle in this E field. We begin the simulation by placing a particle at 10 nm. Most of its energy is PE. In fact, we see that PE = 0.15 eV, in keeping with its location on the potential slope. After

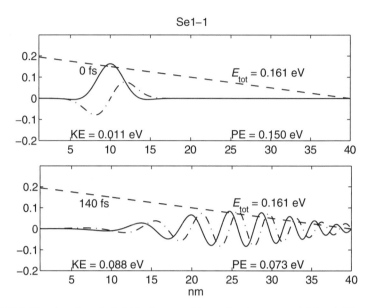

FIGURE 1.11 An electric field is simulated by a slanting potential (top). The particle is initialized at 10 nm. After 140 fs the particle has moved down the potential, acquiring more KE.

140 fs, the particle has started sliding down the potential. It has lost much of its PE and exchanged it for KE. Again, the total energy remains constant.

Note that the simulation of a particle in an E field was accomplished by adding the term $-eV_0$ to the Schrödinger equation of Equation (1.25). But the simulation illustrated in Figure 1.11 looks as if we just have a particle rolling down a graded potential. This illustrates the fact that all phenomena incorporated into the Schrödinger equation must be in terms of energy.

1.5 PROPAGATING THROUGH POTENTIAL BARRIERS

The state variable ψ is a function of both space and time. In fact, it can often be written in separate space and time variables

$$\psi(x, t) = \psi(x)\theta(t). \tag{1.28}$$

Recall that one of our early observations was that the energy of a photon was related to its frequency by

$$E = hf.$$

In quantum mechanics, it is usually written as:

$$E = (2\pi f)\left(\frac{h}{2\pi}\right) = \hbar\omega.$$

The Schrödinger equation is first order in time so we can assume that the time-dependent parameter $\theta(t)$ is in a time-harmonic form,

$$\theta(t) = e^{-i\omega t} = e^{-i(E/\hbar)t}. \tag{1.29}$$

When we put this in the time-dependent Schrödinger equation,

$$i\hbar\frac{\partial}{\partial t}\{\psi(x)e^{-i(E/\hbar)t}\} = -\frac{\hbar^2}{2m}\frac{\partial^2}{\partial x^2}\{\psi(x)e^{-i(E/\hbar)t}\} + V(x)\{\psi(x)e^{-i(E/\hbar)t}\},$$

the left side becomes

$$i\hbar\frac{\partial}{\partial t}\{\psi(x)e^{-i(E/\hbar)t}\} = i\hbar\left(-i\frac{E}{\hbar}\right)\{\psi(x)e^{-i(E/\hbar)t}\}$$
$$= E\{\psi(x)e^{-i(E/\hbar)t}\}.$$

If we substitute this back into the Schrödinger equation, there are no remaining time operators, so we can divide out the term $e^{-i(E/\hbar)t}$, which leaves us with the *time-independent Schrödinger* equation

$$E\psi(x) = -\frac{\hbar^2}{2m_e}\frac{\partial^2\psi(x)}{\partial x^2} + V(x)\psi(x). \tag{1.30}$$

We might find it more convenient to write it as:

$$\frac{\partial^2\psi(x)}{\partial x^2} + \frac{2m}{\hbar^2}[E - V(x)]\psi(x) = 0,$$

or even

$$\frac{\partial^2\psi(x)}{\partial x^2} + k^2\psi(x) = 0, \tag{1.31a}$$

with

$$k = \frac{1}{\hbar}\sqrt{2m_e(E - V(x))}. \tag{1.31b}$$

Equation (1.31a) now looks like the classic Helmholtz equation that one might find in electromagnetics or acoustics. We can write two general types of solutions for Equation (1.31a) based on whether k is real or imaginary. If $E > V$, k will be real and solutions will be of the form

$$\psi(x) = Ae^{ikx} + Be^{-ikx}$$

or

$$\psi(x) = A\cos(kx) + B\sin(kx);$$

that is, the solutions are propagating. Notice that for a given value of E, the value of k changes for different potentials V. This was illustrated in Figure 1.6. If however, $E < V$, k will be imaginary and solutions will be of the form

$$\psi(x) = Ae^{-kx} + Be^{kx}; \tag{1.32}$$

that is, the solutions are decaying. The first term on the right is for a particle moving in the positive x-direction and the second term is for a particle moving in the negative x-direction. Figure 1.12 illustrates the different wave behaviors

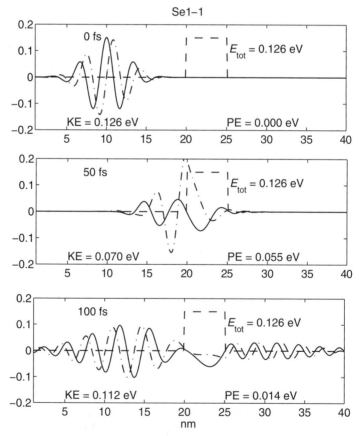

FIGURE 1.12   A propagating pulse hitting a barrier with a PE of 0.15 eV.

for different values of k. A particle propagating from left to right with a KE of 0.126 eV encounters a barrier, which has a potential of 0.15 eV. The particle goes through the barrier, but is attenuated as it does so. The part of the waveform that escapes from the barrier continues propagating at the original frequency. Notice that it is possible for a particle to move through a barrier of higher PE than it has KE. This is a purely quantum mechanical phenomenon called "tunneling."

1.6 SUMMARY

Two specific observations helped motivate the development of quantum mechanics. The photoelectric effect states that energy is related to frequency

$$E = hf. \tag{1.33}$$

The wave–particle duality says that momentum and wavelength are related

$$p = \frac{h}{\lambda}. \tag{1.34}$$

We also made use of two classical equations in this chapter. When dealing with a particle, like an electron, we often used the formula for KE:

$$\mathrm{KE} = \frac{1}{2} m_e v^2, \tag{1.35}$$

where m is the mass of the particle and v is its velocity. When dealing with photons, which are packets of energy, we have to remember that it is electromagnetic energy, and use the equation

$$c_0 = f\lambda, \tag{1.36}$$

where c_0 is the speed of light, f is the frequency, and λ is the wavelength.

We started this chapter stating that quantum mechanics is dictated by the *time-dependent Schrödinger* equation. We subsequently found that each of the terms correspond to energy:

$$i\hbar \frac{\partial}{\partial t} \psi(x,t) = -\frac{\hbar^2}{2m} \nabla^2 \psi(x,t) + V(x)\psi(x,t) \tag{1.37}$$

Total energy   Kinetic energy Potential energy

However, we can also work with the *time-independent Schrödinger* equation:

$$E\psi(x,t) = -\frac{\hbar^2}{2m}\nabla^2\psi(x,t) + V(x)\psi(x,t) \qquad (1.38)$$

Total energy   Kinetic energy   Potential energy

EXERCISES

1.1 Why Quantum Mechanics?

1.1.1 Look at Equation (1.2). Show that h/λ has units of momentum.

1.1.2 Titanium has a work function of 4.33 eV. What is the maximum wavelength of light that I can use to induce photoelectron emission?

1.1.3 An electron with a center wavelength of 10 nm is accelerated through a potential of 0.02 V. What is its wavelength afterward?

1.2 Simulation of the One-Dimensional, Time-Dependent Schrödinger Equation

1.2.1 In Figure 1.6, explain why the wavelength changes as it goes into the barrier. Look at the part of the waveform that is reflected from the barrier. Why does the imaginary part appear slightly to the left of the real, as opposed to the part in the potential?

1.2.2 You have probably heard of the Heisenberg uncertainty principle. This says that we cannot know the position of a particle and its momentum with unlimited accuracy at any given time. Explain this in terms of the waveform in Figure 1.5.

1.2.3 What are the units of $\psi(x)$ in Figure 1.5? (Hint: The "1" in Eq. 1.6 is dimensionless.) What are the units of ψ in two dimensions? In three dimensions?

1.2.4 Suppose an electron is represented by the waveform in Figure 1.13 and you have an instrument that can determine the position to within 5 nm. Approximate the probability that a measurement will find the particle: (a) between 15 and 20 nm, (b) between 20 and 25 nm, and (c) between 25 and 30 nm. Hint: Approximate the magnitude in each region and remember that the magnitude

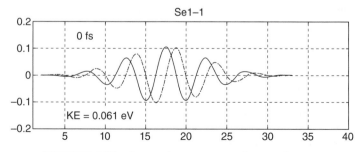

FIGURE 1.13 A waveform representing an electron.

squared gives the probability that the particle is there and that the total probability of it being somewhere must be 1.

1.2.5 Use the program se1_1.m and initialize the wave in the middle (set nc = 200). Run the program with a wavelength of 10 and then a wavelength of 20. Which propagates faster? Why? Change the wavelength to −10. What is the difference? Why does this happen?

1.3 Physical Parameters: The Observables

1.3.1 Add the calculation of the expectation value of position $\langle x \rangle$ to the program se1_1.m. It should print out on the plots, like KE and PE expectation values. Show how this value varies as the particle propagates. Now let the particle hit a barrier as in Figure 1.7. What happens to the calculation of $\langle x \rangle$? Why?

1.4 The Potential $V(x)$

1.4.1 Simulate a particle in an electric field of strength $E = 5 \times 10^6$ V/m. Initialize a particle 10 nm left of center with a wavelength of 4 nm and $\sigma = 4$ nm. (Sigma represents the width of the Gaussian shape.) Run the simulation until the particle reaches 10 nm right of center. What has changed and why?

1.4.2 Explain how you would simulate the following problem: A particle is moving along in free space and then encounters a potential of −0.1 eV.

1.5 Propagation through Barriers

1.5.1 Look at the example in Figure 1.12. What percentage of the amplitude is attenuated as the wave crosses through the barrier? Simulate this using se1_1.m and calculate the probability that the particle made it through the barrier using a calculation similar to Equation (1.15). Is your calculation of the transmitted amplitude in qualitative agreement with this?

REFERENCES

1. R. P. Feynman, R. B. Leighton, and M. Sands, *The Feynman Lectures on Physics*, Reading, MA: Addison-Wesley, 1965.
2. S. Borowitz, *Fundamentals of Quantum Mechanics*, New, York: W. A. Benjamin, 1969.
3. D. M. Sullivan, *Electromagnetic Simulation Using the FDTD Method*, New York: IEEE Press, 2000.
4. D. M. Sullivan and D. S. Citrin, "Time-domain simulation of two electrons in a quantum dot," *J. Appl. Phys.*, Vol. 89, pp. 3841–3846, 2001.
5. D. A. Neamen, *Semiconductor Physics and Devices—Basic Principles*, 3rd ed., New York: McGraw-Hill, 2003.
6. D. K. Cheng, *Field and Wave Electromagnetics*, Menlo Park, CA: Addison-Wesley, 1989.

2

STATIONARY STATES

In Chapter 1, we showed that the Schrödinger equation is the basis of quantum mechanics and can be written in two forms: the time-dependent or time-independent Schrödinger equations. In that first chapter, we primarily use the time-dependent version and talk about particles propagating in space. In the first section of this chapter, we begin by discussing a particle confined to a limited space called an infinite well. By using the time-independent Schrödinger equation, we show that a particle in a confined area can only be in certain states, called eigenstates, and only be at certain energies, called eigenergies. In Section 2.2 we demonstrate that any particle moving within a structure can be written as a superposition of the eigenstates of that structure. This is an important concept in quantum mechanics. Section 2.3 describes the concept of the periodic boundary condition that is widely used to characterize spaces in semiconductors. In Section 2.4 we describe how to find the eigenfunctions and eigenenergies of arbitrary structures using MATLAB, at least for one-dimensional structures. Section 2.5 illustrates the importance of eigenstates in quantum transport by a simulation program. Section 2.6 describes the bra-ket notation used in quantum mechanics.

Quantum Mechanics for Electrical Engineers, First Edition. Dennis M. Sullivan.
© 2012 The Institute of Electrical and Electronics Engineers, Inc.
Published 2012 by John Wiley & Sons, Inc.

2.1 THE INFINITE WELL

One of the important canonical problems of quantum mechanics is the infinite well [1], illustrated in Figure 2.1. In this section, we will determine what an electron would look like if it were trapped in such a structure.

To analyze the infinite well, we will again use the Schrödinger equation. Recall that the time-independent Schrödinger equation is:

$$E\psi(x) = -\frac{\hbar^2}{2m_e}\frac{\partial^2\psi(x)}{\partial x^2} + V(x)\psi(x). \tag{2.1}$$

When we are discussing particles confined to a structure, the time-independent version of the Schrödinger equation in Equation (2.1) is often the best choice. We will rewrite Equation (2.1) in the following manner:

$$\frac{\partial^2\psi(x)}{\partial x^2} + \frac{2m_e}{\hbar^2}(E - V(x))\psi(x) = 0. \tag{2.2}$$

Obviously, we will not be able to solve Equation (2.2) for any instance where $V = \infty$. So we will limit our attention to the area between $x = 0$ and $x = a$. In this region, we have $V = 0$, so we can just leave the $V(x)$ term out. We have now reduced the problem to a simple second-order differential equation:

$$\frac{\partial^2\psi(x)}{\partial x^2} + \frac{2m_e}{\hbar^2}E\psi(x) = 0.$$

In fact, it might be even more convenient to write it in the form

$$\frac{\partial^2\psi(x)}{\partial x^2} + k^2\psi(x) = 0, \tag{2.3}$$

with

$$k = \frac{\sqrt{2m_e E}}{\hbar}. \tag{2.4}$$

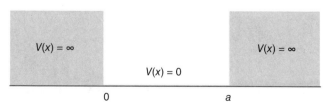

FIGURE 2.1 An infinite well.

Now we have the kind of problem we are used to solving. We still need two boundary conditions. However, we have already defined them as: $\psi(0) = \psi(a) = 0$. Thus, a general solution to Equation (2.3) can be written in either exponential or trigonometric form

$$\begin{aligned}\psi(x) &= Ae^{jkx} + Be^{-jkx} \\ &= A\sin kx + B\cos kx.\end{aligned}$$

It will turn out that the trigonometric form is more convenient. The boundary conditions eliminate the cosine function, leaving

$$\psi(x) = A\sin(kx).$$

The fact that $\psi(a) = 0$ means this is a valid solution wherever

$$k_n = \frac{n\pi}{a} \quad n = 1, 2, 3, \dots .$$

Using Equation (2.4) for values of k, we get

$$k_n = \frac{\sqrt{2m_e E}}{\hbar} = \frac{n\pi}{a}. \tag{2.5}$$

All of the terms in Equation (2.5) are constants, except E; so we will solve for E:

$$E = \varepsilon_n = \frac{\hbar^2}{2m_e}\left(\frac{n\pi}{a}\right)^2 = \frac{\hbar^2\pi^2}{2m_e a^2}n^2 \quad n = 1, 2, 3, \dots . \tag{2.6}$$

Notice that there are only discrete values of E for which we have a solution.

Let us calculate the energy levels for a well that is 10 nm wide:

$$\begin{aligned}\varepsilon_n &= \frac{\hbar^2\pi^2}{2m_e a^2}n^2 = \frac{(1.054\times10^{-34}\text{ J·s})^2(3.14159)^2}{2(9.11\times10^{-31}\text{ kg})(10^{-8}\text{ m})^2}n^2 \\ &= \frac{(1.11\times10^{-68})9.87}{18.22\times10^{-47}}n^2\left[\frac{\text{J}^2\text{s}^2}{\text{kg·m}^2}\right] \\ &= 0.6\times10^{-21}\text{ J}\left(\frac{eV}{1.6\times10^{-19}\text{ J}}\right)n^2 = 0.00375\,n^2\text{ eV}.\end{aligned} \tag{2.7}$$

It appears that the lowest energy state in which an electron can exist in this infinite well is 0.00375 eV, which we can write as 3.75 meV.

But we're not done yet. We know that our solutions are of the form

$$\psi(x) = A\sin\left(\frac{n\pi}{a}x\right),\tag{2.8}$$

but we still do not know the value of A. This is where we use normalization. Any solution to the Schrödinger equation must satisfy

$$\int_{-\infty}^{\infty} \psi^*(x)\psi(x)dx = 1.$$

Since the wave only exists in the interval $0 \le x \le a$, this can be rewritten as

$$\int_0^a \left[A\sin\left(\frac{n\pi}{a}x\right)\right]^* \cdot \left[A\sin\left(\frac{n\pi}{a}x\right)\right] A\, dx = 1.$$

The sines are both real, so we can write this as

$$1 = A^*A\int_0^a \sin^2\left(\frac{n\pi}{a}x\right)dx = A^*A\int_0^a\left(\frac{1}{2}-\frac{1}{2}\sin\left(2\frac{n\pi}{a}x\right)\right)dx$$
$$= \frac{A^*A}{2}a.$$

We are free to choose any A^*, A pair such that

$$A^* \cdot A = 2/a.$$

We choose the simplest, $A = \sqrt{2/a}$. Now we are done. We will rewrite Equation (2.8) as

$$\phi_n(x) = \sqrt{\frac{2}{a}}\sin\left(\frac{n\pi x}{a}\right).\tag{2.9}$$

2.1.1 Eigenstates and Eigenenergies

The set of solutions in Equation (2.9) are referred to as *eigenfunctions* or *eigenstates*. The set of corresponding values Equation (2.7) are the *eigenvalues*, sometimes called the *eigenenergies*. (*Eigen* is a German word meaning *own* or *proper*.) When a state variable is an eigenfunction, we will usually indicate it by ϕ_n instead of ψ, as in Equation (2.9). Similarly, when referring to an eigenenergy, we will usually indicate it by ε_n, as in Equation (2.7). The lowest four eigenstates of the 10 nm infinite well and their corresponding eigenenergies are shown in Figure 2.2. Eigenvalues and eigenfunctions appear throughout

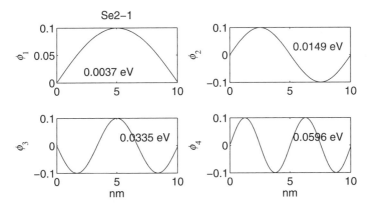

FIGURE 2.2 The first four eigenstates of the 10 nm infinite well and their corresponding eigenenergies.

engineering and science but they play a fundamental role in quantum theory. Any particle in the infinite well must be in an eigenstate or a superposition of eigenstates, as we will see in Section 2.2.

Just suppose, though, that we actually want to see the lowest energy form of this function as it evolves in time. The state corresponding to the lowest energy is called the *ground state*. Equation (2.9) gives us the spatial part, but how do we reinsert the time part? Actually, we have already determined that. If we know the ground state energy is 3.75 meV, we can write the time-dependent state variable as

$$\psi(x,t) = \phi_1(x)e^{-i(E_1/\hbar)t} = \sqrt{\frac{2}{a}}\sin\left(\frac{n\pi}{a}x\right)e^{-i(3.75\,\text{meV}/\hbar)t}. \tag{2.10}$$

Look a little closer at the time exponential

$$\frac{3.75\,\text{meV}}{\hbar} = \frac{3.75\,\text{meV}}{.658\times10^{-15}\,\text{eV}\cdot\text{s}}$$
$$= 5.7\times10^{12}\,\text{s}^{-1}.$$

The units of inverse seconds indicate frequency.

A simulation of the ground state in the infinite well is shown in Figure 2.3.

The simulation program calculated the kinetic energy (KE) as 3.84 meV instead of 3.75, but this is simply due to the finite differencing error. If we used higher resolution, that is, a smaller cell size, we would come closer to the actual value. Notice that after a period of time, 1.08 ps, the state returned to

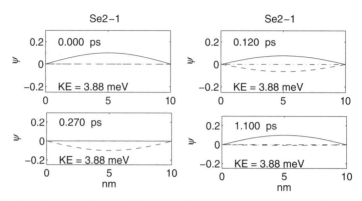

FIGURE 2.3 Time evolution of the ground state of the potential well. The solid line is the real part and the dashed line is the imaginary part.

its original phase. We know the energy of the infinite well, but could we have predicted the time for this "revival"? We know that

$$E = hf = h\left(\frac{1}{T}\right),$$

$$T = \frac{h}{E} = \frac{4.135 \times 10^{-15} \text{ eV} \cdot \text{s}}{3.75 \times 10^{-3} \text{ eV}} = 1.10 \text{ ps}.$$

Let us try another state. Take $n = 4$ state, which has the energy

$$E_4 = 0.00375(4)^2 \text{ eV} = 0.06 \text{ eV},$$

and the corresponding revival time

$$T = \frac{h}{E} = \frac{4.135 \times 10^{-15} \text{ eV} \cdot \text{s}}{60 \times 10^{-3} \text{ eV}} = 0.0689 \text{ ps}.$$

The $n = 4$ state is simulated in Figure 2.4. Notice that in the time period $T = 0.07$ ps, the waveform returns to its original phase.

Example. What is the expected value of the Hamiltonian for the third eigenstate of the 10 nm infinite well?

Solution

$$\langle H \rangle = \langle \phi_3 | H \phi_3 \rangle = \langle \phi_3 | \varepsilon_3 \phi_3 \rangle = \varepsilon_3.$$

The expected value of H is just the expected value of energy, and for an eigenstate that is its corresponding eigenenergy.

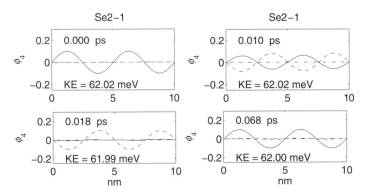

FIGURE 2.4 Time evolution of the $n = 4$ waveform.

TABLE 2.1 The Allowed Wavelengths in the 10 nm Well

n	λ (nm)	$1/\lambda$ (nm^{-1})
1	20	0.05
2	10	0.1
3	6.67	0.15
4	5	0.2
5	4	0.25

2.1.2 Quantization

The examples above illustrate one of the fundamental principles of quantum mechanics: quantization. A particle cannot be just anywhere. Electrons belong to the class of particles called fermions. According to the Pauli exclusion principle, no two fermions can occupy the same state. Because electrons can have different spins, electrons with opposite spins can occupy the same energy level. For now, we will ignore spin and suppose that only one electron can occupy each energy level.

The number of available states is linked to the wavelength. The first eigenstate of the 10 nm well has a wavelength of $\lambda_1 = 2 \times 10$ nm $= 20$ nm. The $n = 2$ function has a wavelength of $\lambda_2 = 10$ nm, and so on. The first few allowed wavelengths are listed in Table 2.1. Perhaps it would be worthwhile to plot available states as a function of wavelength. Actually, we will plot them as $1/\lambda$ (Fig. 2.5).

Physicists prefer to use k-space (Fig. 2.5b). k is related to wavelength by

$$k = \frac{2\pi}{\lambda}.$$

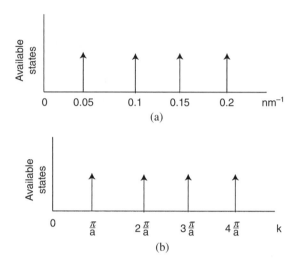

FIGURE 2.5 (a) A plot of the available states as a function of $1/\lambda$; (b) a plot of the available states in k-space. The parameter a is 10 nm.

The parameter k in Fourier-transformed space is similar to *omega* in Fourier-transformed time. However, it is easier just to use λ^{-1}, because it relates to the wavelength.

2.2   EIGENFUNCTION DECOMPOSITION

Let us look again at the eigenstates of the infinite well

$$\varphi_n(x) = \sqrt{\frac{2}{a}} \sin\left(\frac{n\pi x}{a}\right). \tag{2.11}$$

Recall from Fourier theory that if we have a series that repeats with a period T, we can write a Fourier series expansion [2]

$$x(t) = \frac{1}{2}a_0 + \sum_{n=1}^{\infty}[a_n \cos(n\omega_o t) + b_n \sin(n\omega_o t)], \tag{2.12a}$$

where

$$a_n = \frac{2}{T}\int_0^T x(t)\cos(n\omega_o t)\,dt \quad n = 1, 2, 3, \ldots, \tag{2.12b}$$

$$b_n = \frac{2}{T}\int_0^T x(t)\sin(n\omega_o t)\,dt \quad n = 1, 2, 3, \ldots, \tag{2.12c}$$

and

$$\omega_0 = \frac{2\pi}{T}. \tag{2.12d}$$

The infinite well is not periodic, but it does limit our wave function to a specific range, so we can use some of the same methods used in Fourier series. Notice the following differences:

1. We are familiar with signals in the time domain, and now we are dealing with functions of distance.
2. There are no cosine terms because of our restriction $\psi(0) = \psi(10 \text{ nm}) = 0$.
3. We are used to dealing with real functions $x(t)$ in the time domain. But now our space domain function $\psi(x)$ is complex. That means all coefficients will be complex.

If we have a function $\psi(x)$ in the 10 nm infinite well, we can write an expansion similar to Equation (2.12a):

$$\psi(x) = \sum_{n=1}^{N} c_n \phi_n(x). \tag{2.13}$$

In Equation (2.13), N is the number of terms is the expansion. In theory, this number is infinite, but in cases of practical interest it is a manageable finite number. The coefficients are determined individually by the following procedure. To get c_m, for instance, we multiply both sides of Equation (2.13) by $\phi_m^*(x)$ and integrate over the well

$$\int_0^{10 \text{ nm}} \phi_m^*(x)\psi(x)dx = \int_0^{10 \text{ nm}} \phi_m^*(x)\sum_{n=1}^{N} c_n \phi_n(x)dx$$

$$= \sum_{n=1}^{N} c_n \int_0^{10 \text{ nm}} \phi_m^*(x)\phi_n(x)dx = c_m.$$

Therefore,

$$c_m = \int_0^{10 \text{ nm}} \phi_m^*(x)\psi(x)dx. \tag{2.14}$$

In determining c_m we used the following important property:

$$\int_0^{10 \text{ nm}} \phi_m^*(x)\phi_n(x)dx = \begin{cases} 0 & n \neq m \\ 1 & n = m \end{cases}. \tag{2.15}$$

The complex conjugate $\phi_m^*(x)$ in Equation (2.14) is redundant in this case because the $\phi_n(x)$ are all real functions. However, when eigenfunctions are complex the conjugate must be used.

Equation (2.15) shows that the eigenfunctions $\phi_n(x)$ are *orthonormal*, that is, they are normalized and they are orthogonal. You may recall from analytic geometry that a quantity of great interest is the inner product of two vectors: $\vec{a} \cdot \vec{b} = |a||b|\cos(\angle ab)$, where $\angle ab$ is the angle between the two vectors. If two vectors have an angle of 90° between them, then $\vec{a} \cdot \vec{b} = 0$. We say the vectors are orthogonal to each other.

The right side of Equation (2.14) is another type of inner product. This inner product is extremely important in quantum mechanics. The fact that any function can be expressed as a superposition of the eigenfunctions, as in Equation (2.13), means that $\varphi_n(x)$ are a *complete* set. Complete sets of orthonormal eigenfunctions play a key role in quantum mechanics.

Figure 2.6 shows a pulse initiated in the infinite well. The c_n are calculated by Equation (2.14) and plotted as magnitude and phase. Note that the only difference between the pulse at $T = 0$ ps and at $T = 0.008$ ps are the phases of the c_n; the amplitudes did not change.

Remember that each specific eigenfunction φ_n has a specific energy associated with it, which is given by

$$E_n = 0.00375 \, n^2 \text{ eV.}$$

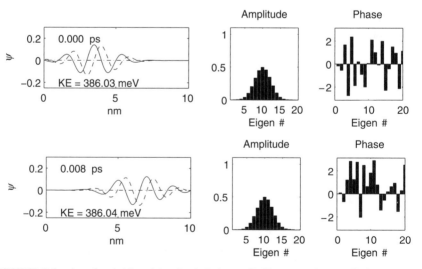

FIGURE 2.6 A pulse initiated in the infinite well. The complex coefficients c_n are determined by Equation (2.14). After the pulse has propagated, the amplitudes of the coefficients are the same but the phases can change.

A specific energy also means that it has a specific frequency associated with it. This frequency is given by

$$\omega_n = 2\pi f_n = 2\pi \frac{E_n}{h} = \frac{2\pi \cdot 0.00375\, n^2\ \text{eV}}{4.135 \times 10^{-15}\ \text{eV} \cdot \text{s}} = 5.7\, n^2 \times 10^{12}\ \text{s}^{-1}. \qquad (2.16)$$

This means each eigenfunction evolves as

$$\varphi_n(x,t) = \varphi_n(x)e^{-i\omega_n t},$$

and we could describe the propagation of a pulse by

$$\psi(x,t) = \sum_{n=1}^{N} c_n e^{-i\omega_n t} \varphi_n(x). \qquad (2.17)$$

This is illustrated in Figure 2.7. A pulse is initialized at $T = 0$ ps. The amplitude and phase of the complex c_n values are shown next to it. The pulse is calculated using Equation (2.17), and the new complex coefficients

$$d_n = c_n e^{-i\omega_n t} \qquad (2.18)$$

are displayed as amplitude and phase.

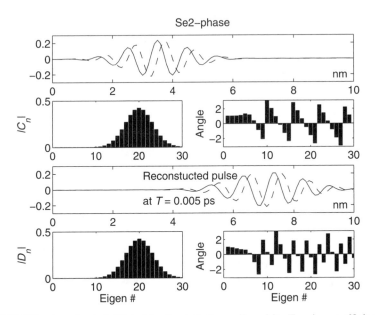

FIGURE 2.7 A pulse is initiated in the infinite well and its Fourier coefficients are calculated. The pulse is propagated by calculated new coefficients for different times.

This observation has enormous implications: If we can find the eigenvalues and corresponding eigenfunctions of a structure, we can predict how a pulse will propagate across it.

2.3 PERIODIC BOUNDARY CONDITIONS

In Sections 2.1 and 2.2, we demonstrated that an orthogonal set of functions can be used to describe any arbitrary function. In this case, we used real sine functions that were the eigenfunctions of an infinite well as our basis function. If the function was wider than 10 nm, we could just make the well larger. However, there is another approach that is much more commonly used in quantum semiconductor theory to limit the discussion to a restricted area. Instead of starting with the boundary conditions of the infinite well, that is, $\psi(0) = \psi(10 \text{ nm}) = 0$, we say

$$\psi(0) = \psi(10 \text{ nm}). \tag{2.19}$$

This is known as the *periodic boundary condition*. It is a very convenient approach to formulating problems in large, homogeneous regions.

The eigenfunctions of a 10 nm infinite well with a periodic boundary condition are

$$\varphi_n(x) = \frac{1}{\sqrt{10}} e^{i\left(\frac{2\pi n x}{10}\right)} \quad n = 0, \pm 1, \pm 2, \pm 3, \ldots. \tag{2.20}$$

The 10 in Equation (2.20) has the unit of nanometers. The factor of $1/\sqrt{10}$ in front is for normalization. Notice that there is an $n = 0$ term. (Electrical engineers would refer to this as the direct current, or "dc" term.) Look at $n = 9$:

$$\varphi_9(x) = \frac{1}{\sqrt{10}} e^{i\left(\frac{2\pi 9 x}{10}\right)} = \frac{1}{\sqrt{10}} e^{i\left(\frac{2\pi(10-1)x}{10}\right)} = \frac{1}{\sqrt{10}} e^{i\left(\frac{2\pi 10 x}{10}\right)} e^{i\left(\frac{-2\pi x}{10}\right)} = \frac{1}{\sqrt{10}} e^{-i\left(\frac{2\pi x}{10}\right)}.$$

It turns out that the higher numbered n are actually the same as the negative n. As before, we can do an eigenfunction decomposition by taking the inner product

$$c_n = \int_0^{10 \text{ nm}} \varphi_n^*(x) \psi(x) dx = \int_0^{10 \text{ nm}} \frac{1}{\sqrt{10}} e^{-i\left(\frac{2\pi n x}{10}\right)} \psi(x) dx.$$

Remember, in quantum mechanics when we take an inner product we use the complex conjugate of the first function. The function can be reconstructed by

$$\psi(x) = \sum_{n=0}^{N-1} c_n \varphi_n(x).$$

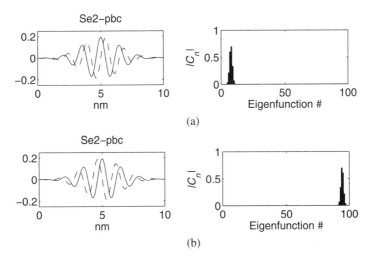

FIGURE 2.8   Decomposition of propagating pulses using the eigenfunctions of the well with the periodic boundary conditions: (a) a pulse propagating left to right; (b) a pulse propagating right to left.

This is demonstrated in Figure 2.8.

In actuality, an eigenfunction decomposition using periodic boundary conditions is equivalent to taking the Fourier transform.

2.4   EIGENFUNCTIONS FOR ARBITRARILY SHAPED POTENTIALS

Suppose we had a potential similar to the one in Figure 2.9a. This potential has its own set of eigenfunctions, the first four of which are shown in Figure 2.9b. (These were calculated with a simple MATLAB function called **eig**, which will be explained later). We can also initiate the ground state in the well and watch how it evolves in time, as shown in Figure 2.10. Notice that the time for one revival, that is, the time to complete one cycle, can be predicted from the total energy of the particle:

$$f = \frac{E}{h} = \frac{0.0158\,\text{eV}}{4.135 \times 10^{-15}\,\text{eV}\cdot\text{s}},$$

$$T_{\text{revival}} = \frac{1}{f} = \frac{h}{E} = \frac{4.135 \times 10^{-15}\,\text{eV}\cdot\text{s}}{0.0158\,\text{eV}} = 262 \times 10^{-15}\,\text{s}.$$

Figure 2.11 shows a particle initialized left of center in the V-shaped well. It can be described by various percentages of the eigenfunctions, as shown in the bar graph to the right. As time evolves, the particle moves right toward the

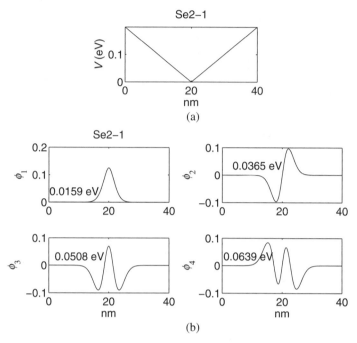

FIGURE 2.9   (a) A V-shaped potential in a 40 nm well and (b) the first four eigen-functions and their corresponding eigenenergies.

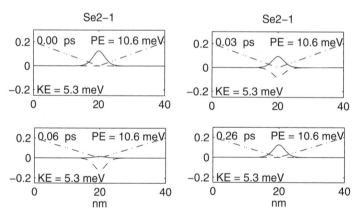

FIGURE 2.10   Time evolution of the ground state function of the V-shaped potential.

lower potential. Potential energy (PE) is lost; KE is gained; but the total energy of the particle remains the same. This is to be expected, because no energy is put into or taken out of the system. The particle reaches maximum KE at the bottom of the well, and then starts up the right side of the well. Notice that throughout, the percentages of the eigenfunctions remain the same.

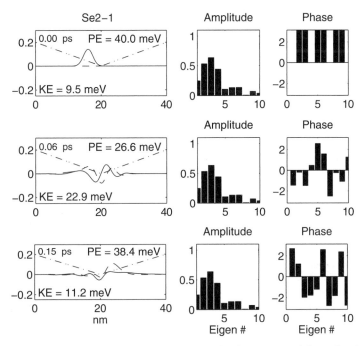

FIGURE 2.11 A pulse initialized left of center in the V potential can be described by a superposition of eigenstates. In time the pulse starts moving right, exchanging PE for KE. It continues oscillating back and forth but the total energy remains constant. The relative amplitudes of the eigenfunctions remains constant.

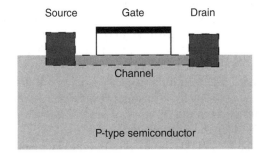

FIGURE 2.12 Diagram of an N-channel MOSFET.

2.5 COUPLED WELLS

Figure 2.12 illustrates an N-channel metal–oxide–semiconductor field-effect transistor, or MOSFET. The size of MOSFETs are constantly shrinking, and as the dimensions get smaller, quantum mechanics plays a more important role. We will try to illustrate this with a very simplistic example.

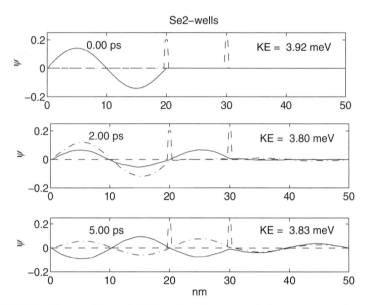

FIGURE 2.13 The particle is initialized in the left well. It can tunnel through the middle well to the right well because the energy, 3.8 meV, has a corresponding eigenstate in the middle well.

In Figure 2.13 an infinite well with two partitions is shown, effectively giving three wells. The partitions are not infinite, but are 0.2 eV, which is a higher energy than we expect the first few eigenenergies to be. Notice that the middle well is 10 nm, similar to the infinite well of Section 2.1. The right and left well are twice as wide, so we can assume that the eigenenergies will be much smaller. Think of the far left well as the "source," the far right well as the "drain," and the middle well as the "channel."

We are interested in determining which particles will flow from the source to the drain, that is, current flow through the device. We begin by assuming that the eigenstates correspond to those of an infinite well of the same size. A particle is initialized in the second eigenstate of the left well, as shown in Figure 2.13. In time, it is able to tunnel into the middle well. Notice the middle well appears to be a ground state eigenfunction. From the middle well, it eventually tunnels to the right well.

Figure 2.14 shows the same three-compartment infinite well. This time, the particle is initialized in the ground state of the left well. The particle is never able to tunnel to the right well, even though the right well will have the same eigenstates as the left well. This is because the particle cannot tunnel through the middle well since 1 meV is not an eigenenergy of this 10 nm well.

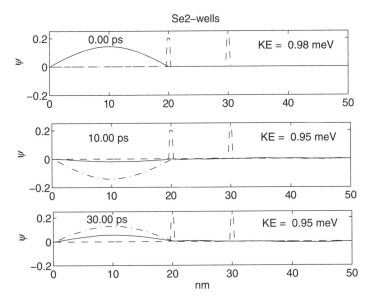

FIGURE 2.14 A particle initialized in the right well cannot tunnel to the left well because it does not have an eigenstate in the middle well.

Suppose we were to add a potential of 0.003 V to the middle well so that the ground state energy is $E_1' = 0.00375\,\text{eV} - 0.003\,\text{eV} = 0.00075\,\text{eV}$. In this case, the 0.98 meV particle might be able to find a state in the channel. This is the situation shown in Figure 2.15. (The decrease in the voltage in the middle well is too small to see on the diagram.) This is an example of how a gate voltage controls the current flow between the source and the drain.

We have been assuming that the left and right wells had about the same eigenstates as a 20 nm infinite well, and that the middle well has approximately those of a 10 nm well. Figure 2.16 illustrates this situation. It also illustrates what we observed above, that is, the left well has a state at about 1 meV. However, there is no available state in the middle available for transport.

These are very simplistic models of a MOSFET. In a real MOSFET, the source and the drain are called contacts, or reservoirs. They can be thought of as large infinite wells containing many eigenstates at closely spaced intervals.

Our philosophy will be the following: we will analyze the channel by finding its eigenstates. Then we look upon the source and drain as external stimuli to the channel. Determining the eigenstates of the channel is analogous to finding the transfer function of a circuit. Once we have the transfer function, we can determine the output for any input. Finding the response to the quantum channel will prove far more difficult.

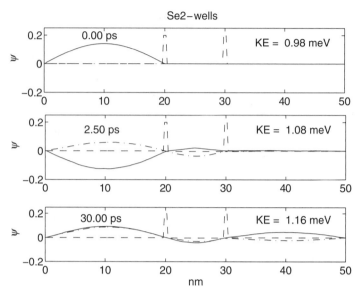

FIGURE 2.15 The potential in the middle channel is lowered by 3 meV, the difference between the first and second eigenenergies.

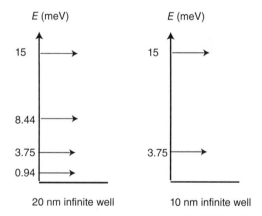

FIGURE 2.16 Available energy states of a 20 nm infinite well (left) and a 10 nm infinite well (right).

2.6   BRA-KET NOTATION

So far we have been developing eigenfunctions for simple one-dimensional structures, such as the infinite well. We have been using notation similar to Equation (2.11)

$$\phi_n(x) = \sqrt{\frac{2}{a}} \sin\left(\frac{n\pi x}{a}\right), \tag{2.21}$$

for the nth eigenstate of an infinite well with length a. A notation widely used in quantum mechanics is the "ket," which is written $|\phi_n\rangle$ or sometimes just $|n\rangle$ to refer to the nth eigenstate of a system [3]. Obviously, $|n\rangle$ will mean one particular function for the infinite well and something different for the V-shaped well or the infinite well with the barrier. However, in the context of one particular system, this does not usually cause any confusion. If we wanted to talk about a state in an infinite well, that is, a superposition of many eigenstates, we can write

$$\psi(x) = \sum_{n=1}^{N} c_n |\phi_n\rangle. \tag{2.22}$$

It turns out that there is a complimentary group of functions, which are the complex conjugates of the original kets. They are called "bras." For the infinite well eigenfunctions of Equation (2.21), bras and kets are the same because the $|\phi_n\rangle$s are real. (This is not always the case.) However, look back at the well with periodic boundary conditions of Section 2.3. Its eigenfunctions (Eq. 2.20) would result in the following kets and bras:

$$|\phi_n\rangle = \frac{1}{\sqrt{a}} e^{i(2\pi nx/a)}, \tag{2.23a}$$

$$\langle\phi_n| = \frac{1}{\sqrt{a}} e^{-i(2\pi nx/a)}. \tag{2.23b}$$

When we multiply them together we get "bra-kets":

$$\langle\phi_m|\phi_n\rangle = \begin{cases} 1 & n = m \\ 0 & n \neq m \end{cases}. \tag{2.24}$$

This of course is just the inner product that we spoke of earlier. For instance, suppose we had a pulse that was the superposition of the $|\phi_n\rangle$ of Equation (2.23a). We could start by expressing this as Equation (2.22). But how do we find the complex c_ns? We take the inner product with each eigenfunction:

$$\langle m|\psi(x)\rangle = \langle m|\sum_{n=1}^{N} c_n |n\rangle$$
$$= \sum_{n=1}^{N} c_n \langle m|n\rangle = c_n. \tag{2.25}$$

The last step results from Equation (2.24). Note that this is nothing but the procedure we use to find the Fourier expansion of a waveform. We have just given it a more compact notation.

Actually, we can take this notation a bit further. Suppose we want to calculate some expected values of observables, as in Section 1.2. Recall how we calculated the expected value of position in Equation (1.16):

$$\langle x \rangle = \int_0^{10\,nm} \psi^*(x) x \psi(x) dx.$$

This is expressed in bra-ket notation as

$$\langle x \rangle = \langle \psi(x)|x|\psi(x) \rangle, \qquad (2.26a)$$

or simply

$$\langle x \rangle = \langle \psi(x)|x\psi(x) \rangle. \qquad (2.26b)$$

Both of these notations appear in the literature and they mean the same thing. Note that the value of $\psi(x)$ on the bra side actually means $\psi^*(x)$. What if we want to determine the expectation value of the KE, as in Equation (1.8)? This can be written as:

$$\langle KE \rangle = \langle \psi(x)|KE\psi(x) \rangle. \qquad (2.27)$$

Remember, this actually means

$$\langle \psi(x)|KE\psi(x) \rangle = \int_0^{10\,nm} \psi^*(x) \cdot -\frac{\hbar^2}{2m}\frac{\partial^2}{dx^2} \psi(x) dx.$$

We have to admit that this is a more concise notation. We can calculate the expected value of KE for the pulse that is already written as the superposition of eigenfunctions as in Equation (2.22):

$$\langle \psi(x)|KE\psi(x) \rangle = \left\{ \sum_{n=1}^{N} c_n^* \langle \phi_n| \right\} \cdot -\frac{\hbar^2}{2m}\frac{\partial^2}{dx^2} \left\{ \sum_{n=1}^{N} c_n |\phi_n\rangle \right\}.$$

If we assume that our eigenfunctions are those of Equation (2.23), then

$$-\frac{\hbar^2}{2m}\frac{\partial^2}{dx^2}|\phi_n\rangle = -\frac{\hbar^2}{2m}\left(i\frac{2\pi n}{a}\right)^2|\phi_n\rangle = \frac{\hbar^2 2\pi^2 n^2}{ma^2}|\phi_n\rangle,$$

and from this:

$$\langle \psi(x)|K\psi(x) \rangle = \sum_{n=1}^{N} |c_n|^2 \frac{\hbar^2 2\pi^2 n^2}{ma^2}.$$

2.7 SUMMARY

In this chapter the extremely important concept of eigenstates and eigenenergies were introduced. Using an infinite well as the primary example, we learned that any quantum state can be decomposed into the eigenstates of a system, and that the mathematics was similar to Fourier series decomposition.

EXERCISES

2.1 Stationary States

2.1.1 If we had used a well that was just 10 Å at its base, what would be the energy levels? (Hint: Do not do a lot of math! Look at the calculation for 100 Å and ask yourself how they will change.)

2.1.2 For the $a = 100$ Å well, how different would our answers be if, instead of an infinite potential, we had a potential of 3 eV?

2.1.3 What can you say about where a particle is in the quantum well? Take the $n = 1$ state. How about $n = 4$? If you had to give a quantitative answer, how would you determine it? Remember the expected value of position?

2.1.4 If we measured the KE of the particle in the $n = 1$ state, what would we get?

2.1.5 What other values of A, the normalization constant used in Equation (2.8), would give an acceptable solution to the infinite square well?

2.1.6 Suppose the floor of an infinite well is set at 3 eV instead of 0 eV. If we repeat the simulation of Figure 2.3 how will the results differ?

2.2 Eigenfunction Decomposition

2.2.1 Suppose we decided to use triangular functions for the infinite well. Is this possible? What would be the advantages and disadvantages?

2.2.2 A system is characterized by the eigenfunctions ϕ_n and corresponding eigenenergies ε_n. The state variable $\psi(x)$ describes a particle in the system. Write the energy of $\psi(x)$ as a function of the eigenenergies.

2.2.3 A system is described by two different complete sets of eigenstates ϕ_n with corresponding energies ε_n and θ_n with corresponding energies δ_n.

Since both sets are complete, any state vector can be written as:

$$\psi = \sum_n c_n \phi_n$$

or

$$\psi = \sum_n d_n \theta_n.$$

A state ψ_0 can be written as:

$$\psi_0 = c_1 \phi_1 + c_2 \phi_2.$$

(a) What is the energy of the state corresponding to ψ_0?

(b) Write ψ_0 as a function of the θ_n eigenstates.

2.3 Periodic Boundary Conditions

2.3.1 Instead of calculating the eigenvalue decomposition for periodic boundary conditions directly, as in Figure 2.8, use the MATLAB Fourier transform and plot the magnitude and phase. Do you get similar results?

Helpful hints:

The MATLAB command for the "fast Fourier transform" is **fft**. The magnitude is obtained by **abs**; the angle can be obtained by **imag(log(. . .))**.

2.5 Coupled Wells

2.5.1 Referring to the three-well MOSFET of Figure 2.13, name three things that will determine the current flow from left to right.

2.6 Bra-ket Notation

2.6.1 Write out a mathematical expression to calculate the following quantities, assuming the state variable is $\psi(x)$:

(a) $\langle p \rangle$

(b) $\langle H \rangle$

2.6.2 Suppose we have a state variable $\psi(x)$ written as the expansion in the basis functions $\phi_n(x)$:

$$\psi(x) = \sum_n c_n \phi_n(x),$$

and we would like to change the basis functions to another known set $\gamma_m(x)$, that is,

$$\psi(x) = \sum_m d_m \gamma_m(x).$$

Determine the d_m as a function of c_n, ϕ_n, γ_m.

REFERENCES

1. D. J. Griffiths, *Introduction to Quantum Mechanics*, Englewood Cliffs, NJ: Prentice Hall, 1994.
2. Z. Gajic, *Linear Dynamic Systems and Signals*, Englewood Cliffs, NJ: Prentice Hall, 2003.
3. J. J. Sakurai, *Modern Quantum Mechanics*, Menlo Park, CA: Addison-Wesley, 1994.

3

FOURIER THEORY IN QUANTUM MECHANICS

Fourier theory is among the most important mathematics in electrical engineering [1]. In Section 3.1, we will review the basic theory while pointing out the two differences in using Fourier theory in quantum mechanics. The first difference is merely a sign convention. The second more significant difference is the fact that Fourier transforms are often taken in the spatial domain instead of the time domain in quantum mechanics. In Section 3.2 we will use Fourier theory to explain the concept of a "broadened" state. Section 3.3 uses Fourier theory to explain the famous Heisenberg uncertainty relation, and in Section 3.4, the spatial Fourier transform theory explains the semiconductor analog of the transfer function, called "transmission."

Throughout this chapter, the MATLAB programs use the fast Fourier transform (FFT) to calculate the Fourier transform of various waveforms. The FFT is reviewed in Appendix B.

3.1 THE FOURIER TRANSFORM

In physics, the time-dependent portion of a propagating sinusoidal wave is described by

$$f(t) = e^{-i\omega_0 t}. \tag{3.1}$$

Quantum Mechanics for Electrical Engineers, First Edition. Dennis M. Sullivan.
© 2012 The Institute of Electrical and Electronics Engineers, Inc.
Published 2012 by John Wiley & Sons, Inc.

(In electrical engineering we usually have a positive quantity in the exponential.) Accordingly, physicists define the forward and inverse Fourier transforms as follows:

$$F(\omega) = \mathcal{F}\{f(t)\} = \int_{-\infty}^{\infty} f(t)e^{i\omega t}dt, \tag{3.2a}$$

$$f(t) = \mathcal{F}^{-1}\{F(\omega)\} = \frac{1}{2\pi}\int_{-\infty}^{\infty} F(\omega)e^{-i\omega t}d\omega. \tag{3.2b}$$

For instance, in the frequency domain, a single frequency is represented by $2\pi\delta(\omega - \omega_0)$, so the corresponding time-domain function is

$$f(t) = \frac{1}{2\pi}\int_{-\infty}^{\infty} 2\pi\delta(\omega - \omega_0)e^{-i\omega t}dt = e^{-i\omega_0 t},$$

as we expect. This is the first entry in Table 3.1, a list of Fourier transform pairs. Another function often of interest is the decaying exponential multiplied by the step function, $e^{-\alpha t}u(t)$. It is a causal function, meaning its value is zero before $t = 0$. The Fourier transform is

$$F(\omega) = \int_{0}^{\infty} e^{-\alpha t}e^{i\omega t}dt = \frac{1}{-\alpha + i\omega}e^{(-\alpha + i\omega)t}\Big|_{0}^{\infty} = \frac{1}{\alpha - i\omega}.$$

TABLE 3.1 Fourier Transform Pairs

	$f(t)$	$F(\omega)$
1.	$e^{-i\omega_0 t}$	$2\pi\delta(\omega - \omega_0)$
2.a	$e^{-\alpha t}u(t)$	$\dfrac{1}{\alpha - i\omega}$
2.b	$e^{-\alpha t}u(-t)$	$\dfrac{1}{\alpha + i\omega}$
3.a	$e^{-\alpha t}e^{-i\omega_0 t}u(t)$	$\dfrac{1}{\alpha - i(\omega - \omega_0)}$
3.b	$e^{\alpha t}e^{-i\omega_0 t}u(-t)$	$\dfrac{1}{\alpha + i(\omega - \omega_0)}$
4.a	$e^{-\alpha|t|}.$	$\dfrac{2\alpha}{\alpha^2 + \omega^2}$
4.b	$e^{-\alpha|t|}e^{-i\omega_0 t}$	$\dfrac{2\alpha}{\alpha^2 + (\omega - \omega_0)^2}$
5.a	$\dfrac{1}{\sqrt{2\pi}\sigma}e^{-\frac{1}{2}\left(\frac{t}{\sigma}\right)^2}$	$e^{-\sigma^2\omega^2/2}$
5.b	$\dfrac{1}{\sqrt{2\pi}\sigma}e^{-\frac{1}{2}\left(\frac{t}{\sigma}\right)^2}e^{i\omega_0 t}$	$e^{-\sigma^2(\omega - \omega_0)^2/2}$
6.	$p_{t_0}(t) = u(t + t_0/2) - u(t - t_0/2)$	$t_0\dfrac{\sin(\omega t_0/2)}{(\omega t_0/2)} = t_0\,\mathrm{sinc}\left(\dfrac{\omega t_0}{2}\right)$

TABLE 3.2   Fourier Transform Properties

1.	Frequency shifting	$F\{x(t)e^{i\omega_0 t}\} = X(\omega + \omega_0)$
2.	Time shifting	$F\{x(t - t_0)\} = X(\omega)e^{i\omega t_0}$
3.	Time reversal	$F\{x(-t)\} = X(-\omega)$

The "anticausal" version of this is $e^{-\alpha t}u(-t)$ and its Fourier transform is

$$F(\omega) = \int_{-\infty}^{0} e^{\alpha t}e^{i\omega t}dt = \frac{1}{\alpha + i\omega}e^{(\alpha + i\omega)t}\Big|_{-\infty}^{0} = \frac{1}{\alpha + i\omega}.$$

Related to these functions are the decaying complex exponentials. The causal single-frequency function is $e^{-\alpha t}e^{-i\omega_0 t}u(t)$, and it has the transform

$$F(\omega) = \int_{0}^{\infty} e^{-\alpha t}e^{-i\omega_0 t}e^{i\omega t}dt = \frac{1}{-\alpha + i(\omega - \omega_0)}e^{(-\alpha - i\omega_0 + i\omega)t}\Big|_{0}^{\infty} = \frac{1}{\alpha - i(\omega - \omega_0)}.$$

Similarly, the transform of the anticausal single-frequency function is

$$F(\omega) = \int_{-\infty}^{0} e^{\alpha t}e^{-i\omega_0 t}e^{i\omega t}dt = \frac{1}{\alpha - i\omega_0 + i\omega}e^{(\alpha - i\omega_0 + i\omega)t}\Big|_{0}^{\infty} = \frac{1}{\alpha + i(\omega - \omega_0)}.$$

These two functions illustrate a very important theorem, usually referred to as the *frequency shifting theorem*. The theorem states that if $x(t)$, $X(\omega)$ is a Fourier transform pair, then so is $x(t)e^{-i\omega_0 t}$, $X(\omega - \omega_0)$ (see Table 3.2). The proof is left as an exercise. We could also look at a function that is a two-sided decaying exponential at a single frequency

$$e^{-|\alpha|t}e^{-i\omega_0 t} = e^{-\alpha t}e^{-i\omega_0 t}u(t) + e^{\alpha t}e^{-i\omega_0 t}u(-t). \tag{3.3a}$$

The Fourier transform of this function is the sum of the Fourier transform of the two terms:

$$\begin{aligned} F\{e^{-|\alpha|t}e^{i\omega_0 t}\} &= \frac{1}{\alpha + i(\omega_0 - \omega)} + \frac{1}{\alpha - i(\omega_0 - \omega)} \\ &= \frac{2\alpha}{\alpha^2 + (\omega_0 - \omega)^2}. \end{aligned} \tag{3.3b}$$

By comparing entries 1 and 4.b in Table 3.1, we see that

$$\lim_{\alpha \to 0} \frac{2\alpha}{\alpha^2 + (\omega_0 - \omega)} = 2\pi\delta(\omega_0 - \omega).$$

The Gaussian function is a very important function in engineering, science, and mathematics:

$$f_\sigma(t) = \frac{1}{\sqrt{2\pi}\sigma} e^{-\frac{1}{2}\left(\frac{t}{\sigma}\right)^2}. \qquad (3.4a)$$

Its Fourier transform is not so easily calculated, so it will just be stated as

$$F_\sigma(\omega) = e^{-\sigma^2\omega^2/2}. \qquad (3.4b)$$

Both the time and frequency domain functions are Gaussian waveforms. The Gaussian function is the only function that Fourier transforms to a function of the same shape. Notice the following very important characteristic: σ appears in the term $(t/\sigma)^2$ in the time domain, but as $(\sigma\omega)^2$ in the frequency domain. The narrower the waveform is in the time domain, the broader it is in the frequency domain. We will see later that this is an indication of the "uncertainty" of the pulse.

We know that the Schrödinger equation is both time and space dependent. It turns out that we will want to look at the Fourier transform in both the time domain and the spatial domain. Electrical engineers are familiar with the time-domain transform in which we go from time t to frequency f, that is, from seconds to Hertz. Since there is a 2π in the Fourier transform, we usually like to use radians per second ω. We are not as used to the spatial-domain trans-form as we are with the time-domain transform. The spatial-domain transform takes us from meters to wavelengths, or inverse meters. The spatial part of a propagating wave is

$$g(x) = e^{ikx}.$$

Therefore, the Fourier transform pair is

$$G(k) = \int_{-\infty}^{\infty} g(x)e^{-ikx}dx, \qquad (3.5a)$$

$$g(x) = \frac{1}{2\pi}\int_{-\infty}^{\infty} G(k)e^{ikx}dk. \qquad (3.5b)$$

Once again, because of the 2π in the Fourier transform, we use the units $k = 2\pi/\lambda$ in the Fourier domain. To summarize

$$t \;\Leftarrow \text{Fourier transform} \Rightarrow\; \omega,$$

$$x \;\Leftarrow \text{Fourier transform} \Rightarrow\; k.$$

How do we get the spatial transforms? Simply go through Table 3.1 and replace every t with x and every ω with $-k$.

There is an added complication to the spatial transform. In time we have one dimension; however, in space, we have three dimensions. Transforms involving x and k can be generalized to $\mathbf{r} = (x, y, z)$ and $\mathbf{k} = k_x, k_y, k_z)$. The Fourier transform pair becomes

$$G(\mathbf{k}) = \int_{-\infty}^{\infty} d\mathbf{r} \cdot g(\mathbf{r}) e^{-i\mathbf{k}\cdot\mathbf{r}}, \tag{3.6a}$$

$$g(\mathbf{r}) = \frac{1}{(2\pi)^3} \int_{-\infty}^{\infty} d\mathbf{k} \cdot G(\mathbf{k}) e^{i\mathbf{k}\cdot\mathbf{r}}. \tag{3.6b}$$

Chapter 1 points out two observations that led to the development of quantum mechanics:

$$E = \hbar\omega, \tag{3.7}$$

$$p = \frac{h}{\lambda} = \hbar k. \tag{3.8}$$

This means that we can think of a temporal Fourier transform as a transform from the time to the energy domain, and we can think of the spatial transform as a transform from the space to the momentum domain.

3.2   FOURIER ANALYSIS AND AVAILABLE STATES

Suppose we start with the following function:

$$\psi(t, x) = \varphi(x) \frac{1}{2\pi} e^{-i\omega_1 t}. \tag{3.9}$$

In this section, the $1/2\pi$ is often added to the time-domain functions for mathematical consistency because of the $1/2\pi$ in the inverse Fourier transform, Equation (3.2b). We will let $\varphi(x)$ be some normalized spatial function because we want to concentrate on the time-dependent part. The Fourier transform of the time-dependent part is

$$\Psi(\omega) = \delta(\omega - \omega_1).$$

If we want to count how many "frequency states" there are, we just integrate

$$\int_{-\infty}^{\infty} \delta(\omega - \omega_1) d\omega = 1.$$

We already knew that—we had one time-harmonic function in the time domain and it corresponded to one state in the frequency domain. Of course, we are more interested in the states as a function of energy. So we can write it as a delta function in the energy domain

$$\Psi(E) = \delta(E - \varepsilon_1).$$

Look back at our example of the 10 nm well in Chapter 2. We know that the allowed states are

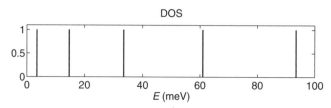

FIGURE 3.1   The available states for the 10 nm infinite well.

$$\varepsilon_n = 3.75\, n^2 \text{ meV},$$

that is, a group of discrete energy values. Mathematically, we express this as the *density of states* (DOS):

$$\text{DOS} = \sum_{n=1}^{\infty} \delta(E - \varepsilon_n). \qquad (3.10)$$

A graph of the available states should look like Figure 3.1.

Now let us look at the case illustrated in Figure 3.2. A particle is initialized in a 10 nm well confined by two 0.2 eV barriers. Similar to the particle in the infinite well, the real and imaginary parts oscillate in time and after a revival time they return to the original position, with one very important difference: the amplitude is decreasing. This occurs because the waveform can tunnel out the sides of the barriers. This will be true of any well that is connected to the "outside world," so we can probably regard this as a more realistic picture than the infinite well. Therefore, a more realistic wave function is

$$\psi(t, x) = \varphi(x) \frac{1}{2\pi} e^{-i\omega_1 t} e^{-\alpha|t|}. \qquad (3.11)$$

This function is a more realistic description of the time-domain characteristics of a quantum eigenstate because a term like $\exp(-\alpha t)$ allows for the possibility of decay of the state due to coupling or some other reasons. We choose to use the two-sided decay $\exp(-\alpha|t|)$ for the following mathematical reasons. First, we can write the time-domain part of Equation (3.11) as

$$e^{-i\omega_1 t} e^{-\alpha|t|} = e^{-i\omega_1 t} e^{\alpha t} u(-t) + e^{-i\omega_1 t} e^{-\alpha t} u(t).$$

If we take the limit as alpha goes to zero

$$\underset{\lim \alpha \to 0}{e^{-i\omega_1 t} e^{-\alpha|t|}} = e^{-i\omega_1 t} u(-t) + e^{-i\omega_1 t} u(t) = e^{-i\omega_1 t},$$

it goes back to original signal.

The second more important reason for using a two-sided exponential is that its Fourier transform is a function that is a good description of a state in the

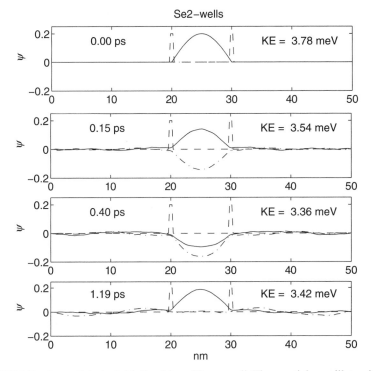

FIGURE 3.2 A particle in initialized in a 10 nm well. The particle oscillates in time similar to the 10 nm infinite well, but the amplitude decays because the waveform can tunnel out the sides. KE, kinetic energy.

frequency domain. From Table 3.1, the Fourier transform of the time-domain part of Equation (3.11) is

$$\Psi(\omega) = \frac{1}{2\pi} \frac{2\alpha}{\alpha^2 + (\omega - \omega_1)^2}.$$

Since this represents one state, its integral should be 1:

$$\int_{-\infty}^{\infty} \frac{1}{2\pi} \frac{2\alpha}{\alpha^2 + (\omega - \omega_1)^2} d\omega = 1.$$

Therefore, the density of the state is

$$D_{\omega_1}(\omega) = \frac{1}{2\pi} \frac{2\alpha}{\alpha^2 + (\omega - \omega_1)^2}.$$

If we want to change the integral to energy, we replace ω with $E / \hbar$:

$$\int_{-\infty}^{\infty} \frac{1}{2\pi} \frac{2\alpha}{\alpha^2 + \frac{1}{\hbar^2}(E-\varepsilon_1)^2} \frac{dE}{\hbar} = \int_{-\infty}^{\infty} \frac{1}{2\pi} \frac{2\alpha\hbar^2}{(\alpha\hbar)^2 + (E-\varepsilon_1)^2} \frac{dE}{\hbar}.$$

We now replace $\alpha\hbar = \gamma/2$:

$$\int_{-\infty}^{\infty} \frac{1}{2\pi} \frac{2(\gamma/2)\hbar}{(\gamma/2)^2 + (E-\varepsilon_1)^2} \frac{dE}{\hbar} = \int_{-\infty}^{\infty} \frac{\gamma/2\pi}{(\gamma/2)^2 + (E-\varepsilon_1)^2} dE.$$

The density is now

$$D_{\varepsilon_1}(E) = \frac{\gamma/2\pi}{(\gamma/2)^2 + (E-\varepsilon_1)^2}.$$

Therefore, the *broadened* DOS for the well can be written as

$$\text{DOS} = \sum_{n=1}^{\infty} \frac{\gamma_n/2\pi}{(\gamma_n/2)^2 + (E-\varepsilon_n)^2}. \tag{3.12}$$

We achieved the broadening by adding an exponential decay in the time domain. Figure 3.3 shows the broadened DOS for the 10 nm well.

We should understand the reason behind the replacement $\alpha\hbar = \gamma/2$. If we use this in Equation (3.11), we get an equation of the form

$$\psi(t,x) = \varphi(x)\frac{1}{2\pi}e^{-i\omega_1 t}e^{-\left(\frac{\gamma}{2\hbar}\right)|t|},$$

which has the modulus

$$|\psi(t,x)|^2 = \frac{1}{(2\pi)^2}|\varphi(x)|^2 e^{-\left(\frac{\gamma}{\hbar}\right)|t|}. \tag{3.13}$$

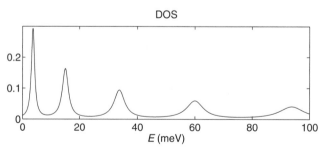

FIGURE 3.3 The broadened DOS for the 10 nm well. Notice that broadening is greater for higher energy states.

that is, a function that decays as $\gamma / \hbar$. Since γ has the units of energy, $\gamma / \hbar$ has the units "per second," as it should.

3.3 UNCERTAINTY

The famed Heisenberg uncertainty relation is part of the mystique of quantum mechanics [2]. Uncertainty comes in two forms: (1) both the position and momentum of a particle cannot be known exactly and (2) both the time and energy of a particle cannot be known exactly. These are expressed as

$$\Delta p \Delta x \geq \hbar / 2, \tag{3.14a}$$

and

$$\Delta E \cdot \Delta t \geq \hbar / 2. \tag{3.14b}$$

In quantum mechanics, we know that

$$E = \hbar \omega = hf.$$

So we can go back to Equation (3.14b) and write

$$\Delta E \cdot \Delta t = \Delta(hf) \cdot \Delta t \geq \hbar / 2 = \frac{h}{2\pi} \frac{1}{2}.$$

Dividing h from both sides gives

$$\Delta f \cdot \Delta t \geq \frac{1}{4\pi} = 0.07958. \tag{3.15a}$$

Without the $\hbar$, it is easier to see the uncertainty as just a number. Can we do the same with Equation (3.14a)? We know that

$$p = \frac{h}{\lambda},$$

so we can write

$$\Delta\left(\frac{h}{\lambda}\right)\Delta x \geq \frac{h}{4\pi},$$

or

$$\Delta\left(\frac{1}{\lambda}\right)\Delta x \geq \frac{1}{4\pi}. \tag{3.15b}$$

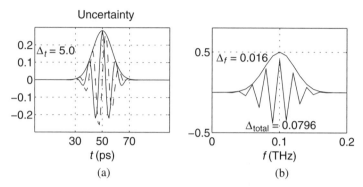

FIGURE 3.4 (a) A sinusoid with a center frequency of 0.1 THz inside a Gaussian envelope defined by $\sigma = 5$. (b) The Fourier transform of the waveform in (a).

Let us first consider the time-frequency uncertainty of Equation (3.15a). We will initialize a time-domain function $\psi(t)$ with a frequency of 0.1 THz inside a Gaussian envelope specified by $\sigma = 5$ ps,

$$\psi(t) = Ae^{-(0.5(t-t_0)/\sigma)^2}e^{-i(2\pi f_0 t)}. \tag{3.16}$$

This is shown in Figure 3.4a. This waveform is centered at 50 ps. If we calculate the "average" time, we get 50 ps.

The uncertainty of quantum mechanics Δ is the standard deviation σ of statistics. It is the mean time from the average time:

$$\langle \Delta_t \rangle = \int_0^{100} \psi^*(t)(t-t_0)^2 \psi(t)dt = 5.0. \tag{3.17}$$

It is not a coincidence that this is the sigma we selected in Equation (3.16).

Now we want to calculate the Δ_f, the uncertainty in the frequency domain. Our first step will be to take the FFT of the waveform. Figure 3.4b shows that it is centered at 0.1 THz. We can use a calculation similar to Equation (3.17) to get the uncertainty in the frequency domain. The program calculated this to be

$$\Delta_f = 0.016,$$

so the total uncertainty is

$$\Delta_t \Delta_f = (5.0)(0.016) = 0.0796 = \frac{1}{4\pi}.$$

In this case, we have found the minimum uncertainty. Luck was not involved. Only Gaussian waveforms Fourier transform with the minimum uncertainty. Figure 3.5 shows a Gaussian pulse that starts with $\sigma_t = 10$ in the time domain and Fourier transforms into $\Delta_f = 0.008$ in the frequency domain. This also gives the minimum uncertainty. But the broader pulse in the time domain (more

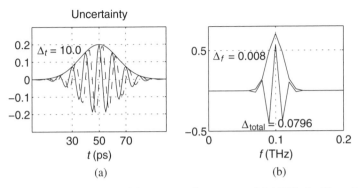

FIGURE 3.5 (a) A sinusoid with a center frequency of 0.1 THz inside a Gaussian envelope defined by $\sigma = 10$. (b) The Fourier transform of the waveform in (a).

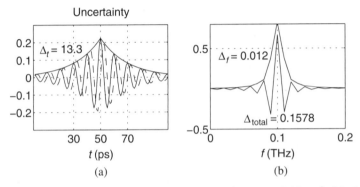

FIGURE 3.6 (a) A decaying exponential (Eq. 3.18) for $\alpha = 0.05$ ps^{-1}; (b) the corresponding FFT. Since the function is not exactly a Gaussian, the total uncertainty is greater than the minimum.

uncertainty) leads to a narrower pulse in the inverse domain (less uncertainty).

If instead of the Gaussian pulse we use

$$\psi(t) = Ae^{-\alpha|t|}e^{-i(2\pi f_0 t)}, \tag{3.18}$$

with $\alpha = 0.05$ ps^{-1}, we get the waveform shown in Figure 3.6a with the time-domain uncertainty of $\Delta_t = 13.3$. If we take the FFT and calculate the uncertainty in the frequency domain, we get $\Delta_f = 0.012$, as shown in Figure 3.6b. When we multiply them together, we get

$$\Delta_t \Delta_f = (13.3)(0.012) = 0.1578 > \frac{1}{4\pi}.$$

This is not an optimum pair from the point of view of uncertainty. However, mathematically, we can often work with this pair easier than with the Gaussian waveforms, as we will see in later chapters.

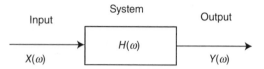

FIGURE 3.7 Linear systems are most easily analyzed by taking the problem into the frequency domain.

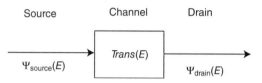

FIGURE 3.8 The transmission function, *Trans*(E), is the probability that a particle of energy E will move from the source to the drain.

3.4 TRANSMISSION VIA FFT

In engineering, if we have a system and we want to know the outputs for various inputs, as in Figure 3.7, the first thing we do is find the transfer function. We do this because we know that for any given input function, $x(t)$, a function of time, we can take the Fourier transform to get $X(\omega)$, and multiply it by the transfer function to get the output in the frequency domain $Y(\omega) = H(\omega) X(\omega)$. We then take the inverse Fourier transform of $Y(\omega)$ to get $y(t)$, the time-domain expression of the output. If we are dealing with causal functions, we usually use Laplace instead of Fourier transforms, but the procedure is the same.

We would like to have a function similar to the transfer function that could tell which particles will go through a channel. In semiconductor engineering this is called *transmission* [3] (Fig. 3.8).

The transmission is a parameter very similar to the frequency response. The main difference is that the transmission is a function of energy instead of frequency,

$$Trans(E) = \frac{|\Psi_{drain}(E)|^2}{|\Psi_{source}(E)|^2}. \tag{3.19}$$

First, we have to explain $\Psi(E)$. As described in Appendix B, we can take a spatial FFT of a waveform and plot it as $1/\lambda$, as shown in Figure 3.9. Figure 3.9a is a waveform with a wavelength of 5 nm enclosed in a Gaussian envelope. Figure 3.9b is an FFT plotted as a function of inverse wavelength. This is analogous to taking a FFT of a time-domain function and plotting it as a function of frequency, which is inverse time.

However, we would prefer a function of energy. We know that inverse wavelength times 2π is k:

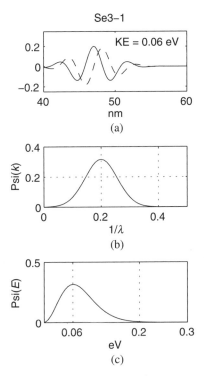

FIGURE 3.9 (a) The plot of a particle with a center wavelength of 5 nm. (b) The spatial Fourier transform. Notice that it is centered at $1/\lambda = 0.2$. (c) The Fourier transform scaled as a function of energy.

$$k = \frac{2\pi}{\lambda}.$$

But we also know that k is related to momentum, and therefore, to kinetic energy.

$$E = \frac{p^2}{2m} = \frac{\hbar^2 k^2}{2m} = \frac{h^2}{2m\lambda^2}.$$

Recall that $\hbar = h/2\pi$. Therefore, we can just as easily plot the k-space data as a function of energy, as shown in Figure 3.9c.

To illustrate the concept of transmission, we will model a channel simply as a potential barrier 0.2 eV high and 1 nm wide, as shown in Figure 3.10. We start by taking the Fourier transform of a waveform moving toward the "channel." Note that the Fourier transform of the input particle is centered at 0.1 eV, but the particle has a fairly broad spectrum. After a simulation time corresponding to 0.2 ps, the waveform has interacted with the barrier; some of the waveform has penetrated the barrier and some has been reflected.

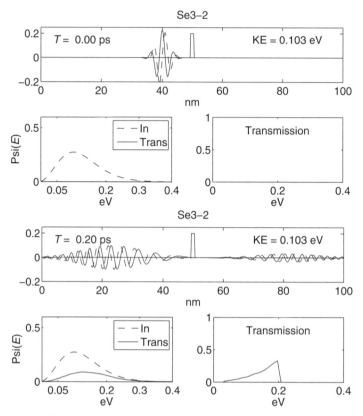

FIGURE 3.10   The calculation of transmission using an input waveform centered at 0.1 eV.

Now we take the Fourier transform of the part of the waveform that penetrated through the barrier. We do this by taking a spatial FFT starting at about 55 nm. We do not want to include the reflected portion of the waveform. This is superimposed on the graph containing the FFT of the input. The graph labeled "Transmission" is the quantity expressed by Equation (3.19). The results do not surprise us. Particles at energies substantially below the 0.2 eV height of the barrier have little chance of making it through. Also notice that we were only able to calculate transmission over the energy range dictated by that of the input waveform. Notice that the magnitude of the input waveform falls off sharply above 0.2 eV. The calculation of Equation (3.19) loses accuracy when the magnitude of the input is too small.

In an effort to calculate transmission over other energy ranges, we repeat the simulation but start with a particle at higher energy as shown in Figure 3.11. This particle's center energy is at about 0.25 eV. After 0.15 ps, a substantially larger portion of the waveform has penetrated the barrier. The transmission once again confirms what we would have guessed, that particles with energies

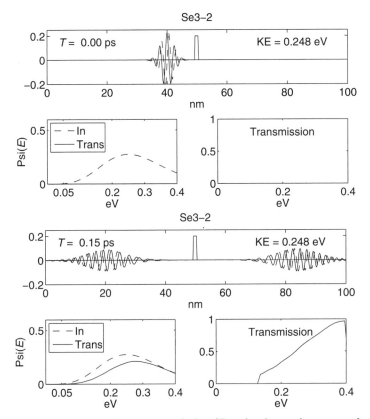

FIGURE 3.11 The calculation of transmission (Trans) using an input waveform centered at 0.25 eV.

above 0.2 eV have a good chance of penetrating the barrier. The transmission calculations of Figures 3.10 and 3.11 together have given us transmission over a range of 0–0.4 eV. This is what we were trying to calculate.

In order to confirm the accuracy of our transmission function, we conduct the simulation with a particle centered at one frequency, but with a broader envelope so its energy spectrum is narrower, as illustrated in Figure 3.12. This pulse is centered at 0.05 eV, and the transmission graph of Figure 3.10 shows us we should only expect about 3% of the waveform to penetrate the barrier. After 0.37 ps, we see that the part that made it through is indeed rather small. It is very easy to calculate this using the following parameter:

$$Trans = \int_{55\,nm}^{100\,nm} |\psi(x)|^2 \, dx. \tag{3.20}$$

Suppose we want to confirm transmission at a higher energy, say 0.3 eV. Figure 3.11 suggests that we should get about 75%. The simulation in Figure 3.13 confirms this.

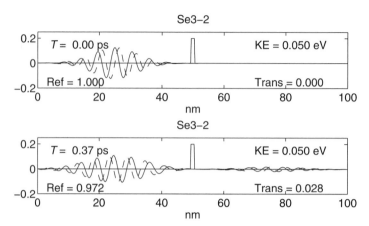

FIGURE 3.12 Transmission (Trans) is determined directly by calculating the fraction of the wave function to the right of the barrier for a particle at 0.05 eV.

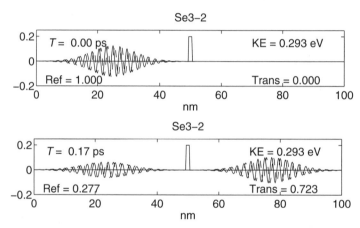

FIGURE 3.13 Transmission (Trans) is determined directly by calculating the fraction of the wave function to the right of the barrier for a particle at 0.3 eV.

A word of caution is in order. In this section, transmission is determined by first taking a spatial FFT and then converting the results to energy. In Section 3.1, we discussed how we could also take the time-domain FFT and convert frequency to energy. We will have occasion to do both. But for now, we are using the spatial transform.

3.5 SUMMARY

This chapter shows the role of Fourier analysis in quantum mechanics. Fourier analysis is the most expedient way of explaining concepts like broadened states and uncertainty. It is also at the heart of one of the key concepts in this book: quantum transport.

EXERCISES

3.1   The Fourier Transform

3.1.1   Prove the time-shifting theorem. Hint: The proof comes about directly from the definition of the Fourier transform in Equation (3.2).

3.1.2   Find the Fourier transforms of the following functions:
   (a) $y(x) = \exp(-3x^2)$
   (b) $f(x) = \sin(10^{-3}x)$
   (c) $g(\mathbf{r}) = \delta(x - x_0, y - y_0, z - z_0)$

3.1.3   Prove the following equation:

$$\lim_{\alpha \to 0} \frac{2\alpha}{\alpha^2 + (\omega - \omega_0)^2} = 2\pi\delta(\omega - \omega_0).$$

Hint: It is probably easier to take the term on the left back into the time domain, look at the limit as α goes to zero, and take this function back to the frequency domain.

3.1.4   Solve the following integral:

$$I = \int_{-\infty}^{\infty} e^{-0.5(t/\sigma)^2} \cos(\omega_o t) dt.$$

Hint: Use Fourier transforms.

3.1.5   Find the Fourier transform of

$$\psi(x) = \frac{1}{\sqrt{2\pi}\sigma} e^{\frac{1}{2}\left(\frac{x}{\sigma}\right)^2} e^{-ik_o x}.$$

3.2   Fourier Analysis and Available States

3.2.1   A one-dimensional well is found to have eigenergies given by

$$\varepsilon_n = 1\, n^2 \text{ eV}.$$

It is also found that after 100 revival periods, there is a 50% chance that the particle has decayed out of the well. Write a mathematical expression for the DOS. Recall that a revival period is the time it takes the phase of the complex waveform to return to its original position.

3.2.2   A particle is initiated in the ground state of the 10 nm well formed by two 0.1 eV barriers, as shown in Figure 3.14. As the simulation proceeds, the time-domain data is saved at the 25 nm point and

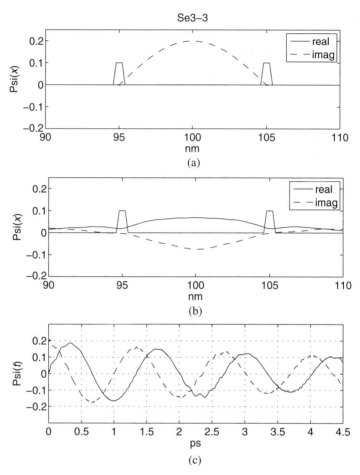

FIGURE 3.14 (a) A particle is initialized between two barriers. (b) The real and imaginary (imag) parts oscillate, similar to the particle in the 10 nm infinite well. (c) The time-domain data from the 100 nm point is saved and plotted after 4.5 ps. (bottom).

then plotted (bottom). For the ground state of the 100 Å infinite well, the DOS would just be

$$D(E) = \delta(E - 3.75 \text{ meV}).$$

What is it for the well in the simulation?

3.3 Uncertainty

3.3.1 Section 3.3 used the MATLAB program **uncer.m** to demonstrate time-frequency uncertainty. Alter this program to demonstrate position-wavelength uncertainty.

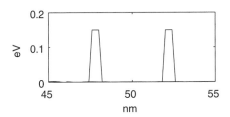

FIGURE 3.15   Each of the barriers is 0.15 eV high and 0.6 nm wide. The barrier centers are 4 nm apart.

3.4   Transmission via FFT

3.4.1   Using the program se3_2.m, calculate the transmission through the barrier in Figure 3.15.

Hints: Try repeating the results of Section 3.4 first to make sure the program is running! To do the simulation for the double barrier, you will need a larger problem space, so set NN = 1000. Once you have the transmission, do a couple verification runs, like Figure 3.11. You will find one very interesting characteristic that you will want to verify.

3.4.2   Describe a completely different approach to solve problems like those described in Exercise 3.4.1, but also using Fourier theory. (Hint: Look at Fig. 3.7 and think like an electrical engineer!)

REFERENCES

1. E. O. Brigham, *The Fast Fourier Transform and Its Applications*, Englewood Cliffs, NJ: Prentice Hall, 1988.
2. D. J. Griffiths, *Introduction to Quantum Mechanics*, Englewood Cliffs, NJ: Prentice Hall, 1994.
3. S. Datta, *Quantum Transport—Atom to Transistor*, Cambridge, UK: Cambridge University Press, 2005.

4

MATRIX ALGEBRA IN QUANTUM MECHANICS

The theoretical basis of quantum mechanics is rooted in linear algebra [1–3]. Therefore, matrices play a key role from both theoretical and practical viewpoints. In Section 4.1, we show that the state vector can be written as a column vector and the operators can be formulated as matrices. In Section 4.2, we explain how the Schrödinger equation can be formulated as a matrix equation. Specifically, once the Hamiltonian operator is written as a matrix, a simple MATLAB command can be used to find the corresponding eigenenergies and eigenstates. Section 4.3 describes the eigenspace representation, an important concept in quantum mechanics that is analogous to the Fourier representation in engineering. Section 4.4 discusses some of the formalism of quantum mechanics. At the end of this chapter, a short appendix reviews some of the basics of matrix algebra that are needed to understand this chapter.

4.1 VECTOR AND MATRIX REPRESENTATION

We begin by writing state variables as vectors and operators as matrices.

4.1.1 State Variables as Vectors

Let us return to our infinite well, but this time we look at a very small one, at least in terms of the number of points used to represent the well. Suppose we use only eight points, as shown in Figure 4.1.

Quantum Mechanics for Electrical Engineers, First Edition. Dennis M. Sullivan.
© 2012 The Institute of Electrical and Electronics Engineers, Inc.
Published 2012 by John Wiley & Sons, Inc.

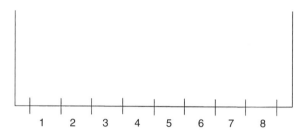

FIGURE 4.1   Infinite well with eight points.

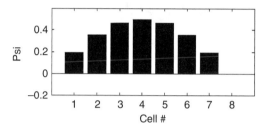

FIGURE 4.2   A function in an eight-point infinite well. The black bars are the real values at the various points, emphasizing that each point in the well can be considered a basis function.

Implicitly, cells 0 and 8 have $\psi = 0$ to form the edge of the well. Figure 4.2 shows a function that has been initialized in the well. This is the ground state eigenfunction, but the crudeness of our model makes it difficult to recognize.

The fact that we are writing this in a discrete form so that it can be computerized means we are actually approximating the function

$$\psi(x) \cong \psi(n \cdot \Delta x) = \sum_{n=1}^{8} \psi_n \delta(x - n \cdot \Delta x), \tag{4.1}$$

where Δx is the cell size. In this case, we are regarding the functions $\delta(x - n \cdot \Delta x)$ as *basis functions* used to construct the function $\psi(n \cdot \Delta x)$. The ψ_n are complex coefficients showing how much of each basis function is used in the construction process. Therefore, we can represent the function of Figure 4.2 as a vector consisting of eight complex elements representing the values. We can do this, because we know that $\psi_1 = 0.19$ is the value at cell number 1, $\psi_4 = 0.5$ is the value at cell number 4, and so on. A waveform like Equation (4.1) becomes a column vector:

$$|\psi\rangle = \begin{bmatrix} \psi_1 \\ \psi_2 \\ \psi_3 \\ \cdots \\ \psi_8 \end{bmatrix}. \tag{4.2}$$

Notice that we have written this as a ket. So the question is, what is a bra? The bra is the *transpose conjugate* of the ket. The term transpose indicates that the column is rewritten as a row and the term conjugate indicates that each element has been replaced with its complex conjugate,

$$\langle \psi | = [\psi_1^* \quad \psi_2^* \quad \cdots \quad \psi_8^*].\tag{4.3}$$

The next logical question is, "What is the inner product?" Obviously it is

$$\langle \psi | \psi \rangle = [\psi_1^* \quad \psi_2^* \quad \cdots \quad \psi_8^*] \begin{bmatrix} \psi_1 \\ \psi_2 \\ \cdots \\ \psi_8 \end{bmatrix} = \psi_1^* \psi_1 + \psi_2^* \psi_2 + \ldots + \psi_8^* \psi_8.$$

We know that this inner product sums to 1 if ψ is normalized. In general, the inner product of two vectors is

$$\langle \psi | \phi \rangle = [\psi_1^* \quad \psi_2^* \quad \cdots \quad \psi_8^*] \begin{bmatrix} \phi_1 \\ \phi_2 \\ \cdots \\ \phi_8 \end{bmatrix} = \psi_1^* \phi_1 + \psi_2^* \phi_2 + \ldots + \psi_8^* \phi_8,\tag{4.4}$$

which produces a complex number. Observe the following important property:

$$\langle \phi | \psi \rangle = \langle \psi | \phi \rangle^*.\tag{4.5}$$

The proof follows from the definition in Equation (4.4):

$$\langle \psi | \phi \rangle^* = \left(\psi_1^* \phi_1 + \psi_2^* \phi_2 + \ldots + \psi_8^* \phi_8 \right)^*$$
$$= \psi_1 \phi_1^* + \psi_2 \phi_2^* + \ldots + \psi_8 \phi_8^* = \langle \phi | \psi \rangle.$$

4.1.2   Operators as Matrices

Suppose that we want to calculate the kinetic energy of the function in Figure 4.2. The kinetic energy operator is

$$T = -\frac{\hbar^2}{2m} \frac{\partial^2}{\partial x^2}.$$

To apply this to our small, finite problem space, we must use the finite-difference approximation

$$T\psi_n = -\frac{\hbar^2}{2m} \left[\frac{\psi_{n-1} - 2\psi_n + \psi_{n+1}}{\Delta x^2} \right].$$

The Δx represents the size of the cells. Let us just suppose 1 nm. The m is the mass of the electron. We will just use the free space value. The parameter we need is

$$\chi_0 = \frac{\hbar^2}{2m \cdot \Delta x^2} = \frac{\left(1.055 \times 10^{-34}\,\text{J} \cdot \text{s}\right)^2}{2 \cdot 9.1 \times 10^{-31} \cdot \left(10^{-9}\right)^2\,\text{kg} \cdot \text{m}^2}$$

$$= \frac{1.113 \times 10^{-68}}{.182 \times 10^{-47}} = 6.115 \times 10^{-21}\,\text{J} \cdot \frac{1\,\text{eV}}{1.6 \times 10^{-19}\,\text{J}}$$

$$= 3.82 \times 10^{-2}\,\text{eV}.$$

Note that χ_0 is closely related to the ra that was used in Chapter 1:

$$ra = \frac{\Delta t}{\hbar}\chi_0.$$

We can write this equation as

$$T\psi_n = -\chi_0 \cdot \left(\psi_{n-1} - 2\psi_n + \psi_{n+1}\right).$$

Now we can write the kinetic energy operator as a matrix

$$\mathbf{T} = -\chi_0 \begin{bmatrix} -2 & 1 & 0 & 0 & 0 & 0 & 0 & 0 \\ 1 & -2 & 1 & 0 & 0 & 0 & 0 & 0 \\ 0 & 1 & -2 & 1 & 0 & 0 & 0 & 0 \\ 0 & 0 & 1 & -2 & 1 & 0 & 0 & 0 \\ 0 & 0 & 0 & 1 & -2 & 1 & 0 & 0 \\ 0 & 0 & 0 & 0 & 1 & -2 & 1 & 0 \\ 0 & 0 & 0 & 0 & 0 & 1 & -2 & 1 \\ 0 & 0 & 0 & 0 & 0 & 0 & 1 & -2 \end{bmatrix}. \tag{4.6}$$

This 8×8 kinetic energy matrix $\mathbf{T}$ can operate on the 1×8 column vector $|\psi\rangle$:

$$\mathbf{T}|\psi\rangle = -\chi_0 \begin{bmatrix} -2 & 1 & 0 & 0 & 0 & 0 & 0 & 0 \\ 1 & -2 & 1 & 0 & 0 & 0 & 0 & 0 \\ 0 & 1 & -2 & 1 & 0 & 0 & 0 & 0 \\ 0 & 0 & 1 & -2 & 1 & 0 & 0 & 0 \\ 0 & 0 & 0 & 1 & -2 & 1 & 0 & 0 \\ 0 & 0 & 0 & 0 & 1 & -2 & 1 & 0 \\ 0 & 0 & 0 & 0 & 0 & 1 & -2 & 1 \\ 0 & 0 & 0 & 0 & 0 & 0 & 1 & -2 \end{bmatrix} \begin{bmatrix} \psi_1 \\ \psi_2 \\ \psi_3 \\ \psi_4 \\ \psi_5 \\ \psi_6 \\ \psi_7 \\ \psi_8 \end{bmatrix} = |\psi'\rangle = \begin{bmatrix} \psi_1^t \\ \psi_2^t \\ \psi_3^t \\ \psi_4^t \\ \psi_5^t \\ \psi_6^t \\ \psi_7^t \\ \psi_8^t \end{bmatrix}.$$

Notice that the $|\psi^t\rangle$ is a 1×8 column vector, just like $|\psi\rangle$. In the parlance of linear algebra, we say that the vector $|\psi\rangle$ has been *transformed* to the vector ψ_n^t using the *transformation* **T**.

Recall that the expectation value of the kinetic energy is calculated by

$$\langle T \rangle = \int_{-\infty}^{\infty} \psi^*(x) \cdot \left(-\frac{\hbar^2}{2m} \frac{\partial^2}{\partial x^2} \right) \psi(x)\, dx,$$

which in our eight-point infinite well we write as

$$\langle T \rangle = \sum_{n=1}^{8} \psi_n^* \cdot T\psi_n. \tag{4.7}$$

This can be calculated in matrix notation by

$$\langle T \rangle = \langle \psi | T\psi \rangle = \langle \psi | T | \psi \rangle$$
$$= \begin{bmatrix} \psi_1^* & \psi_2^* & \cdots & \psi_8^* \end{bmatrix} \begin{bmatrix} \psi_1^t \\ \psi_2^t \\ \cdots \\ \psi_8^t \end{bmatrix} = \begin{bmatrix} \psi_1^* \psi_1^t + \psi_2^* \psi_2^t + \ldots + \psi_8^* \psi_8^t \end{bmatrix}. \tag{4.8}$$

Therefore, $\langle T \rangle$ is the inner product of a row and a column, which gives a number as we would expect. Please see the appendix on matrix algebra at the end of this chapter.

Equation (4.8) can be generalized to

$$\langle \phi | A | \psi \rangle = \begin{bmatrix} \phi_1^* & \phi_2^* & \cdots & \phi_N^* \end{bmatrix} \begin{bmatrix} \psi_1^a \\ \psi_2^a \\ \cdots \\ \psi_N^a \end{bmatrix} = \phi_1^* \psi_1^a + \phi_2^* \psi_2^a + \ldots + \phi_N^* \psi_N^a, \tag{4.9}$$

where **A** is a matrix transformation and ψ_n^a represent the elements of $|\psi\rangle$ that have been transformed by **A**.

Notice the following, which will prove useful:

$$\langle A\psi | \phi \rangle = \begin{bmatrix} \psi_1^{a*} & \psi_2^{a*} & \cdots & \psi_N^{a*} \end{bmatrix} \begin{bmatrix} \phi_1 \\ \phi_2 \\ \cdots \\ \phi_N \end{bmatrix} \tag{4.10}$$
$$= \phi_1 \psi_1^{a*} + \phi_2 \psi_2^{a*} + \ldots + \phi_N \psi_N^{a*} = \langle \phi | A | \psi \rangle^*.$$

4.2 MATRIX REPRESENTATION OF THE HAMILTONIAN

To determine the matrix representation of the Hamiltonian, let us first return to the time-independent Schrödinger equation:

$$
\begin{aligned}
E\psi(x) &= -\frac{\hbar^2}{2m}\frac{\partial^2}{\partial x^2}\psi(x) + V(x)\psi(x) \\
&= H\psi(x).
\end{aligned}
\tag{4.11}
$$

The Hamiltonian consists of a kinetic energy term and potential energy term,

$$
H = T + V = -\frac{\hbar^2}{2m}\frac{\partial^2}{\partial x^2} + V.
\tag{4.12}
$$

The finite-difference implementation is

$$
H\psi_n = -\frac{\hbar^2}{2m}\left[\frac{\psi_{n-1} - 2\psi_n + \psi_{n+1}}{\Delta x^2}\right] + V_n\psi_n.
\tag{4.13}
$$

Once again, we assume Δx equals 1 nm and use the constant

$$
\chi_0 = \frac{\hbar^2}{2m \cdot \Delta x^2} = 3.82 \times 10^{-2} \text{ eV}.
$$

Thus Equation (4.13) becomes

$$
H\psi_n = -\chi_0 \cdot \psi_{n-1} + (2 \cdot \chi_0 + V_n)\psi_n - \chi_0 \cdot \psi_{n+1},
\tag{4.14}
$$

where V_n is the value of the potential in the nth cell. We can write the Hamiltonian matrix as:

$$
\mathbf{H} =
\begin{bmatrix}
2\chi_0 + V_1 & -\chi_0 & 0 & 0 & 0 & 0 & 0 & 0 \\
-\chi_0 & 2\chi_0 + V_2 & -\chi_0 & 0 & 0 & 0 & 0 & 0 \\
0 & -\chi_0 & 2\chi_0 + V_3 & -\chi_0 & 0 & 0 & 0 & 0 \\
0 & 0 & -\chi_0 & 2\chi_0 + V_4 & -\chi_0 & 0 & 0 & 0 \\
0 & 0 & 0 & -\chi_0 & 2\chi_0 + V_5 & -\chi_0 & 0 & 0 \\
0 & 0 & 0 & 0 & -\chi_0 & 2\chi_0 + V_6 & -\chi_0 & 0 \\
0 & 0 & 0 & 0 & 0 & -\chi_0 & 2\chi_0 + V_7 & -\chi_0 \\
0 & 0 & 0 & 0 & 0 & 0 & -\chi_0 & 2\chi_0 + V_8
\end{bmatrix}
\tag{4.15}
$$

4.2.1 Finding the Eigenvalues and Eigenvectors of a Matrix

Equation (4.11) can be written in the matrix notation:

$$H|\psi\rangle = E|\psi\rangle. \tag{4.16}$$

We know that H is the 8×8 matrix given by Equation (4.15) and $|\psi\rangle$ is a 1×8 column vector, similar to Equation (4.2). Does Equation (4.16) look familiar? Suppose we rewrite it as:

$$\mathbf{Hx} = \lambda\mathbf{x}.$$

We recall from the study of linear algebra that if $\mathbf{H}$ is a square matrix, then it has eigenvalues λ, and associated eigenvectors [4, 5]. We have been saying that for each quantum operator there are associated eigenenergies and corresponding eigenfunctions. We see that this is consistent with matrix algebra.

The eigenfunctions and eigenenergies of this matrix can be found from the MATLAB command

$$[\text{phi, D}] = \text{eig}(H). \tag{4.17}$$

"D" is a diagonal matrix with the eigenvalues on the diagonal; "phi" is a matrix that contains the eigenfunctions in the columns. Figure 4.3 displays a very simple potential V, the eigenenergies, and the eigenfunctions.

4.2.2 A Well with Periodic Boundary Conditions

Next we would like to look at the case where the last cell is attached to the first (Fig. 4.4). There is no potential associated with this structure. (This may

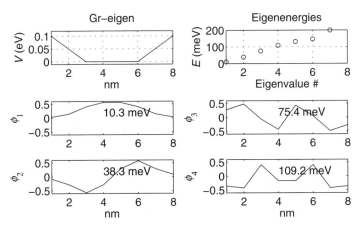

FIGURE 4.3 An infinite well with an added potential, as shown in the upper left plot. The eigenenergies are in the upper right plot, and the first four eigenfunctions are plotted below.

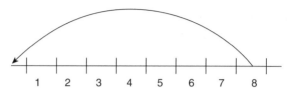

FIGURE 4.4   Eight-point problem space with periodic boundary conditions.

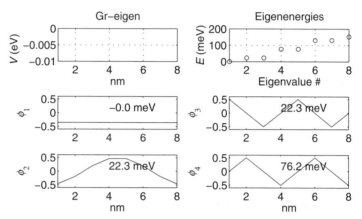

FIGURE 4.5   The eigenfunctions and eigenenergies of the eight-cell structure with periodic boundary conditions. Notice that the lowest eigenenergy, corresponding to the zero energy, is zero, because it does not correspond to a real eigenfunction. The other eigenfunctions come in pairs.

seem like an unusual construction, but it is often used to describe electron propagation in solids leading to Bloch's theorem [1, 3].) For the time being, let us just construct the Hamiltonian matrix, as shown in Equation (4.18). We can now use our MATLAB program to find the eigenvalues and their corresponding eigenfunctions (Fig. 4.5).

$$
\mathbf{H} = \chi_0
\begin{bmatrix}
2 & -1 & 0 & 0 & 0 & 0 & 0 & -1 \\
-1 & 2 & -1 & 0 & 0 & 0 & 0 & 0 \\
0 & -1 & 2 & -1 & 0 & 0 & 0 & 0 \\
0 & 0 & -1 & 2 & -1 & 0 & 0 & 0 \\
0 & 0 & 0 & -1 & 2 & -1 & 0 & 0 \\
0 & 0 & 0 & 0 & -1 & 2 & -1 & 0 \\
0 & 0 & 0 & 0 & 0 & -1 & 2 & -1 \\
-1 & 0 & 0 & 0 & 0 & 0 & -1 & 2
\end{bmatrix}.
\tag{4.18}
$$

Figure 4.5 is slightly different from our previous examples. In Figure 4.5, most of the eigenfunctions come in pairs—one a sine, the other a cosine of the same

energy. When two orthogonal eigenfunctions have the same energy, we refer
to them as *degenerate*. The lowest energy is $E = 0$, and the waveform is flat.
The highest energy at 0.153 eV (not shown) is nondegenerate. However, all of
the eigenfunctions are orthonormal.

Similar to the way we solved the problem of the infinite well, we can come
up with an analytic solution to the well with the periodic boundary condition.
It has to solve the Schrödinger equation

$$E\psi(x) = -\frac{\hbar^2}{2m}\frac{\partial^2}{\partial x^2}\psi(x), \tag{4.19a}$$

subject to the condition

$$\psi(0) = \psi(8). \tag{4.19b}$$

As before, we can rewrite it as

$$\frac{\partial^2}{\partial x^2}\psi(x) = -k^2\psi(x), \tag{4.20a}$$

$$k = \sqrt{\frac{2mE}{\hbar^2}}. \tag{4.20b}$$

Solutions are of the form

$$\psi(x) = \begin{cases} A\cos kx & ka = 0,\pm2\pi,\pm4\pi,\ldots \\ B\sin kx & ka = \pm2\pi,\pm4\pi,\ldots, \end{cases} \tag{4.21}$$

where A and B are constants. The corresponding energies are

$$E_n = \frac{(n2\pi)^2\hbar^2}{2ma^2} = \frac{2}{m}\left(\frac{n\pi\hbar}{a}\right)^2. \tag{4.22}$$

Note that we can construct a new set of orthonormal functions from the functions above:

$$\psi_0 = C_0,$$

$$\psi_{\pm1} = C_1 e^{\pm ik_1 x},$$

$$\psi_{\pm2} = C_2 e^{\pm ik_2 x},$$

$$\psi_{\pm3} = C_3 e^{\pm ik_3 x},$$

$$\psi_4 = B_4\sin k_4 x.$$

Remember that ψ_0 is a constant, and ψ_4 is nondegenerate, so these two functions cannot be made into complex conjugate pairs.

4.2.3 The Harmonic Oscillator

One of the important canonical structures in quantum mechanics is the harmonic oscillator [1, 2], which has a potential given by

$$V(x) = \frac{1}{2}m\left(\frac{E_{ref}}{\hbar}\right)^2 x^2. \tag{4.23}$$

E_{ref} is a reference energy related to the system. If we put this potential into the time-independent Schrödinger equation we get

$$\left[-\frac{\hbar^2}{2m}\frac{\partial^2}{\partial x^2} + \frac{1}{2}m\left(\frac{E_{ref}}{\hbar}\right)^2 x^2\right]\psi(x) = E\psi(x). \tag{4.24}$$

This has analytic solutions that will be described in Chapter 11. Of course, our method is to put this new Hamiltonian operator in a matrix, and use MATLAB to give us the eigenvalues and corresponding eigenfunctions (Fig. 4.6). The Hamiltonian matrix will be similar to Equation (4.15). The V values will contain the potential in Equation (4.23).

The eigenenergies of the harmonic oscillator are

$$E_n = \left(n + \frac{1}{2}\right)E_{ref} \quad n = 0, 1, 2, 3, \dots. \tag{4.25}$$

Therefore, the ground state energy, the lowest energy, is

$$E_0 = \frac{1}{2}E_{ref}. \tag{4.26}$$

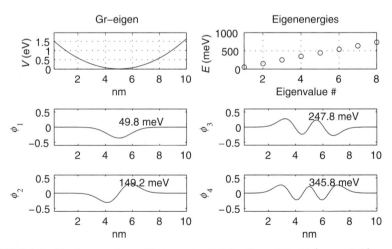

FIGURE 4.6 The harmonic oscillator potential for E_{ref} = 0.1 eV (upper left) and the corresponding eigenenergies (upper right).

Figure 4.6 shows the potential, eigenenergies, and the first six eigenfunctions for a harmonic oscillator potential with $E_0 = 0.1$ eV. This simulation used a 100×100 H matrix and cells of 0.1 nm.

The harmonic oscillator is used as a model for many advanced topics in quantum mechanics [2, 3]. Some of these topics are discussed in Chapter 11.

4.3 THE EIGENSPACE REPRESENTATION

We stated earlier that the *conjugate transpose* of a column vector is a row vector of the same size with the elements replaced with their complex conjugates. Similarly, the conjugate transpose of a matrix means that the rows and column are interchanged, and each element is replaced with its complex conjugate. For example, if $\mathbf{\Phi}$ is a matrix composed of N column vectors ϕ_n

$$\mathbf{\Phi} = [\phi_1 \quad \phi_2 \quad \ldots \quad \phi_N], \tag{4.27a}$$

then the transpose conjugate of $\mathbf{\Phi}$ is

$$\mathbf{\Phi}^+ = \begin{bmatrix} \phi_1^+ \\ \phi_2^+ \\ \ldots \\ \phi_N^+ \end{bmatrix}, \tag{4.27b}$$

where each ϕ_n^+ is a row vector whose elements are the complex conjugates of the corresponding column vector ϕ_n.

We know that the eigenfunctions of an infinite well are sine waves like those shown in Figure 2.2. Note that even though an infinite well has, in theory, an infinite number of eigenfunctions, our eight-point well has only eight.

In Chapter 2, we took the Fourier expansion of waveforms using the sine functions, which were the eigenfunctions of the infinite well, ϕ_n. The coefficients are calculated by

$$c_n = \langle \phi_n | \psi \rangle \tag{4.28a}$$

and the waveform can be reconstructed from:

$$\psi = \sum_{n=1}^{N} c_n \varphi_n. \tag{4.28b}$$

It is desirable to make the calculations of Equation (4.28) using the matrix formalism. To get the c_n values, we put the ψ values in a column vector which we write as a ket:

$$|\psi\rangle = \begin{bmatrix} \psi(1) \\ \psi(2) \\ \cdots \\ \psi(N) \end{bmatrix},$$

and multiply by the transpose conjugate matrix $\mathbf{\Phi}^+$ in Equation (4.27b),

$$\mathbf{\Phi}^+|\psi\rangle = \begin{bmatrix} \phi_1^+ \\ \phi_2^+ \\ \cdots \\ \phi_8^+ \end{bmatrix}\begin{bmatrix} \psi(1) \\ \psi(2) \\ \cdots \\ \psi(N) \end{bmatrix} = \begin{bmatrix} c_1 \\ c_2 \\ \cdots \\ c_N \end{bmatrix} = |c\rangle, \tag{4.29a}$$

which is equivalent to Equation (4.28a). Similarly, if we have the c values in the ket $|c\rangle$, we can reconstruct the function ψ from

$$\mathbf{\Phi}|c\rangle = \begin{bmatrix} \phi_1 & \phi_2 & \cdots & \phi_N \end{bmatrix}\begin{bmatrix} c_1 \\ c_2 \\ \cdots \\ c_N \end{bmatrix} = \begin{bmatrix} \psi(1) \\ \psi(2) \\ \cdots \\ \psi(N) \end{bmatrix} = |\psi\rangle, \tag{4.29b}$$

which is equivalent to Equation (4.28b). If we start with Equation (4.29b) and multiply from the right by $\mathbf{\Phi}^{-1}$, we get

$$\mathbf{\Phi}^{-1}|\psi\rangle = \mathbf{\Phi}^{-1}\mathbf{\Phi}|c\rangle = |c\rangle.$$

Comparing this to Equation (4.29a), we conclude

$$\mathbf{\Phi}^{-1} = \mathbf{\Phi}^+. \tag{4.30}$$

A matrix whose conjugate transpose is equal to its inverse is said to be *unitary*. Since $\mathbf{\Phi}$ consists of orthonormal column vectors, Equation (4.30) is easy to prove:

$$\mathbf{\Phi}^+\mathbf{\Phi} = \begin{bmatrix} \phi_1^+ \\ \phi_2^+ \\ \cdots \\ \phi_N^+ \end{bmatrix}\begin{bmatrix} \phi_1 & \phi_2 & \cdots & \phi_N \end{bmatrix}$$

$$= \begin{bmatrix} \phi_1^+\phi_1 & \phi_1^+\phi_2 & \cdots \\ \phi_1^+\phi_2 & \phi_2^+\phi_2 & \\ \cdots & \cdots & \cdots \\ & & \cdots & \phi_N^+\phi_N \end{bmatrix} = \begin{bmatrix} 1 & 0 & & \\ 0 & 1 & 0 & \\ & 0 & \cdots & 0 \\ & & 0 & 1 \end{bmatrix} = \mathbf{I}.$$

This can only be true if $\mathbf{\Phi}^+ = \mathbf{\Phi}^{-1}$.

In general, we say that

1. $|\psi\rangle$ is the *real space representation*, and
2. $|c\rangle$ is the *eigenfunction* or *eigenspace* representation.

Note that Equation (4.28), or their matrix counterparts in Equation (4.29) forms transform pairs similar to the forward and inverse Fourier transform.

We showed in Section 4.2 that we can write the Hamiltonian as a matrix, as in Equation (4.15). Look at the following matrix operation:

$$
\begin{aligned}
\mathbf{\Phi}^+\mathbf{H}\mathbf{\Phi} &= \mathbf{\Phi}^+\mathbf{H}\begin{bmatrix} \phi_1 & \phi_2 & \cdots & \phi_N \end{bmatrix} \\
&= \mathbf{\Phi}^+\begin{bmatrix} \varepsilon_1\phi_1 & \varepsilon_2\phi_2 & \cdots & \varepsilon_N\phi_N \end{bmatrix} \\
&= \begin{bmatrix} \varepsilon_1 & 0 & & \\ 0 & \varepsilon_2 & 0 & \\ & 0 & \cdots & 0 \\ & & 0 & \varepsilon_N \end{bmatrix} = \mathbf{H}^{\mathbf{e}}.
\end{aligned}
\tag{4.31}
$$

This is the *eigenfunction representation* of the Hamiltonian. We say that the unitary matrix *diagonalizes* the matrix $\mathbf{H}$.

If a matrix H is *diagonalizable*, then its eigenvalues ε_n have the following two important properties:

$$
\det(\mathbf{H}) = \varepsilon_1\varepsilon_2 \ldots \varepsilon_N,
\tag{4.32a}
$$

$$
Tr(\mathbf{H}) = \varepsilon_1 + \varepsilon_2 + \ldots + \varepsilon_N.
\tag{4.32b}
$$

4.4 FORMALISM

The formalism of quantum mechanics is rooted in linear algebra [1–3]. We have chosen to use matrix algebra to introduce these concepts. In this section, we want to state some of the formal definitions and theorems that are at the heart of quantum mechanics.

4.4.1 Hermitian Operators

We know that in quantum mechanics, we calculate properties through operators, like T for kinetic energy and $H = T + V$ for the Hamiltonian. It is particularly desirable that these operators be *Hermitian*. In matrix operators, this means

$$
\mathbf{A}^+ = \mathbf{A}.
$$

The more general definition, in Dirac notation, is

$$\langle \psi_1 | A \psi_2 \rangle = \langle A \psi_1 | \psi_2 \rangle. \tag{4.33}$$

What does this mean? If ψ_1 and ψ_2 are two state variables and A is a Hermitian operator, then Equation (4.33) means

$$\int_{-\infty}^{\infty} \psi_1^*(x) \cdot A \psi_2(x) \, dx = \int_{-\infty}^{\infty} (A \psi_1(x))^* \psi_2(x) \, dx. \tag{4.34}$$

With this definition of Hermitian operators in hand, we state the following three theorems at the heart of quantum mechanics [1]:

1. The eigenvalues of a Hermitian operator are real.
2. The eigenvectors of a Hermitian operator belonging to distinct eigenvalues are orthonormal.
3. The eigenvectors of a Hermitian operator span the space.

The proofs of the above statements are left as exercises.

4.4.2   Function Spaces

A function space is a formal generalization of a vector space [1]. The rectangular coordinate system that we deal with in everyday life has *basis vectors* $\hat{\mathbf{x}}, \hat{\mathbf{y}}, \hat{\mathbf{z}}$ that span the space. This means that every other vector in this space can be written as a superposition of these three basis vectors:

$$\mathbf{w} = \alpha \hat{\mathbf{x}} + \beta \hat{\mathbf{y}} + \gamma \hat{\mathbf{z}}.$$

While we are dealing with finite-dimensional spaces, we say the space is also complete, meaning any vector can be described as a superposition of the basis vectors. Assuming vectors like $\mathbf{w}$ are column vectors, we can define the inner product of any two vectors in the space

$$\mathbf{v}^+ \cdot \mathbf{w} = \chi,$$

where χ is a number.

A *Hilbert space* is a complete inner product space [1]. It is the set of all square-integrable functions, sometimes given the designation L_2. This level of formalism will rarely concern us in this book. Just remember that quantum mechanics is defined in Hilbert space.

We have seen that the eigenfunctions and the corresponding eigenvalues of Hermitian operators are of particular importance. We have stated that eigenvalues of a Hermitian operator are real. The *spectrum* is the set of eigenvalues of a given operator.

APPENDIX: REVIEW OF MATRIX ALGEBRA

There are many different notations that are used in matrix algebra. Throughout this text, the transpose conjugate of a matrix or vector is given by $\mathbf{A}^+$ or $\mathbf{b}^+$, respectively. However, MATLAB uses the commands A' or b'. For a complex conjugate of a number, w, we use w^* in this text. MATLAB also uses w' for the complex conjugate of a number.

The *conjugate* of a matrix replaces each element with its complex conjugate.

If

$$\mathbf{A} = \begin{bmatrix} 2 & 1+i \\ 3-i2 & 4 \end{bmatrix},$$

then

$$conj(\mathbf{A}) = \begin{bmatrix} 2 & 1-i \\ 3+2i & 4 \end{bmatrix}.$$

The *transpose* means the columns and rows are interchanged. $\mathbf{A}^+$ means the *transpose conjugate* of A:

$$\mathbf{A}^+ = \begin{bmatrix} 2 & 3+i2 \\ 1-i & 4 \end{bmatrix}.$$

If x is a column vector, $\mathbf{x}^+$ is the corresponding conjugate row vector.

If

$$\mathbf{x} = \begin{bmatrix} 2 \\ i \\ 1+i2 \end{bmatrix},$$

then

$$\mathbf{x}^+ = \begin{bmatrix} 2 & -i & 1-2i \end{bmatrix}.$$

A very important property is the following:

$$(\mathbf{Ax})^+ = \mathbf{x}^+\mathbf{A}^+.$$

Proof. If $\mathbf{A}$ is made up of a group of row vectors

$$\mathbf{A} = \begin{bmatrix} \mathbf{a}_1 \\ \mathbf{a}_2 \\ \cdots \\ \mathbf{a}_N \end{bmatrix},$$

and **x** is a column vector, then

$$\mathbf{Ax} = \begin{bmatrix} \mathbf{a_1} \\ \mathbf{a_2} \\ \cdots \\ \mathbf{a_N} \end{bmatrix} \mathbf{x} = \begin{bmatrix} \mathbf{a_1 \cdot x} \\ \mathbf{a_2 \cdot x} \\ \cdots \\ \mathbf{a_N \cdot x} \end{bmatrix},$$

where $\mathbf{a_i \cdot x}$ is an inner product. Notice also that

$$(\mathbf{Ax})^* = \begin{bmatrix} (\mathbf{a_1 \cdot x})^* & (\mathbf{a_2 \cdot x})^* & \cdots & (\mathbf{a_N \cdot x})^* \end{bmatrix}.$$

It is not difficult to see that $\mathbf{x}^* \mathbf{A}^*$ gives the same result.

The inverse of a matrix **A**, has the property

$$\mathbf{A} \cdot inv(\mathbf{A}) = \mathbf{I},$$

where *I* is the *identity matrix*. The identity matrix has ones on the diagonal and zeros everywhere else. In MATLAB:

$$eye(3) = \begin{bmatrix} 1 & 0 & 0 \\ 0 & 1 & 0 \\ 0 & 0 & 1 \end{bmatrix}.$$

There are two types of matrices that play a very significant role in quantum mechanics:

1. A matrix is *unitary* if its inverse is equal to its conjugate transpose

$$\mathbf{A}^{-1} = \mathbf{A}^*.$$

   This is significant in using matrices to transform from one representation to another.

2. A matrix is *Hermitian* if it is equal to its conjugate transpose

$$\mathbf{A} = \mathbf{A}^*.$$

This is significant when the matrix is an operator. Hermitian matrices have real eigenvalues and their eigenvectors span the space.

Note the following, which is very helpful for the exercises. If **A** is Hermitian, then for any two vectors **x** and **y**:

$$\mathbf{x}^* \mathbf{Ay} = \mathbf{x}^* \mathbf{A}^* \mathbf{y} = (\mathbf{Ax})^* \mathbf{y}.$$

The *trace* of a matrix is the sum of its diagonal elements, for example,

$$\mathbf{A} = \begin{bmatrix} 1 & 2 \\ 3 & 4 \end{bmatrix},$$
$$Tr(\mathbf{A}) = 1 + 4 = 5.$$

Example. Find the eigenfunctions and eigenvalues of the matrix

$$\mathbf{A} = \begin{bmatrix} 2 & -i \\ i & 2 \end{bmatrix}.$$

Solution. Clearly $\mathbf{A}^+ = \mathbf{A}$, so it is Hermitian. First find the eigenvalues

$$|\mathbf{A} - \lambda\mathbf{I}| = \begin{vmatrix} 2-\lambda & -i \\ i & 2-\lambda \end{vmatrix}$$
$$= (2-\lambda)^2 - 1 = \lambda^2 - 4\lambda + 3 = (\lambda-1)(\lambda-3) = 0.$$

So the eigenvalues are $\lambda_1 = 1$ and $\lambda_2 = 3$. Now find the eigenvectors:
For $\lambda_1 = 1$,

$$\begin{bmatrix} 2 & -i \\ i & 2 \end{bmatrix}\begin{bmatrix} \alpha \\ \beta \end{bmatrix} = (1)\begin{bmatrix} \alpha \\ \beta \end{bmatrix}.$$

One solution is $\alpha = i\beta$. A normalized solution is:

$$\mathbf{x_1} = \begin{bmatrix} i/\sqrt{2} \\ 1/\sqrt{2} \end{bmatrix}.$$

For $\lambda_2 = 3$,

$$\begin{bmatrix} 2 & -i \\ i & 2 \end{bmatrix}\begin{bmatrix} \alpha \\ \beta \end{bmatrix} = (3)\begin{bmatrix} \alpha \\ \beta \end{bmatrix}.$$

One solution is $\alpha = -i\beta$. A normalized solution is:

$$\mathbf{x_2} = \begin{bmatrix} -i/\sqrt{2} \\ 1/\sqrt{2} \end{bmatrix}.$$

The eigenvectors are orthonormal (Remember, in taking the inner product, use the complex transpose of the first vector!):

$$\mathbf{x_1^+ \cdot x_1} = \begin{bmatrix} -i/\sqrt{2} & 1/\sqrt{2} \end{bmatrix}\begin{bmatrix} i/\sqrt{2} \\ 1/\sqrt{2} \end{bmatrix} = \frac{1}{2} + \frac{1}{2} = 1,$$

$$\mathbf{x}_1^{\dagger} \cdot \mathbf{x}_2 = \begin{bmatrix} -i/\sqrt{2} & 1/\sqrt{2} \end{bmatrix} \begin{bmatrix} -i/\sqrt{2} \\ 1/\sqrt{2} \end{bmatrix} = -\frac{1}{2} + \frac{1}{2} = 0.$$

We can now form the matrix

$$\Phi = \begin{bmatrix} x_1 & x_2 \end{bmatrix} = \begin{bmatrix} i/\sqrt{2} & -i/\sqrt{2} \\ 1/\sqrt{2} & 1/\sqrt{2} \end{bmatrix}.$$

We can rewrite $\mathbf{A}$ in the eigenspace representation,

$$\mathbf{A}^{e} = \Phi^{\dagger} \mathbf{A} \Phi$$

$$= \begin{bmatrix} -i/\sqrt{2} & 1/\sqrt{2} \\ i/\sqrt{2} & 1/\sqrt{2} \end{bmatrix} \begin{bmatrix} 2 & -i \\ i & 2 \end{bmatrix} \begin{bmatrix} i/\sqrt{2} & -i/\sqrt{2} \\ 1/\sqrt{2} & 1/\sqrt{2} \end{bmatrix} = \begin{bmatrix} 1 & 0 \\ 0 & 3 \end{bmatrix}.$$

The *trace* of a matrix is the sum of the diagonal elements. Notice that it remains the same in either representation:

$$Tr(\mathbf{A}) = 2 + 2 = 4,$$

$$Tr(\mathbf{A}^{e}) = 1 + 3 = 4.$$

Similarly, the determinant is preserved in going from one representation to another:

$$|\mathbf{A}| = \begin{vmatrix} 2 & -i \\ i & 2 \end{vmatrix} = 4 - 1 = 3,$$

$$|\mathbf{A}^{e}| = \begin{vmatrix} 1 & 0 \\ 0 & 3 \end{vmatrix} = 3 - 0 = 3.$$

This is all calculated very rapidly in MATLAB by

$$[\text{phi, D}] = \text{eig}(A);$$

$$AE = \text{phi'}*A*\text{phi};$$

EXERCISES

4.1　Vector and Matrix Representation

　　4.1.1　Write a momentum operator matrix P similar to the kinetic energy operator matrix T in Equation (4.6) for the eight-point infinite well. Make sure your matrix operator is *Hermitian*.

4.2 Matrix Representation of the Hamiltonian

4.2.1 Show how MATLAB code could use the matrix formulation of the Hamiltonian in Equation (4.15) to implement the finite-difference time-domain (FDTD) formulation of the Schrödinger equation, similar to the program se_1.m.

4.3 The Eigenspace Representation

4.3.1 Find the matrix S that transforms an N-point vector from one representation in one set of eigenfunctions, say v_i, to another set f_i. Assume both are also length N.

4.3.2 For the matrix

$$\sigma_y = \frac{1}{2}\begin{bmatrix} 0 & -i \\ i & 0 \end{bmatrix},$$

do the following by hand, similar to the example given.

(a) Find the eigenvalues and normalized eigenfunctions.

(b) Show that the eigenfunctions are orthonormal.

(c) Find the matrix V that will transform σ_y into the eigenvalue representation.

4.3.3 Look at the matrix A:

$$A = \begin{bmatrix} 2 & 1 & -i \\ 1 & 3 & i \\ i & -i & 2 \end{bmatrix}.$$

(a) Is this matrix Hermitian?

(b) Find the eigenvalues and corresponding eigenvectors. (You would probably want to use MATLAB to do this!)

(c) Find the matrices that will transform A to the eigenfunction representation.

4.3.4 A two-state system can be described by

$$\psi = a_1\phi_1 + a_2\phi_1$$

or

$$\psi = b_1\theta_1 + b_2\theta_1,$$

where the ϕs and θs are eigenstates and as and bs are coefficients. Determine the 2×2 matrix that will take ψ from the ϕ basis to the θ basis, that is, find **M** so

$$\mathbf{b} = \mathbf{Ma},$$

where

$$\mathbf{a} = \begin{bmatrix} a_1 \\ a_2 \end{bmatrix}, \quad \mathbf{b} = \begin{bmatrix} b_1 \\ b_2 \end{bmatrix}.$$

4.4   Formalism

4.4.1   Show that the rows and columns of a unitary matrix are orthonormal sets.

4.4.2   Show that the eigenvalues of a Hermitian matrix must be real.

4.4.3   Show that the eigenvectors of a Hermitian matrix corresponding to distinct eigenvalues are orthogonal.

REFERENCES

1. D. J. Griffiths, *Introduction to Quantum Mechanics*, Englewood Cliffs, NJ; Prentice Hall, 1994.

2. D. A. B. Miller, *Quantum Mechanics for Scientists and Engineers*, Cambridge, UK: Cambridge University Press, 2008.

3. A. Yariv, *Quantum Electronics*, New York: John Wiley and Sons, 1989.

4. D. C. Lay, *Linear Algebra and Its Applications*, New York: Addison-Wesley, 2002.

5. R. C. Penney, *Linear Algebra Ideas and Applications*, 3rd ed., New York: John Wiley and Sons, 2006.

5

A BRIEF INTRODUCTION TO STATISTICAL MECHANICS

Statistical mechanics, the study of the occupation of quantum levels using probability theory [1], plays an essential role in quantum semiconductor theory. Two important quantities, the density of states and the probability distribution, are essential in quantum transport [2]. In Section 5.1, we develop the concept of density of states in one, two, and three dimensions, starting with an understanding of the eigenstates in an infinite well. In Section 5.2, we discuss probability distributions. Because this concept is foreign to most engineers, we begin with some simple heuristic examples to build some intuition for the subject. We eventually describe the Fermi–Dirac distribution that gives the probability of occupation of a state as a function of energy and temperature. This is a fundamental concept in semiconductor physics.

5.1 DENSITY OF STATES

Ultimately, we are going to be interested in how a device conducts electric current. Current is the flow of charge in the form of electrons and holes, although we will limit our discussion to electrons. We have seen that electrons in structures are restricted to certain states, dictated by the physics of the structure. Therefore, before we can discuss current flow, we have to think about the available states for transport.

Quantum Mechanics for Electrical Engineers, First Edition. Dennis M. Sullivan.
© 2012 The Institute of Electrical and Electronics Engineers, Inc.
Published 2012 by John Wiley & Sons, Inc.

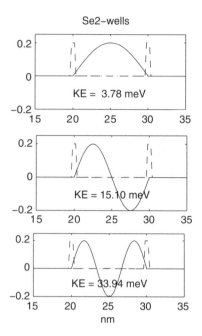

FIGURE 5.1 The three lowest energy eigenfunctions in the 10 nm channel. KE, kinetic energy.

5.1.1 One-Dimensional Density of States

Let us go back to our three-well model of the field-effect transistor (FET) from Chapter 2. If we observe any current flow from the far left side to the far right side, we know that a particle flowing through the channel must be in an eigenstate of the channel. The three lowest energy states of the 10 nm well are illustrated in Figure 5.1.

For our one-dimensional model we will use our old friend the infinite square well. We decided that this had solutions of the form:

$$\phi_n(x) = \sqrt{\frac{2}{a}} \sin\left(\frac{n\pi}{a} x\right) \quad 0 < x < a. \tag{5.1}$$

We have said before that k-space plays a role in quantum analysis similar to the frequency domain in engineering analysis. We also know that the only allowed solutions to Equation (5.1) are at

$$k = \frac{n\pi}{a} \quad n = 1, 2, 3, \dots, \tag{5.2}$$

so we can plot the allowed states on a simple one-dimensional graph, as in Figure 5.2.

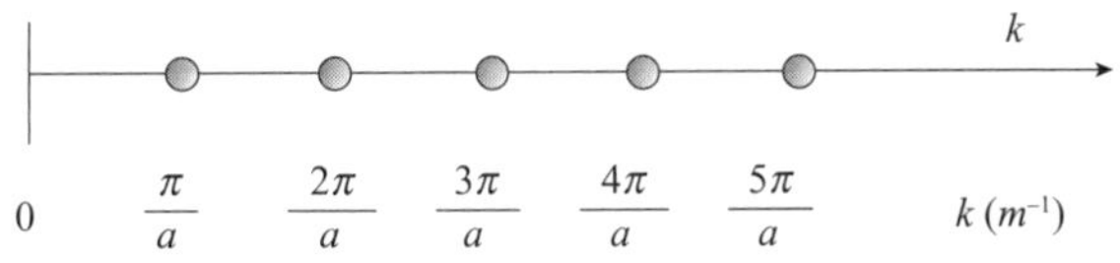

FIGURE 5.2 Plot of the allowed states in the one-dimensional infinite well.

We could think of this in terms of the number of states that are within a given length in k-space. (Here you have to be careful, because k has units of inverse length.) Notice that the well has a finite dimension a, but k extends from zero to plus infinity. Or we could talk about the *density of states* $g_{1D}(k)$. In any given state, there can be two particles, one spin up and the other spin down. Therefore, there are two states for each π/a length in the k-space

$$g_{1D}(k) = \frac{2a}{\pi}. \tag{5.3}$$

In actuality, we are more interested in relating allowed states to energy. So we use

$$E = \frac{\hbar^2 k^2}{2m}$$

or

$$k = \frac{\sqrt{2mE}}{\hbar}. \tag{5.4a}$$

Taking the differential gives

$$dk = \frac{1}{2} \frac{\sqrt{2m}}{\hbar} \frac{dE}{\sqrt{E}}. \tag{5.4b}$$

If we look for the available states up to a certain energy,

$$N = \int_0^{k_{max}} g_{1D}(k)\,dk = \int_0^{k_{max}} \frac{2a}{\pi}\,dk$$

$$= \int_0^{E_{max}} \frac{2a}{\pi} \frac{1}{2} \frac{\sqrt{2m}}{\hbar} \frac{dE}{\sqrt{E}} = \int_0^{E_{max}} \frac{a}{\pi\hbar} \frac{\sqrt{2m}}{\sqrt{E}}\,dE.$$

The term in the integrand is the quantity we want. We divide out the a because we want an expression for the number of available states per unit energy per unit length or the *density of states*, that is, the number of available states per unit energy per unit distance.

$$g_{1D}(E) = \frac{\sqrt{2m}}{\pi \hbar \sqrt{E}}. \tag{5.5}$$

Equation (5.5) has the units:

$$[g_{1D}] = \frac{[\text{kg}]^{1/2}}{[J \cdot s][J]^{1/2}} = \frac{[\text{kg}]^{1/2}}{[J \cdot s]\left[\dfrac{\text{kg}^{1/2} \cdot \text{m}}{\text{s}}\right]} = \frac{1}{[J \cdot \text{m}]}.$$

Remember that the energy of a state in the infinite square well is given by

$$\varepsilon_n = \frac{\hbar^2}{2m}\left(\frac{n\pi}{a}\right)^2 = \frac{\hbar^2 \pi^2}{2ma^2} n^2 \quad n = 1, 2, 3, \ldots$$

It makes sense that as we increase energy, the density of the available states is reduced.

5.1.2   Two-Dimensional Density of States

Suppose now that we have a two-dimensional infinite square well, as shown in Figure 5.3. We will assume both dimensions are of length a.

The separable solution will be of the form

$$\phi_{n,m}(x, y) = \frac{2}{a}\sin\left(\frac{n\pi}{a}x\right)\sin\left(\frac{m\pi}{a}y\right), \quad n = 1, 2, 3 \ldots, m = 1, 2, 3 \tag{5.6}$$

within the region $0 < x < a, 0 < y < a$. The corresponding energies are

$$\varepsilon_{n,m} = \frac{\hbar^2 \pi^2}{2ma^2}\left(n^2 + m^2\right).$$

This gives us the possibility of *degenerate states*: states that have the same energy, but are different wave functions. Obviously, $\varepsilon_{1,2} = \varepsilon_{2,1}$, but the eigen-

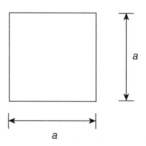

FIGURE 5.3   Diagram of a two-dimensional infinite square well. Inside the rectangle, $V = 0$ and outside the rectangle $V = \infty$.

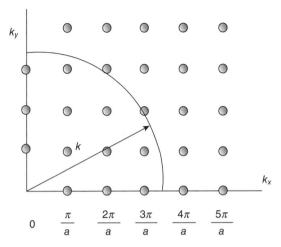

FIGURE 5.4 Diagram of the two-dimensional k-space.

functions correspond to different states that are orthogonal to each other, that is,

$$\int_0^a \int_0^a \phi^*_{1,2}(x, y)\phi_{2,1}(x, y)dxdy = 0.$$

The number of degenerate states increases with increasing n and m.

We can now draw a two-dimensional plot of the available states as shown in Figure 5.4.

Each state occupies an area of $(\pi/a)^2$. For a given value of $k = \sqrt{k_x^2 + k_y^2}$, in the two-dimensional k-space we will encompass an area of

$$A = \frac{1}{4}\pi k^2.$$

The factor 1/4 indicates that we are only using positive values of k. The differential is

$$dA = \frac{1}{2}\pi k \, dk.$$

So if we want to get the number of states up to k_{max}, we calculate

$$N = 2\int_0^{k_{max}} \frac{1}{(\pi/a)^2} dA = 2\int_0^{k_{max}} \frac{1}{(\pi/a)^2}\frac{1}{2}\pi k \, dk. \tag{5.7}$$

The factor 2 in front of the integral represents the two spins per state. Once again, we are more interested in the distribution with respect to energy, so substituting Equation (5.4b) into Equation (5.7) gives:

$$N = \int_0^{E_{max}} \frac{a^2}{\pi} \frac{m}{\hbar^2} dE.$$

The two-dimensional density of states, after dividing out the physical dimension a^2 is

$$g_{2D}(E) = \frac{m}{\pi\hbar^2}. \tag{5.8}$$

The units are

$$[g_{2D}] = \frac{[kg]}{[J \cdot s]^2} = \frac{[kg]}{J\left[\dfrac{kg \cdot m^2}{s^2} \cdot s^2\right]} = \frac{1}{[J \cdot m^2]},$$

that is, states per energy, per area.

5.1.3   Three-Dimensional Density of States

In three dimensions, each state occupies a volume $(\pi/a)^3$. The volume in the three-dimensional k-space enclosed by a distance k is given by

$$V = \frac{1}{8}\frac{4}{3}\pi k^3 = \frac{1}{6}\pi k^3,$$

where we once again consider only the positive values of k_x, k_y, and k_z. The differential is

$$dV = \frac{1}{2}\pi k^2 dk.$$

Now we think of k as the radius in a sphere. The total available number of states up to k_{max} is

$$N = 2\int_0^{k_{max}} \left(\frac{a}{\pi}\right)^3 dV = 2\int_0^{k_{max}} \left(\frac{a}{\pi}\right)^3 \frac{1}{2}\pi k^2 \, dk. \tag{5.9}$$

The number 2 accounts for the two values of spin. Once again, we would prefer to work with energy so using Equation (5.4b), Equation (5.9) becomes

$$N = \int_0^{E_{max}} \left(\frac{a}{\pi}\right)^3 \pi \left(\frac{\sqrt{2mE}}{\hbar}\right)^2 \frac{\sqrt{m}}{\hbar} \frac{dE}{\sqrt{2E}}$$

$$= \frac{\pi a^3 (2m)^{3/2}}{\pi^3 \hbar^3 2} \int_0^{E_{max}} \sqrt{E} dE.$$

If we want to talk about the number per volume, we can divide out the a^3. Also, we can replace $8\pi^3\hbar^3$ with just h^3. Then we have an expression for the three-dimensional density of states as it usually appears in the literature [3]:

$$g_{3D}(E) = \frac{4\pi(2m)^{3/2}}{h^3} \sqrt{E}. \tag{5.10}$$

Here the units are

$$[g_{3D}] = \frac{[kg]^{3/2}}{[J \cdot s]^3} [J]^{1/2} = \frac{[kg]^{3/2}}{J[J^2s^3]} \left[\frac{kg \cdot m^2}{s^2}\right]^{1/2}$$

$$= \frac{[kg]^{3/2}}{J\left[\frac{kg^2m^4}{s}\right]} \left[\frac{kg \cdot m^2}{s^2}\right]^{1/2} = \frac{[kg^2m]}{[J \cdot m^4 \cdot kg^2]} = \frac{1}{[J \cdot m^3]}.$$

5.1.4 The Density of States in the Conduction Band of a Semiconductor

To calculate the states in the conduction band of a semiconductor we must consider two different points:

1. An electron needs a minimum energy E_c just to be in the conduction band, so we will replace E with $E - E_c$.
2. Instead of the free space mass, will use the effective mass m_n^*.

Both of these points are addressed in Chapter 6. Therefore, the density of possible electron states in the conduction band is

$$g_c(E) = \frac{4\pi(2m_n^*)^{3/2}}{h^3} \sqrt{E - E_C}.$$

The same reasoning tells us that the distribution of possible hole states in the valence band is given by

$$g_v(E) = \frac{4\pi(2m_p^*)^{3/2}}{h^3} \sqrt{E_V - E}.$$

5.2 PROBABILITY DISTRIBUTIONS

In Section 5.1, we discussed the density of states, which concerns the number of quantum states that are available for occupation by fermions, either electrons in the conduction band or holes in the valence band. The next step is to think about the particles that are available to fill up these states. However, we cannot just think about the number of particles that are available. We must also consider their energies because this dictates which states they will occupy. In other words, we want to consider distribution of the particles as a function of energy. Because this is quantum mechanics, we have to talk about the *probability* that a state is occupied. Before we do that, we will take a minute to discuss the difference between fermions and classical particles.

5.2.1 Fermions versus Classical Particles

The particle of most interest to us is the electron, and electrons are within the class of particles called *fermions*. Fermions differ from classical particles in two respects [4]:

1. Only two particles can occupy any one system state, one with spin up and the other with spin down.
2. Identical types of fermions are *indistinguishable*.

The second point means that we cannot talk about electron A and electron B. Instead, we have to talk about all the electrons in the universe at once! Actually, if there is an electron in New York and one in Los Angeles, they will not have too much to do with each other. But if we have two particles in the ground state of an infinite well, we cannot deal with them individually. Two simple examples can illustrate how this affects the availability of states.

Suppose we have a billiard table with the usual 15 different balls, each of which has a different number on it. As we rack them up at the beginning of a game, we have 15!, 15 factorial different ways we can put the balls in the rack. We can refer to these 15 factorials as different states. Suppose we rack up 15 snooker balls, all of which are red. How many different states do we have now? The answer is one, because if we change any two balls, the rack looks the same. The rack of indistinguishable snooker balls has only one state, while the rack of billiard balls has 15! states.

Another simple example uses the familiar pair of dice. Each die has six different values, or states. If we have one red die and one green die and we regard each of them as distinguishable, then we have 6×6 or 36 different states, and each of these states is equally likely. However, if both the dice are the same color and we think of the dice as indistinguishable, we have only 11 different states because if we roll two dice, we can get any number between 2 and 12. Furthermore, the distribution of the likelihood of the different states is far different. The numbers 2 and 12 are the least likely to appear, because

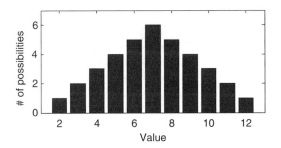

FIGURE 5.5 Number of possibilities of rolling various values with a pair of dice when the dice are indistinguishable.

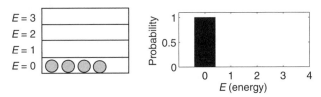

FIGURE 5.6 The total energy $E = 0$ and the total number of balls $N = 4$.

they each have only one possibility: if each die is one, we get a total of two; if each die is six, we get 12. In contrast, there are six different ways to obtain a total of seven. We are six times more likely to roll a seven than a one or 12 because there are six times as many arrangements available. This is illustrated in Figure 5.5.

5.2.2 Probability Distributions as a Function of Energy

Here is a simple example that suggests how energy can be related to the probability distribution. Suppose we have four balls on any of four shelves, each shelf increasing in energy as it goes up. We assume the bottom shelf is zero energy and it requires available energy for the balls to move to the other shelves. We will assume the balls are indistinguishable. We have to conserve two things: the total energy and the total number of balls. If we start with four balls and no energy, there is only one possible configuration, as shown in Figure 5.6. If we randomly select one ball, we know its energy will always be $E = 0$.

Now look at the case where $N = 4$ and $E = 1$ (make up your own units). There is still only one possible configuration as shown on the left side of Figure 5.7. Remember, since the balls are indistinguishable, it does not matter which ball is at the $E = 1$ level. If we randomly select one ball, we have a one in four chance that it has energy $E = 1$, but three chances in four that $E = 0$. This is illustrated in the graph on the right side of Figure 5.7.

If we increase to $N = 4$, $E = 2$, there are two possible configurations, as shown in Figure 5.8. A fundamental postulate of statistical mechanics is that

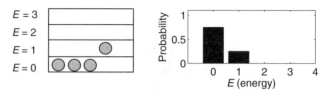

FIGURE 5.7   Four balls, but an energy of $E = 1$.

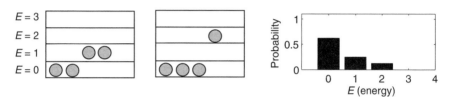

FIGURE 5.8   Four balls with a total energy of $E = 2$.

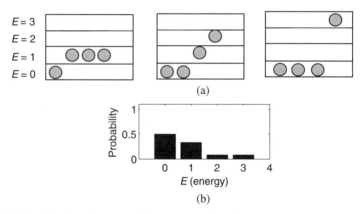

FIGURE 5.9   (a) The three possible configurations for four balls with a total energy of 3 and (b) the probability of the energy of one of the balls selected at random.

in equilibrium, all accessible states are equally probable [1]. This usually deals with microstates, but we will assume it applies to our situation here and say our two configurations are equally likely. Therefore, if we randomly select a ball, we can look at the graph to determine the likelihood that it is at a certain energy. We see there is only one chance in eight of it being at $E = 2$, two chances in eight that it is at $E = 1$, and five chances in eight that it is at $E = 0$. This is shown in the probability distribution on the far right of Figure 5.8.

If we increase the total energy to $E = 3$, we get the three possible configurations shown in Figure 5.9a. Using the same reasoning, we can get the probability distribution shown in Figure 5.9b.

5.2.3 Distribution of Fermion Balls

Now consider the case where the balls are "fermions," that is, no two can occupy the same state. If we have five balls, then the lowest energy configuration is one ball in each of the lowest five states, giving a total energy of $0 + 1 + 2 + 3 + 4 = 10$. If we plot the probability that a given state is occupied, we would find that the lowest five states have a 100% probability of being occupied, and all other states have zero probability of being occupied, as shown in Figure 5.10 below.

If we move to the case where the total energy is 12, we find that there are two possible arrangements of the five balls, as shown on the left side of Figure 5.11. As mentioned earlier, a fundamental assumption of statistical mechanics is that in *thermal equilibrium* every distinct state with the same total energy is equally probable. If we now plot the probability of occupation of each of the energy levels, we get the situation shown on the right side in Figure 5.11.

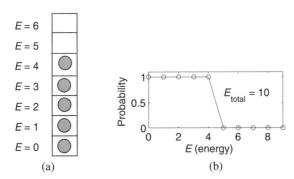

FIGURE 5.10 (a) Occupied states when there are five balls and the total energy is 10; (b) the corresponding probability of occupation of states as a function of energy.

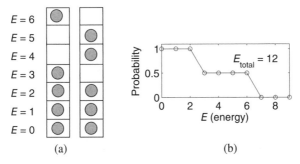

FIGURE 5.11 (a) Occupied states when there are five balls and the total energy is 12; (b) the corresponding probability of occupation of states as a function of energy.

When we have macroscopic numbers of electrons, the probability that a quantum state at energy E is occupied is given by the Fermi–Dirac distribution [3]:

$$f_F(E) = \frac{1}{1 + \exp\left(\dfrac{E - E_F}{k_B T}\right)}, \tag{5.11}$$

where E_F is the Fermi Energy, the value at which the probability of occupation is one-half:

$$f_F(E_F) = \frac{1}{1 + \exp\left(\dfrac{E_F - E_F}{k_B T}\right)} = \frac{1}{1+1} = \frac{1}{2}. \tag{5.12}$$

Equation (5.11) is a probability distribution that has the form shown in Figure 5.12.

In thermodynamics, the Fermi energy is called the *chemical potential*, usually expressed as μ. The chemical potential is defined as the Gibbs free energy required to add another particle to the system at constant temperature and pressure. The Fermi energy is equal the chemical potential at $T = 0$ K [2]. Note that the Fermi energy does not depend on temperature.

The quantity $k_B T$ gives the distribution its shape; the Boltzmann constant is

$$k_B = 1.38 \times 10^{-23} \text{ J/K}$$
$$= 8.62 \times 10^{-5} \text{ eV/K}.$$

T is temperature in degrees Kelvin. The quantity $k_B T$ has units of energy, and therefore $(E - E_F)/k_B T$ is dimensionless.

One set of values will be used so often that it is worth memorizing. At room temperature, $T = 300$ K:

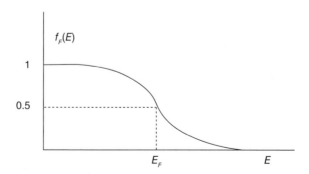

FIGURE 5.12   The Fermi–Dirac distribution.

$$k_BT = (8.62 \times 10^{-5} \text{ eV/K})(300 \text{ K})$$
$$= 0.0259 \text{ eV}.$$

Example. At room temperature, what is the probability that a state 0.03 eV above the Fermi level is occupied?

Solution

$$f_F(0.03) = \frac{1}{1+\exp\left(\dfrac{E-E_F}{k_BT}\right)}$$

$$= \frac{1}{1+\exp\left(\dfrac{0.03}{0.0259}\right)} = \frac{1}{1+3.18} = 0.239.$$

Example. What is the probability that a state 0.3 eV above the Fermi level is occupied?

Solution

$$f_F(0.3) = \frac{1}{1+\exp\left(\dfrac{0.3}{0.0259}\right)} = \frac{1}{1+1.07 \times 10^5} = 9.3 \times 10^{-6}.$$

Example. What is the probability that a state at energy 0.3 eV above the Fermi energy is occupied at a temperature of 400 K?

Solution. We know that $k_BT = 0.0259$ eV at $T = 300$ K. To get it at 400 K, instead of going back and recalculating, use the simple linearity

$$k_B(400) = \frac{k_B(400)}{k_B(300)}k_B(300) = \frac{4}{3}0.0259 \text{ eV} = 0.0345 \text{ eV},$$

$$f_F(0.3) = \frac{1}{1+\exp\left(\dfrac{0.3}{0.0345}\right)} = \frac{1}{1+5,977} = 1.67 \times 10^{-4}.$$

As the temperature approaches 0 K, a situation that indicates a total lack of thermal energy, the distribution approaches the shape shown in Figure 5.13.

What does a distribution like Figure 5.13 mean? It means that every energy state below the Fermi energy level is occupied, and every state above the Fermi level is empty. This is similar to Figure 5.12 where we know every ball is in the lowest possible energy state. We could say that the Fermi energy is $E_F = 4.5$. For that matter, it could be anything between $4 < E_F < 5$ and give the same result.

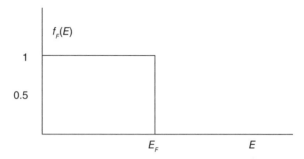

FIGURE 5.13   The Fermi–Dirac distribution at $T = 0$ K.

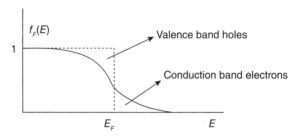

FIGURE 5.14   The Fermi–Dirac distribution as T increases. The number of missing states below the Fermi level equals the number of states above the Fermi level.

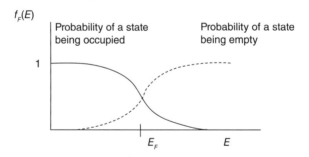

FIGURE 5.15   The probability distributions for electrons and holes.

If a semiconductor were held at 0 K, all electrons would remain in place in their covalent bands. Since no thermal energy is available, none of them can acquire the additional energy need to break a bond and move up to the conduction band. As the temperature increases above $T = 0$ K, we get a distribution that looks more like Figure 5.14.

As thermal energy becomes available, more bonds are broken and more electrons can move into the conduction band. However, that also means that a hole is left behind in the valence band. This is reflected in the Fermi–Dirac distribution by the fact that their respective areas are the same. This might be further illustrated by Figure 5.15, where separate probabilities for electrons and holes are graphed.

5.2.4   Particles in the One-Dimensional Infinite Well

To start with a concrete example we use the one-dimensional infinite well that is 10 nm wide and talk about what happens as we add more particles to the well. In the discussion of this section, we will ignore the Coulomb effects, that is, the change in energy levels that comes from the Coulomb interaction between the particles. The 10 nm infinite well has eigenenergies of

$$\varepsilon_n = 0.00375\ n^2 \text{ eV}.$$

If we put one particle in the well, it will most likely occupy the lowest energy state given by $E_1 = 0.00375$ eV. If the well were at $T = 0$ K, we could be sure that would be the case. In fact, the second particle would enter the same assuming the two particles have different spins. At 0 K, we could proceed in a similar way, adding particles and knowing each will go in the next available state.

Example. If the Fermi energy in the 10 nm well is $E_F = 0.01$ eV, what is the total number of electrons in the well?

Solution. Notice that a temperature has not been specified. Looking at Figure 5.14, one can see that the shape of the distribution is affected by temperature, but not by the number of particles. Therefore, in solving this problem, we will assume $T = 0$ K, as in Figure 5.13. If an energy ε_n is above the Fermi energy, then

$$f_F(E) = \frac{1}{\left(1 + \exp\left[\dfrac{E_n - E_F}{k_B T}\right]\right)} = \frac{1}{\left(1 + \exp\left[\dfrac{E_n - E_F}{0}\right]\right)} = \frac{1}{(1 + e^\infty)} = 0,$$

indicating that the state is empty. If an energy ε_n is below the Fermi energy, then

$$f_F(E) = \frac{1}{\left(1 + \exp\left[\dfrac{E_n - E_F}{0}\right]\right)} = \frac{1}{(1 + e^{-\infty})} = 1,$$

indicating that the state is occupied. The total number of particles is the sum

$$n = 2 \sum_{n=1}^{E_{max}} 1 \quad E_{max} < E_F.$$

In this case, only the ground state energy is below the Fermi energy, $E_F = 0.01$ eV, so $n = 2$.

Example. (a) If we have 10 particles in the 10 nm one-dimensional well at 0 K, what is the Fermi energy? (b) What is the Fermi energy if we have 11 particles in the well?

Solution. (a) Assuming double occupancy, we know the levels up to $n = 5$ are filled. So the Fermi energy will lie between the $n = 5$ and $n = 6$ eigenenergies:

$$E_5 = 0.00375 \times 5^2 \text{ eV} = 0.09375 \text{ eV},$$

$$E_6 = 0.00375 \times 6^2 \text{ eV} = 0.135 \text{ eV}.$$

So

$$0.0937 \text{ eV} \leq E_F \leq 0.135 \text{ eV}.$$

(b) There is one electron in the $n = 6$ state out of the two possible states. By definition, $E_6 = 0.00375 \times 6^2$ eV $= 0.135$ eV is the Fermi energy.

5.2.5   Boltzmann Approximation

Though the Fermi–Dirac distribution is not conceptually difficult, it is not the type of function that we would like to have to deal with in integrals, for instance. Fortunately, there is an approximation that can be very helpful when we are primarily concerned with the distribution at higher energies,

$$f_F(E \gg E_F) = \frac{1}{1 + \exp\left(\dfrac{E - E_F}{k_B T}\right)}$$

$$\cong \frac{1}{\exp\left(\dfrac{E - E_F}{k_B T}\right)} = \exp\left(\frac{E_F - E}{k_B T}\right),$$

referred to as the *Boltzmann approximation*. It is illustrated in Figure 5.16.

Example. What is the probability that a state at an energy 0.2 eV above the Fermi energy is occupied? Assume room temperature.

Solution. Using the Fermi–Dirac distribution we calculate

$$f_F(E) = \frac{1}{1 + \exp\left(\dfrac{0.2}{0.0259}\right)} = \frac{1}{1 + \exp(7.722)} = \frac{1}{1 + 2257} = 4.427 \times 10^{-4}.$$

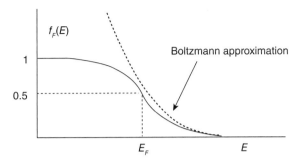

FIGURE 5.16 Region in which the Boltzmann approximation holds.

The Boltzmann approximation gives

$$f_F(E \gg E_F) \cong \exp\left(\frac{E_F - E}{k_B T}\right) = \exp(-7.72) = 4.43 \times 10^{-4}.$$

5.3 THE EQUILIBRIUM DISTRIBUTION OF ELECTRONS AND HOLES

If we are dealing with electrons in the conduction band of a semiconductor, the quantity that primarily concerns us is the number of electrons per unit volume per unit energy in the conduction band [3]. Quantitatively, we can express that as

$$n(E) = g_C(E) \cdot f_E(E). \tag{5.13}$$

The density of electrons in the conduction band is

$$\frac{\text{Total } e^-}{\text{Vol}} = \int_0^\infty n(E) dE. \tag{5.14}$$

Remember that in determining the density of states in three dimensions in a semiconductor, we realized the density was proportional to the square root of the energy above the conduction band.

$$g_C(E) \propto \sqrt{E - E_c}. \tag{5.15}$$

Similarly, the density of holes in the valence band is

$$g_V(E) \propto \sqrt{E_v - E}. \tag{5.16}$$

These are plotted in Figure 5.17. Notice, however, that E is plotted on the vertical axis.

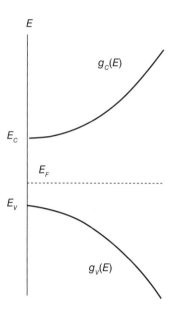

FIGURE 5.17 Density of states diagrams for electrons and holes.

Recall also that the Fermi–Dirac distribution gave us the probability that a state was occupied (Eq. 5.11), as shown in Figure 5.14.

We would like Figure 5.14 to match up with the available states in the conduction and valence band shown in Figure 5.17. To do this, we have to turn Figure 5.14 sideways and be sure that the Fermi levels match, as shown in Figure 5.18. Look back at Equation (5.13). The part that interests us is the part where they overlap, as shown in Figure 5.18. The probability that a state in the conduction band is occupied by an electron is

$$n_C(E) = g_c(E) f_F(E). \tag{5.17a}$$

In the valence band, the probability that a state is *not* occupied is given by

$$p(E) = g_v(E)[1 - f_F(E)]. \tag{5.17b}$$

Therefore, we can also get the density of holes from the diagram.

The position of the Fermi energy, E_F is dictated by the fact that in an intrinsic semiconductor, the number of electrons must equal the number of holes. If

$$m_n^* = m_p^*,$$

then $g_c(E)$ and $g_v(E)$ are symmetric and

$$E_F = \frac{E_C + E_v}{2}.$$ (5.18)

To get the total number of available electrons, we integrate over energy

$$n_0 = \int_0^\infty n(E)dE = \int_0^\infty g_C(E)f_F(E)dE.$$ (5.19)

We have previously determined that

$$g_C(E) = \frac{4\pi(2m_n^*)^{3/2}\sqrt{E - E_c}}{h^3}.$$ (5.20)

Looking at Figure 5.18, we are justified in using the Boltzmann approximation for the Fermi–Dirac distribution, so putting the two together,

$$n_0 = \int_0^\infty \frac{4\pi(2m_n^*)^{3/2}\sqrt{E - E_c}}{h^3} \exp\left(\frac{E_F - E}{k_B T}\right)dE.$$ (5.21)

This is a gamma function integral with the solution

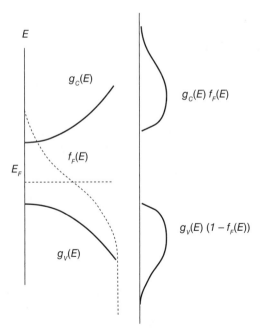

FIGURE 5.18 The density of states and Fermi–Dirac distributions plotted together.

$$n_0 = N_c \exp\left(\frac{E_F - E_c}{k_B T}\right), \tag{5.22}$$

where

$$N_c = 2\left(\frac{2\pi m_n^* k_B T}{h^2}\right)^{3/2}. \tag{5.23}$$

N_c is *the effective density of states in the conduction band*; and n_0 is *the thermal equilibrium density of electrons in the conduction band*. Similarly,

$$p_0 = N_v \exp\left(\frac{E_v - E_F}{kT}\right), \tag{5.24}$$

$$N_v = 2\left(\frac{2\pi m_p^* k_B T}{h^2}\right)^{3/2}, \tag{5.25}$$

where N_v is *the effective density of states in the valance band*; and p_0 is *the thermal equilibrium density holes in the valence band*. N_c and N_v are constants that characterize semiconductor materials. The effective density of states are related topics and covered thoroughly in Neamen [3].

5.4   THE ELECTRON DENSITY AND THE DENSITY MATRIX

We have talked about the Fermi–Dirac distribution and how it affects the number of particles in a system. In this section, we want to look at the spatial distribution of particles for a given Fermi energy and temperature. We will use the 10 nm infinite well as our model.

We will start with the case of $T = 0$ K, because we know that it leads to the simple Fermi–Dirac distribution, that is, every state below the Fermi energy is occupied and every state above it is empty. Two examples are shown in Figure 5.19.

Consider the case where the Fermi energy is 0.01 eV. We know that in the 10 nm well, only the ground state is below this energy. If there is one particle in a well, then the electron density is given by

$$n(x) = \phi_1^*(x) \cdot \phi_1(x), \tag{5.26}$$

where $\phi_1(x)$ is the ground state eigenfunction. The top part of Figure 5.19 is for the case of $E_F = 0.01$. Only the ground state wave function contributes to the electron density. Spin degeneracy means there are actually two particles, and we multiply Equation (5.26) by two to get the $n(x)$ in the figure. Notice that it has the shape of the ground state eigenfunction squared. If we take the

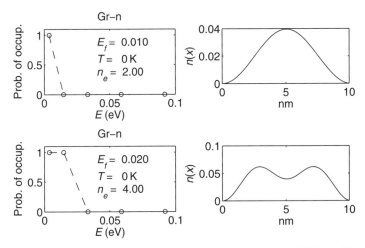

FIGURE 5.19 Particle densities in a 10 nm infinite well for two different Fermi energies at $T = 0$ K.

case of $E_F = 0.02$ eV, there are two states below this energy, giving four particles with energies below E_F, and a $n(x)$ that is a superposition of the first two eigenfunctions squared.

When we have temperatures other than 0 K, the situation is a little different. Now the electron density is

$$n(x) = \sum_n f_F(\varepsilon_n) \cdot \phi_n^*(x) \cdot \phi_n(x). \tag{5.27}$$

Recall that the Fermi–Dirac distribution is

$$f_F(E) = \frac{1}{1 + \exp\left(\dfrac{E - E_F}{k_B T}\right)}.$$

Two examples at $T = 100$ K are given in Figure 5.20.

5.4.1 The Density Matrix

A topic closely related to electron density is the density matrix, ρ. The density matrix at equilibrium can be written as the Fermi function of the Hamiltonian matrix

$$\rho = f_F(H - \mu I). \tag{5.28}$$

This general matrix relation is valid in any representation. However, when we are in the eigenstate representation, we relate it to the electron density. If

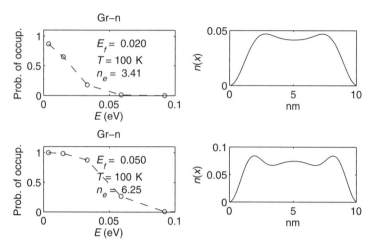

FIGURE 5.20   Particle densities for two values of E_F when $T = 100$ K.

$$H_e = \begin{bmatrix} \varepsilon_1 & 0 & 0 & \cdots \\ 0 & \varepsilon_2 & 0 & \cdots \\ 0 & 0 & \varepsilon_3 & \cdots \\ \cdots & \cdots & \cdots & \cdots \end{bmatrix},$$ (5.29)

then

$$\rho_e = \begin{bmatrix} f_0(\varepsilon_1 - \mu) & 0 & 0 & \cdots \\ 0 & f_0(\varepsilon_2 - \mu) & 0 & \cdots \\ 0 & 0 & f_0(\varepsilon_3 - \mu) & \cdots \\ \cdots & \cdots & \cdots & \cdots \end{bmatrix},$$ (5.30)

and the number of electrons is just

$$N_{\text{elec}} = \sum_{n=1}^{N} f_0(\varepsilon_n - \mu) = \text{Trace}(\rho).$$ (5.31)

Equation (5.31) is the sum of the diagonal elements of ρ, or the *trace* of ρ. But the interesting thing is that Equations (5.28) and (5.30) hold in any representation. The real space and eigenspace representations are linked by the matrix Φ that contains the columns which are the eigenfunctions of the Hamiltonian (Chapter 2):

$$\rho = \Phi \rho_e \Phi'.$$

It may seem silly that we took a simple one-dimensional quantity like the particle density and fabricated a two-dimensional quantity like the density matrix. This has mathematical advantages due to the fact that the density matrix is the same size as our operators, an advantage that will be utilized in later chapters.

EXERCISES

5.1 Density of States

5.1.1 In developing the expression for the density of states in one dimension we used the infinite well as a model. What would be different if we used the well with periodic boundary conditions described in Section 2.3?

5.2 Probability Distributions

5.2.1 For the problem of the balls on the shelves, find the distribution if the number of balls is still four, but the total energy is increased to four.

5.2.2 Similar to Figure 5.11, what is the probability distribution when there are still five fermion balls but the total energy is 12?

5.2.3 Consider the system below which has six fermion particles. We include spin degeneracy, so up to two particles can be in any given state. The example below is for $E = 6$. Find the probability of occupation if $E = 8$. (Hint: Whether or not the state is occupied by one or two particles is irrelevant.)

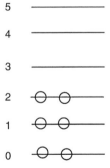

5.2.4 When is the Boltzmann approximation valid? Give your answer in terms of $E - E_F$ versus kT.

5.2.5 In a two-dimensional infinite well that is a Å on each side, what is the Fermi energy if the density of particles is J particles/Å^2? (This is an analytic calculation. Your final answer should be in terms of J and basic constants. Hint: start with Eq. 5.7.)

5.2.6 A two-dimensional well is a meters on one side and b meters on the other side. What is its density of states in units of $J^{-1}m^{-2}$?

5.3 The Equilibrium Distribution of Electrons and Holes

5.3.1   What is the probability of a state 0.2 eV above the conduction band being occupied if the Fermi energy is 0.5 eV below the conduction band? Assume a room temperature of 300 K. What is the probability if the temperature is 310 K?

5.4 The Electron Density and the Density Matrix

5.4.1   Graph the probability of occupation of the states as well as the particle densities, similar to Figure 5.20, for $E_F = 0.01$ eV at $T = 20$ K and $T = 100$ K.

REFERENCES

1. P. L. Hagelstein, S. D. Senturia, and T. P. Orlando, *Introductory Applied Quantum and Statistical Mechanics*, New York: John Wiley and Sons, 2004.

2. S. Datta, *Quantum Transport—Atom to Transistor*, Cambridge, UK: Cambridge University Press, 2005.

3. D. A. Neamen, *Semiconductor Physics and Devices*, 3rd ed., New York: McGraw-Hill, 2003.

4. D. J. Griffiths, *Introduction to Quantum Mechanics*, Englewood Cliffs, NJ: Prentice Hall, 1994.

6

BANDS AND SUBBANDS

Most of the particle transport that interests us takes place in semiconductors. We briefly spoke about the conduction bands of semiconductors and the fact that a particle propagating in a semiconductor must be in a conduction band. In this chapter, we quantify this concept. We begin by using simulation to demonstrate that a particle traveling in a periodic structure is influenced by the spacing of the structure and can only propagate at certain energies dictated by the periodic structure. In Section 6.2, we introduce an important quantity called the effective mass. In Section 6.3, we describe the modes of a two-dimensional structure. The modes describe the influence of the dimension perpendicular to the direction of propagation of a particle. Some two-dimensional simulations are included to illustrate the concept of modes.

6.1 BANDS IN SEMICONDUCTORS

Semiconductors are crystals. This means that the atoms that compose the semiconductors are in very specific patterns. An example is given in Figure 6.1a. When we analyze the behavior of electrons in a semiconductor, more important than the atoms themselves is the location of the atoms, which is indicated by the group of points called the "lattice" (Fig. 6.1b).

Look at the situation in Figure 6.2a. A particle is initialized in a potential that consists of spikes of −1 meV potential spaced at 20 Å intervals. The spikes

Quantum Mechanics for Electrical Engineers, First Edition. Dennis M. Sullivan.
© 2012 The Institute of Electrical and Electronics Engineers, Inc.
Published 2012 by John Wiley & Sons, Inc.

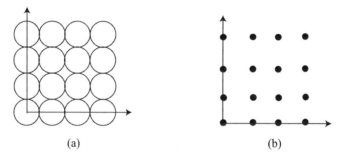

FIGURE 6.1   (a) An illustration of how atoms might be arranged in a crystal; (b) the corresponding lattice.

are intended to mimic a potential created by a one-dimensional lattice. In Figure 6.2b the spatial Fourier transform of the particle waveform shows that its center is at $(1/\lambda) = (1/20\text{ Å}) = 0.05(\text{Å})^{-1}$, as it should be. Classically, the particle would just skip over spikes of the potential. We have seen that in quantum mechanics, any potential will cause some interaction with the waveform. In fact, after the waveform has traveled only a short distance, a large part of the waveform has been reflected and travels in the negative direction, as seen in Figure 6.2c. This is quantified by the spatial Fourier transform that appears in Figure 6.2d. Remember that the fast Fourier transform puts the negative frequencies in the upper part of the array, and the negative frequencies indicate that part of the waveform that is traveling right to left. That is the reason that the negative of $(1/\lambda) = 0.05(\text{Å})^{-1}$ appears at $(1/\lambda) = 0.95(\text{Å})^{-1}$.

To try and understand this, let us think back to the case of the infinite well and ask ourselves the following question: If the walls of the well are not infinite, will there still be energy states that are more likely to constrict a particle based on whether or not the wavelengths of the particle are integer multiples of the length of the well? The answer is yes. Look at Figure 6.3. We know that an infinite well with a base of d only has eigenstates corresponding with wavelengths given by $2d/n$, where n is a positive integer. In trying to propagate through such a well, the wavelengths that are most likely to be restricted are similarly given by

$$\lambda_n^{\text{restricted}} = \frac{2d}{n}. \tag{6.1}$$

For the lattice in Figure 6.2a,

$$\lambda_n^{\text{restricted}} = \frac{40\text{ Å}}{n}.$$

The first few restricted wavelengths are $\lambda_1^{\text{restricted}} = 40\text{ Å}$, $\lambda_2^{\text{restricted}} = 20\text{ Å}$, and $\lambda_3^{\text{restricted}} = 13.3\text{ Å}$. These are the spikes in Figure 6.2b,d that indicate wavelengths that will not propagate in the lattice.

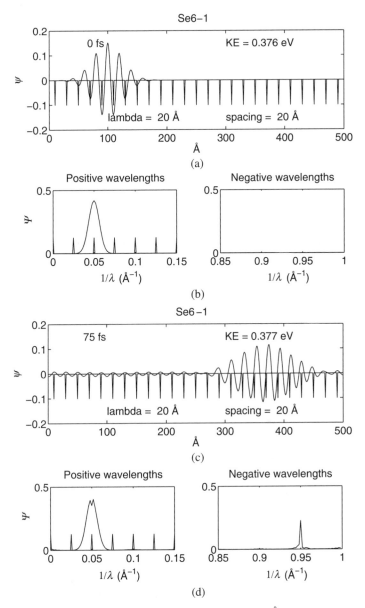

FIGURE 6.2 (a) A particle with a center wavelength of 20 Å is initialized in a potential that consists of −1 meV spikes at intervals of 20 Å. (b) The Fourier transform shows only positive frequencies. (c) As the particle propagates from left to right, a large portion of the waveform is reflected. (d) The Fourier transform has lost some amplitude at the corresponding positive wavelength and acquired some at the negative wavelength. KE, kinetic energy.

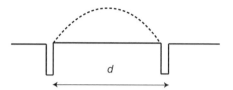

FIGURE 6.3 Even a small, negative potential will try to hold states whose wave-lengths are given by $2d/n$.

Figure 6.4a is similar to Figure 6.2a except that the center wavelength is 16 Å. After 60 fs, the waveform continues to propagate with virtually no loss in amplitude.

The two examples illustrated in Figures 6.2 and 6.4 had fairly narrow wave-length spectra because they started with wide Gaussian envelopes in real space. It is interesting to start with a narrow Gaussian envelope and observe the filtering action that the lattice exerts on the waveform as it tries to propa-gate. This is illustrated in Figure 6.5.

6.2 THE EFFECTIVE MASS

Even for the particles with wavelengths that can propagate through the 20 Å lattice, that is, at $\lambda = 30$, 15, and 8 Å, their behavior is still affected by the lattice. In the following discussion, we will assume that we are talking about particles that are within one of the allowed propagating bands. Once again, look at the Schrödinger equation:

$$i\hbar \frac{\partial}{\partial t} \psi(x,t) = \left(-\frac{\hbar^2}{2m_e} \frac{\partial^2}{\partial x^2} + V(x) \right) \psi(x,t). \tag{6.2}$$

Notice that within the operator

$$H = \left(-\frac{\hbar^2}{2m_e} \frac{\partial^2}{\partial x^2} + V(x) \right), \tag{6.3}$$

there are two parameters, m_e and V. Since $V(x)$ consists of a constant pattern over a relatively long distance, we could simply consolidate the two parameters and write

$$H = \left(-\frac{\hbar^2}{2m_e^*} \frac{\partial^2}{\partial x^2} \right). \tag{6.4}$$

The new parameter m_e^* is the *effective mass* [1]. Each semiconductor material has a different effective mass (The effective masses for the three most common semiconductors are given in Table A.2 in Appendix A).

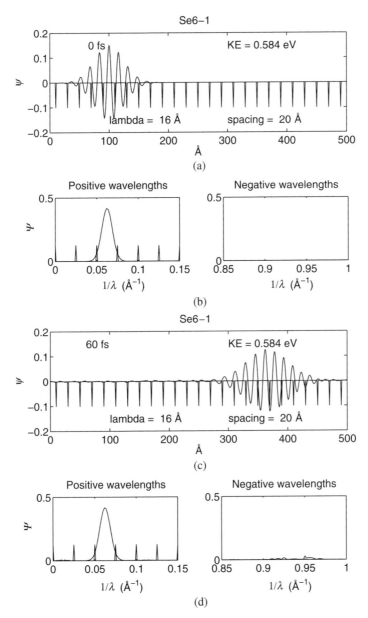

FIGURE 6.4 Simulation in a lattice similar to Figure 6.2, except that the particle has a center wavelength of 16 Å. After 60 fs, it has almost no loss. (a) A particle with a wavelength of 16 Å initialized in a lattice potential. The particle is moving from left to right. Only the real part of ψ is shown. (b) The spatial Fourier transform of the waveform in (a). (c) The particle in (a) after 60 fs. (d) The spatial Fourier transform of the waveform in (c).

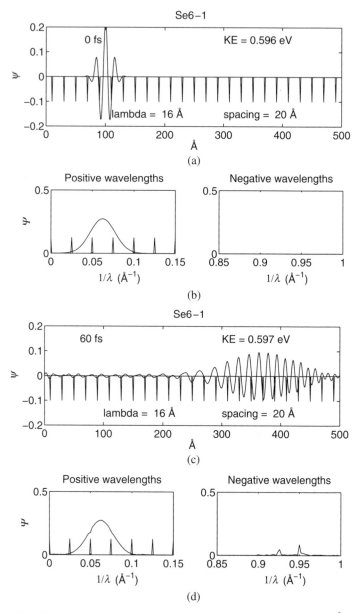

FIGURE 6.5 The simulation of a particle with a center wavelength of 16 Å in a lattice of 20 Å. The waveform describing the particle is fairly narrow in the space domain, leading to a broad spectrum in the Fourier domain. The spikes in the positive wavelengths graph indicate forbidden regions. Even though 16 Å is not a forbidden wavelength, the parts of the spectrum at 20 and 13.33 Å are attenuated. (a) A particle with a wavelength of 16 Å initialized in a lattice potential. The particle is moving from left to right. Only the real part of ψ is shown. (b) The spatial Fourier transform of the waveform in (a). (c) The particle in (a) after 60 fs.

If we eliminate the potential, that is, incorporate it into the effective mass, then all the energy is kinetic energy:

$$E = \frac{\hbar^2 k^2}{2m_e^*}. \tag{6.5}$$

Taking the second derivative of each side and dividing by $\hbar^2$,

$$\frac{1}{m_e^*} = \frac{1}{\hbar^2} \frac{d^2 E}{dk^2}. \tag{6.6}$$

This is often taken as the effective mass.

The effect of the lattice is illustrated in an E versus k diagram. If we look at Equation (6.5), we would conclude that this is just a parabolic curve (Fig. 6.6). However, Equation (6.1) tells us that there are certain values of k that are not allowed, so there will be jumps in curves as shown in Figure 6.7. The results are usually folded into the region

$$-\frac{\pi}{a} \le k \le \frac{\pi}{a},$$

which is referred to as the Brillouin zone, as shown in Figure 6.8.

We have been working in one dimension. As is usually the case, working in the real world of three dimensions is more complicated. The lattice that a particle sees is dictated by its direction through the lattice. The two-dimensional case is illustrated by Figure 6.9, showing that x-direction (indicated by [100]) propagation will see different spacing than propagation at a 45° angle [110]. (The notations [100] and [110] are known as *Miller indices* [2]. They are always three integers that indicate the direction of the particle.)

The lattices of real semiconductors are far more complicated and result in complicated conduction and valence bands, as shown in Figure 6.10.

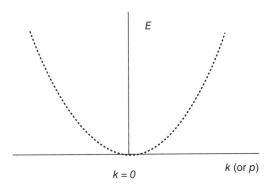

FIGURE 6.6 The E versus k curve for a free electron.

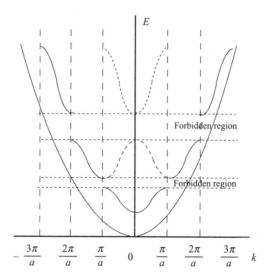

FIGURE 6.7 The E versus k diagram showing the effects of the band gaps.

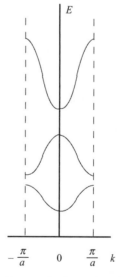

FIGURE 6.8 The E versus k diagram in the reduced-zone representation. The zone between $-\pi/a$ and π/a is referred to as the Brillouin zone.

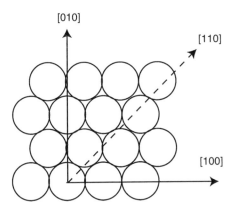

FIGURE 6.9 An illustration showing different directions in a semiconductor, which result in different k versus E curves.

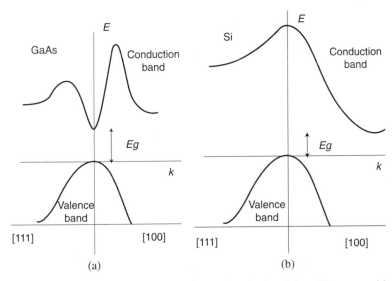

FIGURE 6.10 The conduction and valence bands for (a) gallium arsenide and (b) silicon.

6.3 MODES (SUBBANDS) IN QUANTUM STRUCTURES

We have seen that transport through a channel depends on the particle having an energy that corresponds to the eigenenergy of the channel. In the one-dimensional example in Figure 2.13, the particle in the left well tunnels to the right well because the 10 nm well in the middle has an eigenenergy of about 3.75 meV. The 10 nm well in the middle is not a true "infinite well," because it is connected to the wells on each side. But we did find it useful to use the energies of the infinite well as a starting point to estimate which particles would have the appropriate energy to pass through the channel.

The one-dimensional infinite well is a useful model to describe many concepts in quantum mechanics. However, we wonder what effect the dimensions perpendicular to the direction of propagation will have. In this section we will briefly describe the two-dimensional infinite well and use it to describe the concept of modes.

The two-dimensional, time-independent Schrödinger equation is

$$\frac{\partial^2}{\partial x^2}\psi(x, y)+\frac{\partial^2}{\partial y^2}\psi(x, y)+\frac{2m}{\hbar^2}(E-V(x, y))\psi(x, y)=0. \tag{6.7}$$

For a two-dimensional infinite well, we will have boundary conditions,

$$\psi(0, y)=\psi(a, y)=\psi(x, 0)=\psi(x, b)=0, \tag{6.8}$$

within the well $V(x, y) = 0$. If we follow a process similar to the one we used in Section 2.1, we will get solutions of the form

$$\psi(x, y)=\sqrt{\frac{4}{a\cdot b}}\sin\left(\frac{n\pi}{a}x\right)\sin\left(\frac{m\pi}{b}y\right), \tag{6.9}$$

with corresponding eigenenergies

$$E_{n,m}=\frac{\hbar^2\pi^2}{2m}\left(\frac{n^2}{a^2}+\frac{m^2}{b^2}\right). \tag{6.10}$$

If for instance the well is 10 nm in the x-direction and 50 nm in the y-direction, we can use the mathematics developed for the 10 nm one-dimensional infinite well and write

$$E_{n,m}=\frac{\hbar^2\pi^2}{2m}\left(\frac{n^2}{(10\text{ nm})^2}+\frac{m^2}{(50\text{ nm})^2}\right)=\left(n^2+\frac{m^2}{25}\right)3.75\text{ meV}.$$

The lowest energy state, the $(1, 1)$ state, is

$$E_{1,1}=\left(1^2+\frac{1^2}{25}\right)3.75\text{ meV}=3.9\text{ meV}.$$

The second lowest, the $(1, 2)$ state has energy

$$E_{1,2}=\left(1^2+\frac{2^2}{25}\right)3.75\text{ meV}=4.35\text{ meV}.$$

Some other states are:

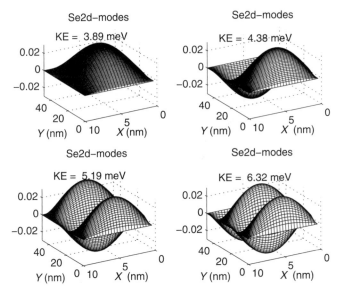

FIGURE 6.11   The first four states of a two-dimensional infinite well.

$$E_{1,3} = \left(1^2 + \frac{3^2}{25} \right) 3.75 \text{ meV} = 5.1 \text{ meV},$$

$$E_{1,4} = \left(1^2 + \frac{4^2}{25} \right) 3.75 \text{ meV} = 6.15 \text{ meV}.$$

The first four eigenstates and their corresponding energies are shown in Figure 6.11.

In contrast, the next highest state in the x-direction is

$$E_{2,1} = \left(2^2 + \frac{1^2}{25} \right) 3.75 \text{ meV} = 15.15 \text{ meV},$$

as seen in Figure 6.12. (The lack of agreement is due to the fact that cells of 0.2 nm are being used in these programs.)

If the channel consisted of a two-dimensional well instead of a one-dimensional well, with x being the direction of current flow, all of the above states, $E_{1,1}$ though $E_{1,4}$ would contribute to transport in the x-direction because they all have the same eigenstate in the x-direction. These states are referred to as *subbands*, or *modes*. Remember that since the x-direction is the direction of propagation, there will certainly be a physical connection on each side in the x-direction, so it won't be a true infinite two-dimensional well. However, we have seen that the eigenenergies of corresponding infinite well provide insight into which particles will propagate through and which won't.

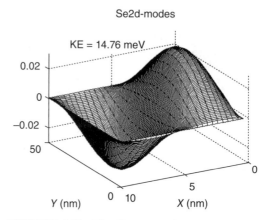

FIGURE 6.12   The $E_{2,1}$ state of the infinite well.

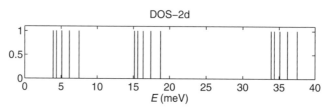

FIGURE 6.13   The first five subbands in the y-direction for each x-direction eigenstate of the 10×50 nm well.

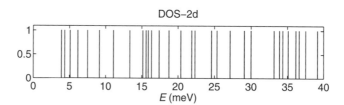

FIGURE 6.14   All the subbands of the 10×50 nm well.

The relationship between the fundamental x-direction eigenenergies and those of the subbands is illustrated in Figure 6.13. The density of states appears as a group of delta functions in the energy domain, similar to Figure 3.1. We know that they are actually broadened due to coupling, but that is not relevant to this discussion. The original eigenenergies at 3.75, 15, and 33.75 meV are there, but the related energies of the subbands are also there. It is clear that these subbands have a substantial influence on the available states for

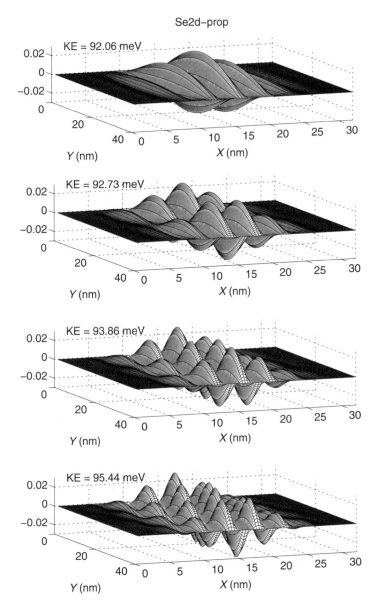

FIGURE 6.15   Particles propagating in the *x*-direction with different wavenumbers in the *y*-direction.

transport. Note that only the five lowest subbands are shown in Figure 6.14 for illustrative purposes. Figure 6.15 shows all the subbands up to 40 meV, but there is still a clustering around the *x*-directed eigenenergies.

We will want to consider particles that propagate in one dimension and be confined in the other. But it is important to realize that while the waveform propagates in only one direction, it must be in one of the accepted states in the other direction. Figure 6.15 shows four particles propagating in the positive x-direction. They each have a much higher energy than the particles in Figure 6.11 because the x-direction wavelength is about 4 nm. Going from top to bottom they have slightly higher energies because their wavelengths are smaller in the y-direction, but that is relatively insignificant. In fact, they all propagate at about the same speed.

Remember that even though the four particles have about the same speed, they are orthogonal to one another by virtue of their y-direction waveforms. These are the different *modes*.

EXERCISES

6.1   Bands

6.1.1   Using the program se6_1.m, change to a periodic potential with spikes at +1 meV. Are the results any different? Run this and show the results.

6.1.2   Using the program se6_1.m, change to a periodic potential with double spikes at 0.5 meV, that is, a spike, a space, and then another spike. Are the results any different? Run this and show the results.

6.1.3   If the lattice constant is 50 Å, what energy range constitutes the conduction band? (This assumes the bottom of the conduction band is 0 eV.) Use se6_1.m to qualitatively verify your result, that is, show that in one region it will propagate and in another it won't.

6.1.4   In this section we used the eigenstates of the infinite well to predict which wavelengths wouldn't propagate in a lattice. But in Exercise 3.4.1 we saw that a dual barrier passes an energy related to the ground state of a well of the same dimension. How can that be?

6.2   The Effective Mass

6.2.1   Figure 6.16 shows the first Brillouin zone of the conduction band of a semiconductor. Estimate the effective mass in terms of the constants $e0$ and a.

6.3   Modes (Subbands) in Quantum Structures

6.3.1   Suppose you have a three-dimensional structure with dimensions $100 \times 1000 \times 1000$Å in the x-, y-, and z-directions, respectively. If a particle is propagating in the x-direction, what is the difference in energy between the lowest and second lowest mode? What is the difference in energy between the second and third mode? What will the density of states look like, compared to Figure 6.14?

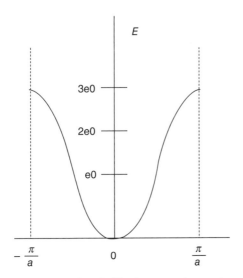

FIGURE 6.16   The first Brillouin zone of a semiconductor.

REFERENCES

1. G. W. Neudec and R. F. Pierret, *Advanced Semiconductor Fundamentals*, Englewood Cliffs, NJ: Prentice Hall, 2003.
2. D. A. Neamen, *Semiconductor Physics and Devices*, 3rd ed., New York: McGraw-Hill, 2003.

7

THE SCHRÖDINGER EQUATION FOR SPIN-1/2 FERMIONS

In this chapter we discuss more elaborate formulations of the Schrödinger equation that allow us to model the influence of electric and magnetic fields on an electron, and even the effects that two electrons in close proximity have on each other. We start by taking a closer look at the spin of an electron. We also describe some of the properties and operators associated with spin. In Section 7.2, we describe how the spin of an electron can be manipulated by an external magnetic field. In Section 7.3, we describe the influence of the Lorentz force on a moving electron and how this influence is incorporated into the Schrödinger equation. This is illustrated with a two-dimensional simulation. In Section 7.4, we describe the Hartree–Fock approximation. This approximation adds terms to the Schrödinger equation that describes the influence that two particles in close proximity have on each other.

7.1 SPIN IN FERMIONS

Classically, there are two kinds of angular momenta: *orbital angular momentum* $L = r \times p$, and *spin angular momentum* $S = I\omega$. The orbiting of the moon around the earth is an example of orbital angular momentum, while the earth's rotation about its axis is spin.

In quantum mechanics, finding the electron orbitals of the hydrogen atom (neglecting spin) provides states of definite orbital angular momentum. This

Quantum Mechanics for Electrical Engineers, First Edition. Dennis M. Sullivan.
© 2012 The Institute of Electrical and Electronics Engineers, Inc.
Published 2012 by John Wiley & Sons, Inc.

is one of the canonical problems in quantum mechanics that is covered in every beginning text [1]. We will not cover orbital angular momentum in this book. The quantum mechanical spin of a particle is analogous to classical spin in only the most superficial way. Every elementary particle has an inherent value s, which we call the *spin* of that particle. Electrons and other Fermions like protons and neutron have spin 1/2; photons have spin 1.

We will concentrate on the $s = 1/2$ case, since we are most interested in electrons. If we restrict ourselves to considering the spin of an electron, only two eigenstates, *spin up* and *spin down*, exist. With only two eigenstates, the state vectors can be represented by a two-component column vector; spin-up and spin-down eigenstates can be expressed as

$$\chi_\uparrow = \begin{pmatrix} 1 \\ 0 \end{pmatrix} \quad \text{and} \quad \chi_\downarrow = \begin{pmatrix} 0 \\ 1 \end{pmatrix}, \tag{7.1}$$

respectively. Any other spin state can be represented as a superposition of these two states:

$$\chi = \alpha\chi_\uparrow + \beta\chi_\downarrow = \begin{pmatrix} \alpha \\ \beta \end{pmatrix}. \tag{7.2}$$

We call these column vectors *spinors*. Remember that α and β can be complex. The coefficients must satisfy the normalization condition $|\alpha|^2 + |\beta|^2 = 1$.

7.1.1   Spinors in Three Dimensions

The three-dimensional spherical coordinates are illustrated in Figure 7.1.

Assuming $r = 1$, a position vector in rectangular coordinates can be described by three real numbers

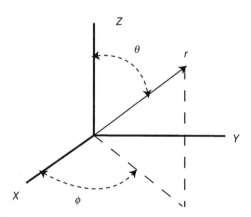

FIGURE 7.1   The three-dimensional spherical coordinates.

$$v(\theta, \phi) = \begin{pmatrix} \sin\theta\cos\phi \\ \sin\theta\sin\phi \\ \cos\theta \end{pmatrix} = \begin{pmatrix} x \\ y \\ z \end{pmatrix}, \tag{7.3}$$

Notice that θ dictates how far the vector is removed from the positive z-axis, and satisfies $0 \le \theta \le \pi$. If the vector is not on the z-axis, then it needs another angle, ϕ, to show how far it is from the x-axis, that is, its projection on to the xy plane.

A spinor can be represented as two complex numbers:

$$\chi(\theta, \phi) = \begin{pmatrix} \cos\dfrac{\theta}{2} \\ e^{i\phi}\sin\dfrac{\theta}{2} \end{pmatrix} = \begin{pmatrix} e^{-i\phi}\cos\dfrac{\theta}{2} \\ \sin\dfrac{\theta}{2} \end{pmatrix}. \tag{7.4}$$

The relative magnitudes of the two complex numbers dictate the angle with the z-axis, while the relative phase between the two numbers dictates the angle with the x-axis. Note that the magnitudes of the two complex numbers satisfy the normalization condition

$$\chi(\theta, \phi)^{+}\chi(\theta, \phi) = \left(\cos\frac{\theta}{2}\right)^{*}\left(\cos\frac{\theta}{2}\right) + \left(e^{i\phi}\sin\frac{\theta}{2}\right)^{*}\left(e^{i\phi}\sin\frac{\theta}{2}\right)$$
$$= \left[\left(\cos\frac{\theta}{2}\right)^{2} + \left(\sin\frac{\theta}{2}\right)^{2}\right] = 1.$$

By convention, as in Equation (7.1) we say

$$\chi(\theta = 0, \phi) = \chi_{\uparrow} = \begin{pmatrix} 1 \\ 0 \end{pmatrix} \tag{7.5a}$$

is spin "up," while

$$\chi(\theta = \pi, \phi) = \chi_{\downarrow} = \begin{pmatrix} 0 \\ 1 \end{pmatrix} \tag{7.5b}$$

is spin "down".

Spinors trace out a sphere of unit size called the *Bloch sphere* [2], as shown in Figure 7.2.

Note that the Cartesian unit vectors can be written as spin vectors, or spinors:

$$X = v(\theta = \pi/2, \phi = 0) = \begin{pmatrix} 1 \\ 0 \\ 0 \end{pmatrix} = \chi(\theta = \pi/2, \phi = 0) = \begin{pmatrix} 0.707 \\ 0.707 \end{pmatrix},$$

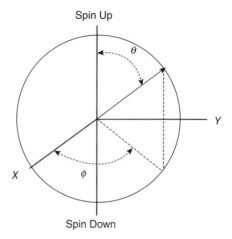

FIGURE 7.2   The Bloch sphere.

$$Y = v(\theta = \pi/2, \phi = \pi/2) = \begin{pmatrix} 0 \\ 1 \\ 0 \end{pmatrix} = \chi(\theta = \pi/2, \phi = \pi/2) = \begin{pmatrix} 0.707 \\ i0.707 \end{pmatrix}.$$

Note also that $\theta = 0$ means spin up

$$Z = v(\theta = 0) = \begin{pmatrix} 0 \\ 0 \\ 1 \end{pmatrix} = \chi(\theta = 0) = \begin{pmatrix} 1 \\ 0 \end{pmatrix},$$

and $\theta = \pi$ means spin down

$$-Z = v(\theta = \pi) = \begin{pmatrix} 0 \\ 0 \\ -1 \end{pmatrix} = \chi(\theta = \pi) = \begin{pmatrix} 0 \\ 1 \end{pmatrix}.$$

Example. In what direction does the following spinor point?

$$\chi = \begin{pmatrix} -i0.707 \\ 0.707 \end{pmatrix}$$

Solution. Remember that it is the relative phase between the upper and lower parts that determines ϕ, so we can write:

$$\chi = \begin{pmatrix} -i0.707 \\ 0.707 \end{pmatrix} = \begin{pmatrix} 0.707 \\ i0.707 \end{pmatrix} = \begin{pmatrix} 0.707 \\ 0.707e^{i90°} \end{pmatrix},$$

From Equation (7.4), we know $\phi = 90°$:

$$\cos\left(\frac{\theta}{2}\right) = 0.707 \Rightarrow \theta = 90°,$$

that is, the spinor is directed along the y-axis.

Example. In what direction does the following spinor point?

$$\chi = \frac{1}{5}\begin{pmatrix} 1+i\sqrt{8} \\ 2\sqrt{2}\,(1+i) \end{pmatrix}$$

Solution

$$\chi = \frac{1}{5}\begin{pmatrix} 1+i\sqrt{8} \\ 2\sqrt{2}\,(1+i) \end{pmatrix} = \frac{1}{5}\begin{pmatrix} 3e^{i70°} \\ 4e^{i45°} \end{pmatrix} = \begin{pmatrix} 0.6e^{i70°} \\ 0.8e^{i45°} \end{pmatrix},$$

$$\cos\left(\frac{\theta}{2}\right) = 0.6 \Rightarrow \theta = 106°,$$

$$\phi = 45° - 70° = -25°.$$

7.1.2   The Pauli Spin Matrices

Since our spin states have only two elements, the spin operators will be represented by 2×2 matrices. One very important group of operators is the Pauli spin matrices:

$$\sigma_x = \begin{bmatrix} 0 & 1 \\ 1 & 0 \end{bmatrix}, \quad \sigma_y = \begin{bmatrix} 0 & -i \\ i & 0 \end{bmatrix}, \quad \sigma_z = \begin{bmatrix} 1 & 0 \\ 0 & -1 \end{bmatrix}. \tag{7.6}$$

If we operate on the up spinor with the σ_z operator,

$$\sigma_z \chi_\uparrow = \begin{bmatrix} 1 & 0 \\ 0 & -1 \end{bmatrix}\begin{pmatrix} 1 \\ 0 \end{pmatrix} = \begin{pmatrix} 1 \\ 0 \end{pmatrix},$$

we get the same spinor back. However, if we operate on the up spinor with σ_x:

$$\sigma_x \chi_\uparrow = \begin{bmatrix} 0 & 1 \\ 1 & 0 \end{bmatrix}\begin{pmatrix} 1 \\ 0 \end{pmatrix} = \begin{pmatrix} 0 \\ 1 \end{pmatrix} = \chi_\downarrow.$$

What if we operate on a Y spinor with σ_z?

$$\sigma_z \chi_Y = \begin{bmatrix} 1 & 0 \\ 0 & -1 \end{bmatrix} \frac{1}{\sqrt{2}} \begin{pmatrix} 1 \\ i \end{pmatrix} = \frac{1}{\sqrt{2}} \begin{pmatrix} 1 \\ -i \end{pmatrix}.$$

The result points in the $-y$-direction. Each of the Pauli matrices spins the spinor 180° around its own axis. If a Pauli matrix operates on a spin in its own direction, it leaves it unchanged.

What if we take the expectation values of the Pauli matrices in a given state? We will start with a spinor located at $\theta = 45°$, $\phi = 90°$, that is, 45°between the y- and z-axes:

$$\chi = \begin{pmatrix} \cos(45°/2) \\ \sin(45°/2)e^{i90°} \end{pmatrix} = \begin{pmatrix} 0.924 \\ i0.383 \end{pmatrix}.$$

Consider the expectation value of σ_y:

$$\langle \sigma_y \rangle = [0.924 \quad -i0.383] \begin{bmatrix} 0 & -i \\ i & 0 \end{bmatrix} \begin{pmatrix} 0.924 \\ i0.383 \end{pmatrix}$$

$$= [0.924 \quad -i0.383] \begin{pmatrix} 0.383 \\ i0.924 \end{pmatrix} = 0.354 + 0.354 = 0.707.$$

This is the expectation value of finding the spin along the y-axis. In a similar manner we would calculate $\langle \sigma_x \rangle = 0$ and $\langle \sigma_z \rangle = 0.707$. This tells us that the expectation value of the spin along the x-direction is zero and is equal along the y- and z-directions. Note that the Pauli matrices satisfy

$$\langle \sigma_x \rangle^2 + \langle \sigma_y \rangle^2 + \langle \sigma_z \rangle^2 = 1.$$

7.1.3 Simulation of Spin

If we want to include spin into the state vector, ψ, we will have to represent it as a two-component column vector [3],

$$\psi = \begin{pmatrix} \psi_U \\ \psi_D \end{pmatrix}, \tag{7.7}$$

where ψ_U is the spin-up part and ψ_D is the spin-down part.

Let us take as our first example the standard one-dimensional quantum well. Suppose

$$\psi_U(x) = \sqrt{\frac{2}{a}} \sin\left(\frac{\pi x}{a}\right), \quad \psi_D(x) = 0.$$

Clearly this particle is "spin up." If it were

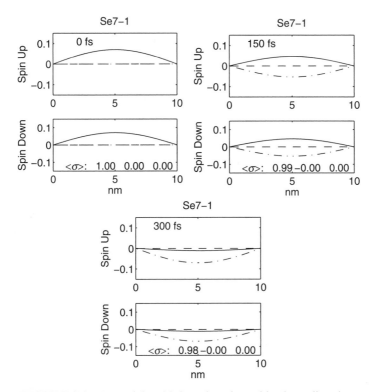

FIGURE 7.3   A particle with its spin oriented in the x-direction.

$$\psi_U(x) = 0, \quad \psi_D(x) = \sqrt{\frac{2}{a}} \sin\left(\frac{\pi x}{a}\right),$$

we would say it is spin down. How about

$$\psi_U(x) = 0.707\sqrt{\frac{2}{a}} \sin\left(\frac{\pi x}{a}\right), \quad \psi_D(x) = 0.707\sqrt{\frac{2}{a}} \sin\left(\frac{\pi x}{a}\right)?$$

This spinor lies in the x-direction. Figure 7.3 shows a simulation of this particle, including both how it changes in real space and in spin space. Notice that as time progresses, the spin-up and spin-down parts oscillate in phase, which maintains the overall spin orientation.

If

$$\psi_U(x) = 0.707\sqrt{\frac{2}{a}} \sin\left(\frac{\pi x}{a}\right), \quad \psi_D(x) = i0.707\sqrt{\frac{2}{a}} \sin\left(\frac{\pi x}{a}\right),$$

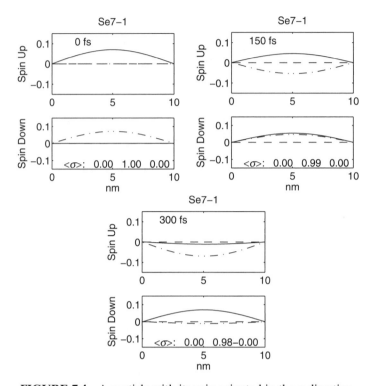

FIGURE 7.4   A particle with its spin oriented in the *y*-direction.

then we would say the spin is in the *y*-direction. A simulation of this is shown in Figure 7.4.

In the above, we simply carried out independent calculations for the evolution of ψ_U and ψ_D; what happens with ψ_U does not affect what happens with ψ_D, and vice versa. We shall see shortly, however, that there are important situations where the evolution of the two components of the spinors are coupled.

The following spin angular momentum operators are closely related to the Pauli matrices:

$$S_x = \frac{\hbar}{2}\sigma_x, \quad S_y = \frac{\hbar}{2}\sigma_y, \quad S_z = \frac{\hbar}{2}\sigma_z.$$

This can be written as:

$$\mathbf{S} = \frac{\hbar}{2}\boldsymbol{\sigma}.$$

For example:

$$S_z = \frac{\hbar}{2}\begin{bmatrix} 1 & 0 \\ 0 & -1 \end{bmatrix}. \tag{7.8}$$

If we operate on a pure spin-up state with S_z we get

$$S_z\chi_\uparrow = S_z\begin{pmatrix} 1 \\ 0 \end{pmatrix} = \frac{\hbar}{2}\begin{bmatrix} 1 & 0 \\ 0 & -1 \end{bmatrix}\begin{pmatrix} 1 \\ 0 \end{pmatrix} = \frac{\hbar}{2}\chi_\uparrow.$$

Since the operator S_z returned the spinor $\chi\uparrow$ times a value $\hbar/2$, spin up must be an *eigenspinor* of S_z with corresponding eigenvalue $\hbar/2$. In other words, $\chi\uparrow$ is a state of definite z-component spin angular momentum with eigenvalue $\hbar/2$. If S_z operates on a spin-down state, then

$$S_z\chi_\downarrow = S_z\begin{pmatrix} 0 \\ 1 \end{pmatrix} = \frac{\hbar}{2}\begin{bmatrix} 1 & 0 \\ 0 & -1 \end{bmatrix}\begin{pmatrix} 0 \\ 1 \end{pmatrix} = -\frac{\hbar}{2}\begin{pmatrix} 0 \\ 1 \end{pmatrix} = -\frac{\hbar}{2}\chi_\downarrow.$$

So spin down is also an eigenspinor but with eigenvalue $-\hbar/2$. If we take the expectation value of S_z with respect to the spin-down eigenspinor, then

$$\langle S_z \rangle = \langle \chi_\downarrow | S_z\chi_\downarrow \rangle = \left\langle \chi_\downarrow \left| -\frac{\hbar}{2}\chi_\downarrow \right. \right\rangle = -\frac{\hbar}{2}.$$

Needless to say the expectation value of the z-component of spin angular momentum of the spin-up eigenspinor is $\hbar/2$. Let us see if that is also true of S_x operating on a spinor in the x-direction:

$$S_x\chi_x = \frac{\hbar}{2}\sigma_x\frac{1}{\sqrt{2}}\begin{bmatrix} 1 \\ 1 \end{bmatrix} = \frac{\hbar}{2}\frac{1}{\sqrt{2}}\begin{bmatrix} 0 & 1 \\ 1 & 0 \end{bmatrix}\begin{bmatrix} 1 \\ 1 \end{bmatrix} = \frac{\hbar}{2}\begin{bmatrix} 1 \\ 1 \end{bmatrix} = \frac{\hbar}{2}\chi_x,$$

$$\langle S_x \rangle = \langle \chi_x | S_x\chi_x \rangle = \frac{\hbar}{2}.$$

Thus, χ_x is an eigenspinor of S_x. Let us start with a spin-up spinor, and operate on it with S_x and S_y:

$$S_x\chi_\uparrow = \frac{\hbar}{2}\sigma_x\begin{bmatrix} 1 \\ 0 \end{bmatrix} = \frac{\hbar}{2}\begin{bmatrix} 0 & 1 \\ 1 & 0 \end{bmatrix}\begin{bmatrix} 1 \\ 0 \end{bmatrix} = \frac{\hbar}{2}\begin{bmatrix} 0 \\ 1 \end{bmatrix} = \frac{\hbar}{2}\chi_\downarrow,$$

$$S_y\chi_\uparrow = \frac{\hbar}{2}\sigma_y\begin{bmatrix} 1 \\ 0 \end{bmatrix} = \frac{\hbar}{2}\begin{bmatrix} 0 & -i \\ i & 0 \end{bmatrix}\begin{bmatrix} 1 \\ 0 \end{bmatrix} = \frac{\hbar}{2}\begin{bmatrix} 0 \\ i \end{bmatrix} = i\frac{\hbar}{2}\chi_\downarrow.$$

We can set

$$\begin{bmatrix} 0 \\ i \end{bmatrix} = \begin{bmatrix} 0 \\ 1 \end{bmatrix}$$

because only the relative phase between 1 and 0 is relevant. Thus, S_x or S_y flips a spin-up spinor to spin down (along with the introduction of an additional coefficient). What does the following do?

$$S_z \chi_X = \frac{\hbar}{2}\begin{bmatrix} 1 & 0 \\ 0 & -1 \end{bmatrix}\begin{bmatrix} .707 \\ .707 \end{bmatrix} = \frac{\hbar}{2}\begin{bmatrix} .707 \\ -.707 \end{bmatrix}$$

S_z spins the X spinor around the z-axis so the spinor is pointing in the $-x$-direction as well as multiplying the result by $(\hbar/2)$.

What is the expectation value of S_z on an x-directed spinor?

$$\langle S_z \rangle = \begin{pmatrix} 0.707 \\ 0.707 \end{pmatrix}^+ \frac{\hbar}{2}\begin{bmatrix} 1 & 0 \\ 0 & -1 \end{bmatrix}\begin{pmatrix} 0.707 \\ 0.707 \end{pmatrix} = \frac{\hbar}{2}(0.707 \quad 0.707)\begin{pmatrix} 0.707 \\ -0.707 \end{pmatrix} = 0.$$

We might summarize what we have seen by the following observations:

1. If one of the S operators operates on a spinor in its direction, the operator returns the same spinor times the value $\pm\hbar/2$. The expected value of this operation is $\pm\hbar/2$.
2. If one of the S operators operates on a spinor perpendicular to its direction, the operator returns the spinor to the opposite direction that it started times the value $\pm\hbar/2$. The expected value of this operation is 0.

Example. For the following spinor,

$$\chi = \frac{1}{\sqrt{3}}\begin{pmatrix} 1+i \\ 1 \end{pmatrix},$$

calculate $\langle S_x \rangle$, $\langle S_y \rangle$, and $\langle S_z \rangle$.

Solution. We calculate the expectation values by the following operations:

$$\langle S_x \rangle = \frac{1}{3}\begin{pmatrix} 1+i \\ 1 \end{pmatrix}^+ \frac{\hbar}{2}\begin{bmatrix} 0 & 1 \\ 1 & 0 \end{bmatrix}\begin{pmatrix} 1+i \\ 1 \end{pmatrix}$$

$$= \frac{1}{3}\frac{\hbar}{2}(1-i \quad 1)\begin{pmatrix} 1 \\ 1+i \end{pmatrix} = \frac{1}{3}\frac{\hbar}{2}[1-i-1+i] = \frac{\hbar}{2}\frac{2}{3} = \frac{\hbar}{3},$$

$$\langle S_y \rangle = \frac{1}{3}\begin{pmatrix} 1+i \\ 1 \end{pmatrix}^+ \frac{\hbar}{2}\begin{bmatrix} 0 & -i \\ i & 0 \end{bmatrix}\begin{pmatrix} 1+i \\ 1 \end{pmatrix}$$

$$= \frac{1}{3}\frac{\hbar}{2}(1-i \quad 1)\begin{pmatrix} -i \\ -1+i \end{pmatrix} = \frac{1}{3}\frac{\hbar}{2}[-1-i-1+i] = -\frac{\hbar}{2}\frac{2}{3} = -\frac{\hbar}{3},$$

$$\langle S_z \rangle = \frac{1}{3}\begin{pmatrix} 1+i \\ 1 \end{pmatrix}^\dagger \frac{\hbar}{2}\begin{bmatrix} 1 & 0 \\ 0 & -1 \end{bmatrix}\begin{pmatrix} 1+i \\ 1 \end{pmatrix}$$

$$= \frac{1}{3}\frac{\hbar}{2}(1-i \quad 1)\begin{pmatrix} 1+i \\ -1 \end{pmatrix} = \frac{\hbar}{2}\frac{1}{3} = \frac{\hbar}{6}.$$

Note that:

$$\langle S \rangle^2 = \langle S_x \rangle^2 + \langle S_y \rangle^2 + \langle S_z \rangle^2$$

$$= \left(\frac{\hbar}{2}\right)^2 \left(\frac{1}{3^2} + \frac{2^2}{3^2} + \frac{2^2}{3^2}\right) = \left(\frac{\hbar}{2}\right)^2.$$

The S matrices are Hermitian. Therefore, we can determine eigenvalues and eigenstates using standard matrix algebra. Start with the S_z operator. We know it has an eigenergy ε, so

$$\varepsilon\psi = S_z\psi,$$

$$|\varepsilon I - S_z| = 0,$$

$$\begin{vmatrix} \varepsilon - \hbar/2 & 0 \\ 0 & \varepsilon + \hbar/2 \end{vmatrix} = 0,$$

$$(\varepsilon - \hbar/2)(\varepsilon + \hbar/2) = 0.$$

There are two solutions:

$$\varepsilon_1 = \frac{\hbar}{2}, \quad \varepsilon_2 = -\frac{\hbar}{2}.$$

What are the corresponding eigenfunctions? With little difficultly we can show that they are $\chi\uparrow$ and $\chi\downarrow$, respectively.

If we have a spinor pointing in the x-direction and make a measurement with S_z, what are the possible outcomes?

$$S_z\chi_x = \frac{\hbar}{2}\begin{bmatrix} 1 & 0 \\ 0 & -1 \end{bmatrix}\begin{pmatrix} 0.707 \\ 0.707 \end{pmatrix} = \frac{\hbar}{2}\begin{pmatrix} 0.707 \\ -0.707 \end{pmatrix}.$$

We could write this as a superposition of the up and down spinors:

$$S_z\chi_x = \frac{\hbar}{2}0.707(\chi_\uparrow - \chi_\downarrow).$$

So we would say the outcome of the measurement has a 50% chance of being spin up and a 50% chance of being spin down.

7.2 AN ELECTRON IN A MAGNETIC FIELD

A spinning charged particle is a magnetic dipole. Similarly, a particle with spin produces a magnetic dipole moment. One can define the magnetic dipole moment [1] as:

$$\mu = \gamma S, \qquad (7.9)$$

where γ is a constant called the gyromagnetic ratio, and it is given by

$$\gamma = \frac{q_{electron}}{m_{electron}} = \frac{1.6 \times 10^{-19} \text{ C}}{9.1 \times 10^{-31} \text{ kg}} = 0.176 \times 10^{12} \text{ C/kg}.$$

Putting a magnetic dipole in a magnetic field B produces a torque

$$T = \mu \times B.$$

The energy is $-\mu \cdot B$. Therefore, we can write the Hamiltonian of a spin at rest as [4]:

$$H_s = -\mu \cdot B = -\gamma B \cdot S. \qquad (7.10)$$

Check the units (Remember the $\hbar$ in the S operator):

$$[H_s] = \left[\frac{\text{C}}{\text{kg}} \frac{\text{Wb}}{\text{m}^2} \cdot \text{J} \cdot \text{s} \right] = \left[\frac{\text{C}}{\text{kg}} \frac{1}{\text{m}^2} \frac{\text{kg} \cdot \text{m}^2}{\text{C} \cdot \text{s}} \text{J} \cdot \text{s} \right] = [\text{J}].$$

(B, the magnetic flux density, has units of Webers per meter squared, or Tesla.) Note that γS results in the following:

$$-\gamma \frac{\hbar}{2} = 0.5 \left(.176 \times 10^{12} \text{ C/kg} \right) \left(6.56 \times 10^{-14} \text{ eV} \times \text{s} \right)$$

$$= -0.00577 \text{ eV/Tesla}.$$

Equation (7.10) is a term that must be added to the Hamitonian when describing an electron *at rest* in a magnetic field.

Up until now, we have used the state variable ψ to describe the wave function that represents a particle, usually an electron. Now it will be necessary to include spin. So we might describe the particle by

$$\psi = \psi(x) \chi, \qquad (7.11)$$

where χ is the spinor. The full Hamiltonian for a particle with spin in a magnetic field is

$$H_{\text{total}} = H_0 + H_s = -\frac{\hbar^2}{2m}\frac{\partial^2}{\partial x^2} + V(x) - \gamma \boldsymbol{B} \cdot \boldsymbol{S}. \qquad (7.12)$$

The first two terms on the right in Equation (7.12) will act on the spatial wave form $\psi(x)$ while the third term on the right will act on the spinor ($\boldsymbol{B}$ is spatially uniform). We want to concentrate on the term acting on the spin, so we start by assuming a particle is in the ground state of a 10 nm infinite well. Now we can write Equation (7.11) as:

$$\psi = \phi_1(x)\chi. \qquad (7.13)$$

The time-dependent Schrödinger equation is

$$i\hbar \frac{\partial}{\partial t}\psi = H\psi, \qquad (7.14)$$

where H is given by Equation (7.12) and $V(x) = 0$. First look at the time derivative

$$i\hbar \frac{\partial}{\partial t}\psi = i\hbar \frac{\partial \phi_1}{\partial t}\chi + i\hbar \phi_1 \frac{\partial \chi}{\partial t}.$$

H_0 only operates on ϕ_1 while the H_s only operates on χ, so Equation (7.14) is

$$i\hbar \chi \frac{\partial \phi_1}{\partial t} + i\hbar \phi_1 \frac{\partial \chi}{\partial t} = \chi H_0 \phi_1 + \phi_1 H_s \chi,$$

and rearranging gives:

$$(\varepsilon_1 \phi_1 - H_0 \phi_1)\chi + i\hbar \phi_1 \frac{\partial \chi}{\partial t} = \phi_1 H_s \chi.$$

The term in parentheses goes to zero because ϕ_1 is in an eigenstate. Dividing out ϕ_1 leaves

$$i\hbar \frac{\partial \chi}{\partial t} = H_s \chi = -\gamma \boldsymbol{B} \cdot \boldsymbol{S}\chi. \qquad (7.15)$$

Note that this separation of the spatial and spin degrees of freedom only takes place because we assumed a wave function where these degrees of freedom factored.

Example. A particle is in the ground state of a 10 nm infinite well. Suppose the spin of the particle is in the x-direction, that is,

$$\chi = \frac{1}{\sqrt{2}}(\chi_\uparrow + \chi_\downarrow).$$

What happens when a magnetic field of B_0 Tesla is applied in the z-direction?

Solution. We are looking for a solution to Equation (7.15). If the B is directed in the z-direction, then

$$-\gamma \boldsymbol{B} \cdot \boldsymbol{S} = -\gamma B_o S_z.$$

Equation (7.15) becomes

$$i\hbar \frac{\partial}{\partial t}\chi = H_s \chi = -\gamma \frac{\hbar}{2} B_0 \sigma_z \chi. \tag{7.16}$$

Start by looking for the eigenstates. Since

$$H_s = -\gamma \frac{\hbar}{2} B_0 \begin{bmatrix} 1 & 0 \\ 0 & -1 \end{bmatrix},$$

the eigenenergies are

$$\varepsilon_1 = -\gamma \frac{\hbar}{2} B_0, \quad \varepsilon_2 = \gamma \frac{\hbar}{2} B_0,$$

and the corresponding eigenstates are $\chi_\uparrow$ and $\chi_\downarrow$. Therefore, any solution to the time-domain Schrödinger equation must be a superposition of these two solutions:

$$\begin{aligned} \chi(t) &= c_1 \chi_\uparrow e^{-i\varepsilon_1 t/\hbar} + c_2 \chi_\downarrow e^{-i\varepsilon_2 t/\hbar} \\ &= c_1 \chi_\uparrow e^{i(\gamma B_0/2)t} + c_2 \chi_\downarrow e^{-i(\gamma B_0/2)t}, \end{aligned} \tag{7.17}$$

where $|c_1|^2 + |c_2|^2 = 1$. The c values are determined by the initial conditions. It was stated that the spin is pointing in the x-direction, so

$$c_1 = c_2 = 0.707.$$

To follow the particle under the influence of the magnetic field, we look at the expectation values of the Pauli matrices:

$$\langle \sigma_x \rangle = \langle \chi | \sigma_x \chi \rangle = \begin{bmatrix} c_1^* e^{-i(\gamma B_0/2)t} & c_2^* e^{i(\gamma B_0/2)t} \end{bmatrix} \begin{bmatrix} 0 & 1 \\ 1 & 0 \end{bmatrix} \begin{bmatrix} c_1 e^{i(\gamma B_0/2)t} \\ c_2 e^{-i(\gamma B_0/2)t} \end{bmatrix}$$

$$= \begin{bmatrix} c_1^* e^{-i(\gamma B_0/2)t} & c_2^* e^{i(\gamma B_0/2)t} \end{bmatrix} \begin{bmatrix} c_2 e^{-i(\gamma B_0/2)t} \\ c_1 e^{i(\gamma B_0/2)t} \end{bmatrix} = \left(e^{i2(\gamma B_0/2)t} + e^{-i2(\gamma B_0/2)t} \right) = \cos(\omega_L t).$$

The frequency

$$2\left(\frac{\gamma}{2}B_0\right) = \gamma B_0 = \omega_L$$

is called the *Larmor frequency*; this is the frequency at which the spin precesses in the magnetic field. Similarly,

$$\langle\sigma_y\rangle = \langle\chi|\sigma_y\chi\rangle = \begin{bmatrix} c_1 e^{-i(\gamma B_0/2)t} & -ic_2 e^{i(\gamma B_0/2)t} \end{bmatrix}\begin{bmatrix} 0 & -i \\ i & 0 \end{bmatrix}\begin{bmatrix} c_1 e^{i(\gamma B_0/2)t} \\ ic_2 e^{-i(\gamma B_0/2)t} \end{bmatrix}$$

$$= \begin{bmatrix} c_1 e^{-i(\gamma B_0/2)t} & -ic_2 e^{i(\gamma B_0/2)t} \end{bmatrix}\begin{bmatrix} c_2 e^{-i(\gamma B_0/2)t} \\ ic_1 e^{i(\gamma B_0/2)t} \end{bmatrix} = -\sin(\omega_L t).$$

If we calculate the expectation value in the z-direction, then

$$\langle\sigma_z\rangle = \langle\chi|\sigma_z\chi\rangle = \begin{bmatrix} c_1 e^{i\omega_0 t} & c_2 e^{-i\omega_0 t} \end{bmatrix}\begin{bmatrix} 1 & 0 \\ 0 & -1 \end{bmatrix}\begin{bmatrix} c_1 e^{-i\omega_0 t} \\ c_2 e^{i\omega_0 t} \end{bmatrix}$$

$$= \begin{bmatrix} c_1 e^{i\omega_0 t} & c_2 e^{-i\omega_0 t} \end{bmatrix}\begin{bmatrix} c_1 e^{-i\omega_0 t} \\ -c_2 e^{i\omega_0 t} \end{bmatrix} = c_1 c_2 (1-1) = 0.$$

Note the units:

$$[\gamma B] = \left[\frac{C}{kg}\right]\left[\frac{kg \cdot m^2}{C \cdot s}\frac{1}{m^2}\right] = [s^{-1}].$$

The period is related to the Larmor frequency ω_L by

$$T_1 = \frac{2\pi}{\omega_L} = \frac{2\pi}{(0.176\times10^{12})B_0} = \frac{35.7 \text{ ps}}{B_0(\text{in Tesla})}.$$

Remember that we said a spinor could be written in a manner that related it to the three-dimensional representation of Figure 7.1:

$$\chi(\theta, \phi) = \begin{pmatrix} \cos\dfrac{\theta}{2} \\ e^{i\phi}\sin\dfrac{\theta}{2} \end{pmatrix}.$$

Therefore, we can take

$$c_1 = \cos\frac{\theta_0}{2}, c_2 = e^{i\phi_0}\sin\frac{\theta_0}{2},$$

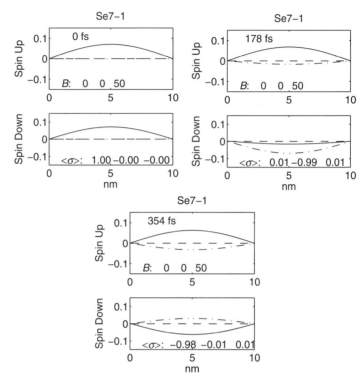

FIGURE 7.5 Simulation of a particle in a B_z field of 50 T that starts with its spin in the x-direction and precesses around the z-axis.

and we can write the solution, Equation (7.17). Basically, this says that if we start at an angle θ, we stay at that angle. But the angle ϕ is dependent on the relative phase between the spin-up and spin-down portion, and that it is changing in time. This says that it is precessing around the z-axis, which is what we expected because we applied a B field in the z-direction—precisely the meaning of the Larmor frequency.

A simulation of a particle in a magnetic field with a 50 T magnetic flux density in the z-direction is shown in Figure 7.5. If $B_0 = 50$ T, then the time period is $T = 0.714$ ps. This simulation shows that the spin of the particle starts in the x-direction. It is in the $-y$-direction after T/4, and is in the $-x$-direction after T/2.

7.3 A CHARGED PARTICLE MOVING IN COMBINED *E* AND *B* FIELDS

In Section 7.2, we talked about a particle at rest, that is, not moving, while it is in a uniform magnetic field. If a charged particle is moving, relativity dictates that the forces acting on it are far more complicated! The well-known Lorentz force law is

$$F = q(E + v \times B).$$

To represent this force in the Schrödinger equation, we need the Hamiltonian of a particle in the respective fields. Remember that we represented the electric field as the gradient of the scalar potential φ, that is, $E = -\nabla \varphi$. To represent the B field we use the vector potential A, which is related to the B field by $B = \nabla \times A$ [4],

$$A = -\frac{1}{2} r \times B = -\frac{1}{2} \begin{vmatrix} \hat{i} & \hat{j} & \hat{k} \\ x & y & z \\ B_x & B_y & B_z \end{vmatrix}$$

$$= \frac{1}{2} \left[-\hat{i}(yB_z - zB_y) + \hat{j}(xB_z - zB_x) - \hat{k}(xB_y - yB_x) \right].$$

(7.18)

The Hamiltonian [3] is:

$$H = \frac{1}{2m} \left(\frac{\hbar}{i} \nabla - q \cdot A \right)^2$$

$$= -\frac{\hbar^2}{2m} \nabla^2 - \frac{\hbar q}{im} \nabla \cdot A + \frac{q^2}{2m} A \cdot A.$$

(7.19)

Before we insert Equation (7.18) into Equation (7.19), let us make some simplifying assumptions. We will restrict ourselves to a two-dimensional problem in the xy plane, and assume our B field is a constant in the z-direction,

$$B = B_0 \hat{k}.$$

(7.20)

Consequently, A becomes

$$A = \frac{1}{2} \left[-\hat{i} y B_0 + \hat{j} x B_0 \right],$$

and

$$\nabla \cdot A = \frac{1}{2} \left[-\hat{i} y B_0 + \hat{j} x B_0 \right] = \frac{1}{2} B_0 \left(-y \frac{\partial}{\partial x} + x \frac{\partial}{\partial y} \right),$$

$$A \cdot A = \frac{1}{4} \left[-\hat{i} y B_0 + \hat{j} x B_0 \right] = \frac{B_0^2}{4} (x^2 + y^2).$$

Equation (7.19) now becomes

$$H = -\frac{\hbar^2}{2m}\left(\frac{\partial^2}{\partial x^2} + \frac{\partial^2}{\partial y^2}\right) + i\frac{\hbar B_0 q}{2m}\left(-y\frac{\partial}{\partial x} + x\frac{\partial}{\partial y}\right) + \frac{q^2 B_0^2}{8m}\left(x^2 + y^2\right). \quad (7.21)$$

Figure 7.6 is a two-dimensional simulation of a particle in a z-directed magnetic field, without considering spin. Figure 7.6a is the potential due to a **B** field of 25 T. This is simulated in the program by adding the potential

$$V_B(x, y) = \frac{q^2 B_0^2}{8m}\left[(x - 60\ \text{nm})^2 + (y - 60\ \text{nm})^2\right]. \quad (7.22)$$

Equation (7.22) is the third term on the right side of Equation (7.21). It has been found that the second term on the right side of Equation (7.21) plays a far less significant role for this kind of simulation, and so it has been omitted. In Figure 7.6b we see the wave function of an electron that is initially traveling in the positive x-direction. It is directed inward by the **B** field and tends to move around the inside of the potential like a marble moving around a bowl.

7.4 THE HARTREE–FOCK APPROXIMATION

Up until now, we have solved the Schrödinger equation for one electron. In this section, we will discuss the Schrödinger equation for two electrons in close proximity. There are two forces that come into play. There is the Coulomb force that says two particles with the same charge will be repelled from one another. This will result in an additional integral in the Schrödinger equation that we refer to as the Hartree term. There is another purely quantum mechanical term that stems from the Pauli exclusion principle that says no two particles can occupy the same state. This is often called the exchange force [1]. It also results in an integral in the Schrödinger equation that is called the Fock term.

7.4.1 The Hartree Term

Consider the situation shown in Figure 7.7. There are two electrons, one at position r_1 and one at position r_2. The potential energy (PE) due to the Coulomb force is

$$\text{PE}_{\text{Coul}} = \frac{e^2}{4\pi\varepsilon_0 |r_1 - r_2|}. \quad (7.23)$$

Now look at the situation in Figure 7.8. In an approximate sense that can be obtained rigorously [5], electron 1 effectively sees the potential of the charge density $-e|\Psi_2(r_2)|^2$ of electron 2, giving the effective PE

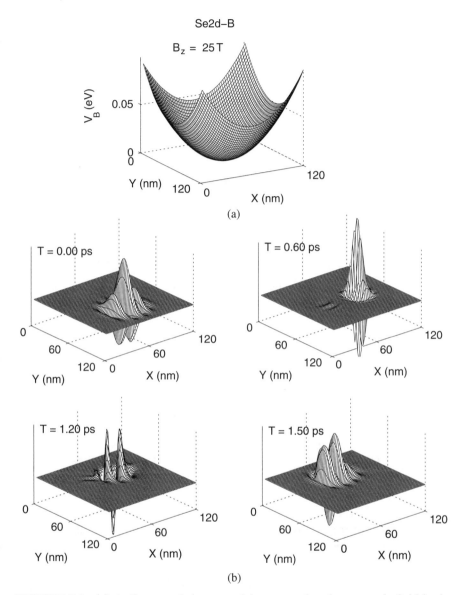

FIGURE 7.6 (a) A diagram of the potential representing the magnetic field in the z-direction, Equation (7.22). (b) Wave function of an electron moving in a magnetic field. At time $T = 0$, the particle is moving in the x-direction. At $T = 0.6$ ps, it has been turned inward by the magnetic field and is moving in the negative y-direction. The particle continues turning under the influence of the magnetic field, and by $T = 1.5$ ps, it is moving in the negative x-direction. KE, kinetic energy; PE, potential energy.

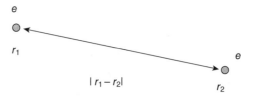

FIGURE 7.7 Two electrons at r_1 and r_2.

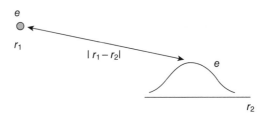

FIGURE 7.8 Electron 1 interacting with the effective charge distribution associated with electron 2.

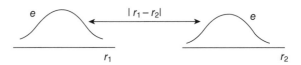

FIGURE 7.9 Effective charge distribution of electron1 interacting with the effective charge distribution associated with electron 2.

$$\text{PE}_{\text{Coul}} = \frac{e^2}{4\pi\varepsilon_0} \int \frac{1}{|r_1 - r_2|} |\psi_2(r_2)|^2 dr_2. \tag{7.24}$$

In fact, each electron behaves in some sense as a charge cloud, as shown in Figure 7.9, giving the effective PE

$$\text{PE}_{\text{Coul}} = \frac{e^2}{4\pi\varepsilon_0} \int dr_1 \int dr_2 |\psi_1(r_1)|^2 \frac{1}{|r_1 - r_2|} |\psi_2(r_2)|^2. \tag{7.25}$$

How would this enter the Schrödinger equation? We start by considering electron 1 with wave function ψ_1. The Schrödinger equation governing electron 1 under the influence of electron 2 is Equation (7.26a). Similarly, Equation (7.26b) is the Schrödinger equation for electron 2 as influenced by electron 1:

$$i\hbar \frac{\partial \psi_1(r_1)}{\partial t} = -\frac{\hbar^2}{2m} \nabla_{r_1}^2 \psi_1(r_1) + \frac{e^2}{4\pi\varepsilon_0} \int dr_2 \frac{|\psi_2(r_2)|^2}{|r_1 - r_2|} \psi_1(r_1), \tag{7.26a}$$

$$i\hbar\frac{\partial\psi_2(r_2)}{\partial t} = -\frac{\hbar^2}{2m}\nabla_{r_2}^2\psi_2(r_2) + \frac{e^2}{4\pi\varepsilon_0}\int dr_1 \frac{|\psi_1(r_1)|^2}{|r_1 - r_2|}\psi_2(r_2). \qquad (7.26b)$$

The term on the far right of each equation is called the *Hartree term*. Note that when we calculate the expectation value of the PE for Equation (7.26a), we get the integral of Equation (7.25). Equation (7.26a,b) shows two coupled *nonlinear* differential equations for ψ_1 and ψ_2. How such approximate nonlinear equations arise from the original Schrödinger equation is described in Ashcroft and Mermin [5].

We would like to be able to simulate the interaction of two particles under the influence of the Coulomb potential with our MATLAB programs. On the surface, that seems straightforward. However, take a closer look at the Hartree term of Equation (7.26a). This says that as we calculate the updated values of ψ_1, we have to stop at every point r_1 and perform the integral over the entire r_2 length of ψ_2! This would easily overload the simulation program. Suppose we look at the Hartree term with a different perspective. Let us rewrite Equation (7.26a) as:

$$h(r) = \frac{e^2}{4\pi\varepsilon_0}\frac{1}{|r|}, \qquad (7.27a)$$

$$i\hbar\frac{\partial\psi_1(r_1)}{\partial t} = -\frac{\hbar^2}{2m}\nabla_{r_1}^2\psi_1(r_1) + \left[\int h(r_1 - r_2)|\psi_2(r_2)|^2 \, dr_2\right]\psi_1(r_1). \qquad (7.27b)$$

The integral is a convolution. Fourier theory tells us that we can take the Fourier transforms of h and ψ_1, multiply them together, and take the inverse Fourier transform to get the Hartree term. We can take the Fourier transform of h once at the beginning of the simulation, but we have to do the Fourier transform of ψ_2 at every iteration as well as the inverse Fourier transform. Using the fast Fourier transform (FFT) makes this process efficient. The integral is still the most computationally intensive part of the calculation, but it becomes more manageable using this method.

One remaining trick speeds up the above process. The FFT algorithm uses complex numbers for its input and output. But $|\Psi_1(r_1)|^2$ and $|\Psi_2(r_2)|^2$ are both real functions. Therefore, by using the function $|\Psi_1(r_1)|^2 + i|\Psi_2(r_2)|^2$ as the input, after performing the convolution using FFT, the real part is used in Equation (7.26a) and the imaginary part in Equation (7.26b).

As an example, we start with two particles in the 10 nm infinite well and for now we will assume they have opposite spins; one is in the ground state and the other is in the first excited state, as shown in the far left column of Figure 7.10. The Hartree terms of Equation (7.26) must be added gradually during the simulation. A Hanning window is applied to the Hartee term over a period of 10,000 iterations. (See Appendix A for an explanation of the Hanning window.) The right column of Figure 7.10 shows the simulation at

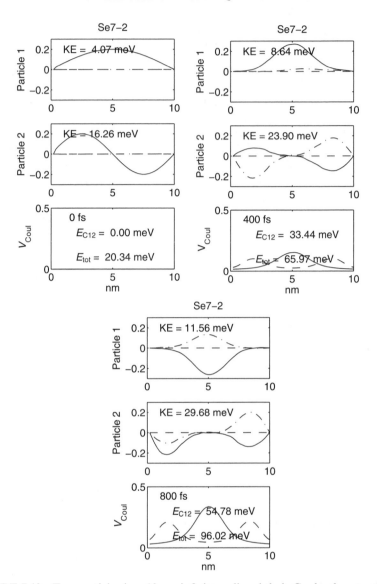

FIGURE 7.10 Two particles in a 10 nm infinite well and their Coulomb potentials at three different time periods as the Coulomb potential is gradually added. KE, kinetic energy.

the halfway point. The bottom column shows the simulation with the full Hartree term. The paramener E_{C12} is the total Coulomb energy between the particles. Not unexpectedly, the particle initially in the ground state has been squeezed toward the middle and the second particle's wave function has largely split into two pieces.

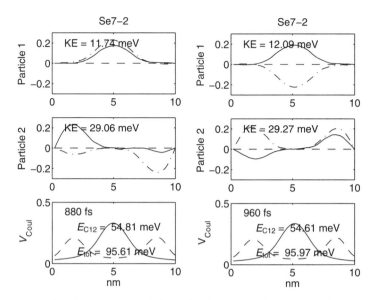

FIGURE 7.11 The two particles of Figure 7.10 after the Hartree term has been added. Notice that each particle remains in a stable state as dictated by the Coulomb potential presented by the other particle. KE, kinetic energy.

If we let the simulation continue (see Fig. 7.11), the energy of each particle as well as the total energy remains the same.

7.4.2 The Fock Term

An examination of Equation (7.26a,b) shows that there is nothing to prevent two electrons from being in the same state. In other words, the Hartree approximation does not obey the Pauli exclusion principle. The Fock term will correct this. To arrive at a consistent view, consider more carefully the *two-electron* wave function. We might be tempted to say that it is a simple product: $\Psi(r_1, r_2) = \Psi_1(r_1)\Psi_2(r_2)$. This assumes we can tell the particles apart: the first electron is in state ψ_1 and the second electron is in state ψ_2. We have already said that in quantum mechanics, we cannot talk about this electron or that electron. All electrons must be *indistinguishable*. This is addressed through the *Slater determinant*, which for two electrons is:

$$\psi(r_1, r_2) = \frac{1}{\sqrt{2}} \begin{vmatrix} \psi_1(r_1) & \psi_2(r_1) \\ \psi_1(r_2) & \psi_2(r_2) \end{vmatrix}$$

$$= \frac{1}{\sqrt{2}}[\psi_1(r_1)\psi_2(r_2) - \psi_1(r_2)\psi_2(r_1)]. \tag{7.28}$$

The $1/\sqrt{2}$ is the normalization factor. If ψ_1 and ψ_2 are both the same, then $\Psi(r_1, r_2)$ goes to zero. The Slater determinant automatically embodies the *Pauli exclusion principle*. Let us see what happens when the coordinates are exchanged:

$$\psi(r_2, r_1) = \frac{1}{\sqrt{2}}[\psi_1(r_2)\psi_2(r_1) - \psi_1(r_1)\psi_2(r_2)] = -\psi(r_1, r_2).$$

Thus, we arrive at the original two-particle wave function but with the overall sign reversed. We say that this wave function is odd under the exchange parity operation. This is also a characteristic of identical Fermions. It will be interesting to determine the modulus of this function for the two-particle system:

$$
\begin{aligned}
|\psi(r_1, r_2)|^2 &= \frac{1}{\sqrt{2}}[\psi_1(r_1)\psi_2(r_2) - \psi_1(r_2)\psi_2(r_1)] \\
&\times \frac{1}{\sqrt{2}}[\psi_1^*(r_1)\psi_2^*(r_2) - \psi_1^*(r_2)\psi_2^*(r_1)] \\
&= \frac{1}{2}[|\psi_1(r_1)|^2 |\psi_2(r_2)|^2 + |\psi_1(r_2)|^2 |\psi_2(r_1)|^2 \\
&\quad - \psi_1(r_1)\psi_2(r_2)\psi_1^*(r_2)\psi_2^*(r_1) \\
&\quad - \psi_1(r_2)\psi_2(r_1)\psi_1^*(r_1)\psi_2^*(r_2)].
\end{aligned}
\tag{7.29}
$$

Now we write the Coulomb interaction for our two-fermion system, similar to Equation (7.25),

$$
\begin{aligned}
\text{PE} &= \frac{e^2}{4\pi\varepsilon_0}\int dr_1 \int dr_2 |\psi(r_1, r_2)|^2 \frac{1}{|r_1 - r_2|} \\
&= \frac{e^2}{4\pi\varepsilon_0}\int dr_1 \int dr_2 |\psi_1(r_1)|^2 \frac{1}{|r_1 - r_2|}|\psi_2(r_2)|^2 \\
&\quad + \frac{e^2}{4\pi\varepsilon_0}\int dr_1 \int dr_2 |\psi_1(r_2)|^2 \frac{1}{|r_1 - r_2|}|\psi_2(r_1)|^2 \\
&\quad - \frac{e^2}{4\pi\varepsilon_0}\int dr_1 \int dr_2 \psi_1(r_1)\psi_2(r_2)\frac{1}{|r_1 - r_2|}\psi_1^*(r_2)\psi_2^*(r_1)\delta_{s_1,s_2} \\
&\quad - \frac{e^2}{4\pi\varepsilon_0}\int dr_1 \int dr_2 \psi_1(r_2)\psi_2(r_1)\frac{1}{|r_1 - r_2|}\psi_1^*(r_1)\psi_2^*(r_2)\delta_{s_1,s_2}.
\end{aligned}
$$

The first two integrals are just the Coulomb integral of Equation (7.25). The last two integrals are a purely quantum mechanical terms. The function δ_{s_1,s_2} is the Kronecker delta function. The s_1 and s_2 refer to spin up and spin down, respectively. It has the value

$$\delta_{s_1,s_2} = \begin{cases} 1 & \text{if } s_1 = s_2, \\ 0 & \text{if } s_1 \neq s_2 \end{cases}.$$

When we include these PE terms in the Schrödinger equation [5], we get

$$i\hbar \frac{\partial \psi_1(r_1)}{\partial t} = -\frac{\hbar^2}{2m}\nabla_{r_1}^2 \psi_1(r_1) + \frac{e^2}{4\pi\varepsilon_0}\int dr_2 \frac{|\psi_2(r_2)|^2}{|r_1-r_2|}\psi_1(r_1)$$
$$-\frac{e^2}{4\pi\varepsilon_0}\int dr_2 \frac{\psi_2^*(r_2)\psi_1(r_2)}{|r_1-r_2|}\psi_2(r_1)\delta_{s_1,s_2},$$

(7.30a)

$$i\hbar \frac{\partial \psi_2(r_2)}{\partial t} = -\frac{\hbar^2}{2m}\nabla_{r_2}^2 \psi_2(r_2) + \frac{e^2}{4\pi\varepsilon_0}\int dr_1 \frac{|\psi_1(r_1)|^2}{|r_1-r_2|}\psi_2(r_2)$$
$$-\frac{e^2}{4\pi\varepsilon_0}\int dr_1 \frac{\psi_1^*(r_1)\psi_2(r_1)}{|r_1-r_2|}\psi_1(r_2)\delta_{s_1,s_2}.$$

(7.30b)

The last term on the right in Equation (7.30a,b) is called the *exchange term* or the *Fock term*. Equation (7.30) is termed the *Hartree–Fock* approximation.

The implementation of the Fock term into the finite-difference simulation is similar to that of the Hartree term [6, 7]. The Fock term can be regarded as a convolution and the FFT can be used to calculate the integral. Note that the inputs to the convolution integrals, $\psi_2^*(r_2)\psi_1(r_2)$ in Equation (7.30a) and $\psi_1^*(r_1)\psi_2(r_1)$ in Equation (7.30b), are complex conjugates of each other. Therefore, it is only necessary to calculate one additional convolution.

EXERCISES

7.1 Spin in Fermions

7.1.1 Write the X and Y spinors, $\chi^{(X)}$, $\chi^{(Y)}$ as superpositions of the up and down spinors, χ_+ and χ_-.

7.1.2 Find a spinor that has angles $\theta = 45°$ and $\phi = 30°$ on the Bloch sphere. (Remember, it must be normalized.) Calculate the expectation values of S_x, S_y, and S_z.

7.1.3 A spinor is located $45°$ between the y- and z-axis.

(a) What is this spinor (represented by a two-component column vector)?

(b) What is the resulting spinor if it is rotated around the y-axis? (Use the appropriate operator; do not just do it graphically.)

7.1.4 An important quantity in quantum mechanics is the commutator

$$[\mathbf{A}, \mathbf{B}] = \mathbf{AB} - \mathbf{BA},$$

where **A** and **B** are both operators. Show that

$$[S_x, S_y] = i\hbar S_z.$$

The following are also true: $[S_y, S_z] = i\hbar S_x$ and $[S_z, S_x] = i\hbar S_y$. Make an argument that you do not have to go through the explicit algebra to know that the second two statements are true if the first one is true.

A note on commutators. When $[A, B] = 0$, we say that these operators *commute*, and A and B are called *compatible observables*. Compatible observables share a complete set of eigenvectors. If they are matrices, their matrices can be simultaneously diagonalized.

7.1.5 Suppose a spinor is halfway between the x- and z-axis, that is,

$$\chi = \begin{pmatrix} \cos\left(\dfrac{45°}{2}\right) \\ \sin\left(\dfrac{45°}{2}\right) \end{pmatrix} = \begin{pmatrix} 0.924 \\ 0.195 \end{pmatrix}.$$

What **B** fields would you apply for what length of time to rotate the spin to end up along the x-axis? Assume the **B** fields always have a magnitude of 100 T. (Hint: Look at the example at the end of Section 7.2.)

7.2 An Electron in a Magnetic Field

7.2.1 A particle is in the ground state of a 10 nm infinite well. A magnetic flux density of 0.1 T in the z-direction is turned on. What is the difference in energy if the particle is spin up or spin down?

7.2.2 In the development that led up to Equation (7.15), Equation (7.13) was used because we assumed that the particle was in the ground state. Show that Equation (7.15) is still valid for any wave function $\psi(x)$ in the infinite well.

7.3 A Charged Particle Moving in Combined E and B Fields

7.3.1 Show that A of Equation (7.18) is adequate to represent the **B** field.

7.4 The Hartree–Fock Approximation

7.4.1 Using the program se7_2.m, begin with two particles in the 10 nm well that are separated spatially and calculate their total energy after the Hartree term has been added (Fig. 7.12).

7.4.2 Take the inner product of Equation (7.30a) with $\psi_1(r_1)$ and the inner product of Equation (7.30b) with $\psi_2(r_2)$ and show that the

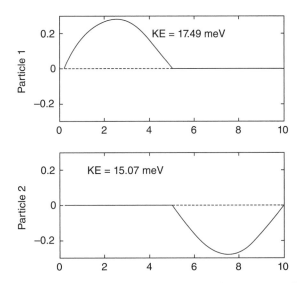

FIGURE 7.12 Two separated particles in a 10 nm well.

energy due to the Hartree–Fock terms are the same as when we start with the Slater determinant and take the inner product with itself.

REFERENCES

1. D. J. Griffiths, *Introduction to Quantum Mechanics*, Englewood Cliffs, NJ: Prentice Hall, 1995.
2. M. A. Nielsen and I. L. Chuang, *Quantum Computation and Quantum Information*, Cambridge, UK: Cambridge University Press, 2000.
3. D. M. Sullivan and D. S. Citrin, "Time-domain simulation of quantum spin," *J. Appl. Phys.*, Vol. 94, p. 6518, 2003.
4. C. Cohen-Tannoudji, B. Diu, and F. Laloe, *Quantum Mechanics*, New York: Wiley, 1977.
5. N. W. Ashcroft and N. D. Mermin, *Solid State Physics*, Orlando, FL: Saunders College, 1976.
6. D. M. Sullivan and D. S. Citrin, "Time-domain simulation of two electrons in a quantum dot," *J. Appl. Phys.*, Vol. 89, p. 3841, 2001.
7. D. M. Sullivan and D. S. Citrin, "Time-domain simulation of a universal quantum gate," *J. Appl. Phys.*, Vol. 96, p. 1450, 2004.

8

THE GREEN'S FUNCTION FORMULATION

Central to circuit analysis is the impulse response. If we have the impulse response of a circuit, we can determine the output for any input by convolving the impulse response and the input. Or we can work in the frequency domain, in which case the Fourier transform of the impulse response, the transfer function, characterizes the circuit in the frequency domain.

A concept similar to the impulse response or the transfer function is the Green's function. The Green's function is a somewhat broader concept because it can deal with space as well as time. In this chapter, we show how the Green's function can be used to determine electron density in a channel. Readers not familiar with the Green's function may want to read Appendix C.

In the first section of this chapter, we will discuss how the Green's function relates to what we have done so far and what we still need to do to determine how electrons flow through a channel. In Section 8.2, we will learn about the density matrix and the spectral matrix, which are essential to the Green's function formulation. Section 8.3 describes the Green's function matrix, first in the eigenspace representation, and then in the real space representation. Section 8.4 introduces the concept of the self-energy, which links the channel to the outside world.

The material of this chapter and the next are based on the development of the Green's function formulation by Supriyo Datta [1–3]. This pioneering work on Green's function for semiconductor devices is part of an overall effort to develop a new approach to electronic devices [4].

Quantum Mechanics for Electrical Engineers, First Edition. Dennis M. Sullivan.
© 2012 The Institute of Electrical and Electronics Engineers, Inc.
Published 2012 by John Wiley & Sons, Inc.

8.1 INTRODUCTION

So far, we have been looking for solutions of the Schrödinger equation in the following general form:

$$i\hbar \frac{\partial \psi(x,t)}{\partial t} - H\psi(x,t) = 0.$$

In simulations, we usually started with an initial condition, that is, a value of $\psi(x,0)$. This was often an eigenstate of the infinite well or a Gaussian wavepacket. We always started with some value of ψ initialized in the problem space.

What happens when we start with an empty channel and we want to talk about particles coming into and going out of the channel? We have already talked a little bit about particles leaving the channel through a loss term, but we have not spoken about a source to put particles in the channel. So we probably want to think about a Schrödinger equation that looks like

$$i\hbar \frac{\partial \psi(x,t)}{\partial t} - H\psi(x,t) = f(x,t). \tag{8.1}$$

The term on the right, $f(x,t)$, is a *forcing function*.

Let us start with an analogous problem in circuit analysis, say a simple resistor-inductor (RL) circuit like the one shown in Figure 8.1. This circuit can be described by a differential equation

$$L\frac{d}{dt}i_{\text{out}}(t) + Ri_{\text{out}}(t) = v_{\text{in}}(t).$$

We can take this into the frequency domain

$$-i\omega L I_{\text{out}}(\omega) + R I_{\text{out}}(\omega) = V_{\text{in}}(\omega).$$

Note that the time derivative results in $-i\omega$ because we are using the physics convention. From this equation we develop a transfer function

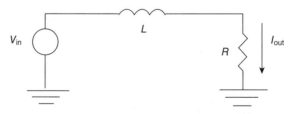

FIGURE 8.1 A simple RL circuit.

$$H(\omega) = \frac{I_{\text{out}}(\omega)}{V_{\text{in}}(\omega)} = \frac{1}{R - i\omega L}.$$

The transfer function characterizes the circuit independent of the input. We can then find the output from the transfer function for any input by multiplying by the Fourier transform of the input and then taking the product back into the time domain.

It would be desirable to have something like this for the Schrödinger equation in Equation (8.1). To do so, we start by taking the Fourier transform, the way we did with the circuit equation. Since ω is related to E in quantum mechanics, we usually prefer to write it as a function of energy:

$$E\psi(x, E) - H\psi(x, E) = f(x, E). \tag{8.2}$$

This looks like the time-independent Schrödinger equation with a forcing function.

We would like to find a function that tells us what to expect in the channel for a given forcing function. Since the Schrödinger equation is a function of space and time, we will not have a simple transfer function like that of the RL circuit. However, there is something analogous. It is called the Green's function. In the development of the Green's function, we will use the matrix formulation of the Hamiltonian, so the Green's function will be an $N \times N$ matrix, even though the original channel is only $1 \times N$. Toward this end, we will rewrite some previous one-dimensional functions as two-dimensional functions in the coming sections. This may seem a little silly at first, but the reasoning will become clear.

8.2 THE DENSITY MATRIX AND THE SPECTRAL MATRIX

In this section we define several new functions that will be crucial for understanding Green's function in Section 8.3. It will seem as if we are deliberately making things difficult by creating two-dimensional functions from simpler one-dimensional functions. Remember that even though our infinite well is one dimensional, the operators related to it, such as the Hamiltonian, are two dimensional. This is the reason for the new definitions. These new two-dimensional functions will be compatible with the matrix operators.

We saw in Chapter 5 that the electron density in an infinite well can be specified by

$$n(x) = \sum_m |\phi_m(x)|^2 f_F(\varepsilon_m - \mu), \tag{8.3}$$

where the Fermi–Dirac distribution is given by

$$f_F(\varepsilon_m - \mu) = \frac{1}{1 + \exp\left(\dfrac{\varepsilon_m - \mu}{kT}\right)}.$$

We are using the chemical potential, μ, instead of the Fermi energy E_F that we used in Chapter 5. For our purpose, they are essentially the same thing, but μ is the parameter most often used in semiconductor theory.

We also saw in Chapter 5 that the one-dimensional electron density of Equation (8.3) can be generalized into a two-dimensional function called the *density matrix*:

$$\rho(x, x') = \sum_m \phi_m(x) f_F(\varepsilon_m - \mu) \phi_m^*(x'). \tag{8.4}$$

The diagonal of the density matrix is the electron density,

$$n(x) = \rho(x, x).$$

Figure 8.2 illustrates the relationships among the Fermi–Dirac distribution, the electron density, and the density matrix in a 10 nm infinite well. Figure 8.2a is the Fermi–Dirac distribution for $k_B T = 0.001$ eV and a chemical potential of $\mu = 0.05$ eV. The low $k_B T$ results in an "in or out" distribution of the states. The probability of occupation of the first three states is 1 and the probability of occupation of all other states is 0. Figure 8.2b shows the electron density. Figure 8.2c shows the corresponding density matrix. Notice that the diagonal of the density matrix is the electron density.

We now define another function called the *spectral function*:

$$A(x, x'; E) = 2\pi \sum_m \phi_m(x) \delta(E - \varepsilon_m) \phi_m^*(x). \tag{8.5}$$

Note that the spectral function is a function of energy that is dependent on the eigenenergies of the structure it is describing. Strictly speaking, Equation (8.5) only has nonzero values at the eigenenergies. Figure 8.3 shows the spectral function of the 10 nm infinite well at the third eigenenergy.

The spectral function is related to the density matrix by the following equation:

$$\rho(x, x') = \int_{-\infty}^{\infty} \frac{dE}{2\pi} f_F(E - \mu) A(x, x'; E). \tag{8.6}$$

Equation (8.6) says that if we have the spectral function, we can get the density matrix for a given Fermi energy.

Equation (8.5) is a real space representation. Similar to the way that the eigenspace representation of the Hamiltonian is just a diagonal matrix with

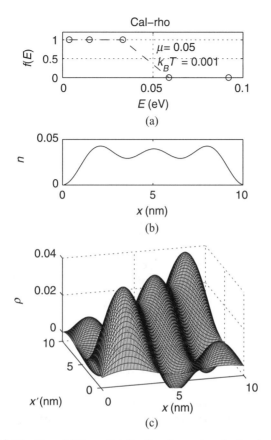

FIGURE 8.2 (a) The Fermi–Dirac distribution for a chemical potential $\mu = 0.05$ eV and $k_B T = 0.001$. The probability of occupation is 1 for the first three eigenenergies and 0 for all other. (b) The electron density in a 10 nm infinite well for the distribution in part (a). (c) The density matrix corresponding to the electron density of (b).

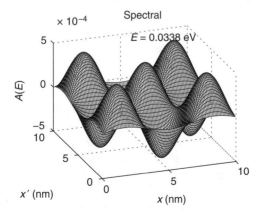

FIGURE 8.3 The spectral function for the 10 nm infinite well at $E = 0.0338$ eV.

the eigenenergies on the diagonal, it can be shown that the eigenspace representation of the spectral matrix is

$$
A_e(E) = 2\pi \begin{bmatrix} \delta(E-\varepsilon_1) & 0 & & \\ 0 & \delta(E-\varepsilon_2) & 0 & \\ & 0 & \ldots & 0 \\ & & 0 & \delta(E-\varepsilon_N) \end{bmatrix}.
\tag{8.7}
$$

In the discussion above, we have assumed that the density of states (DOS) in the channel is given by

$$
\text{DOS} = \sum_m \delta(E-\varepsilon_m).
$$

This tells us that eigenstates exist exactly at the eigenenergies and only at the eigenergies. We saw in Chapter 3 that a more realistic picture is given by the broadened DOS:

$$
\text{DOS} = \sum_m \frac{1}{2\pi} \frac{\gamma_m}{(\gamma_m/2)^2 + (E-\varepsilon_m)^2}.
$$

These broadened terms could be used in place of the delta functions in the spectral function. In particular, we can write the eigenstate representation of the spectral function as:

$$
A_e(E) = 2\pi \begin{bmatrix} \dfrac{\gamma_1/2\pi}{(\gamma_1/2)^2 + (E-\varepsilon_1)^2} & 0 & & \\ 0 & \dfrac{\gamma_2/2\pi}{(\gamma_2/2)^2 + (E-\varepsilon_2)^2} & 0 & \\ & 0 & \ldots & 0 \\ & & 0 & \ldots \end{bmatrix}.
\tag{8.8}
$$

8.3   THE MATRIX VERSION OF THE GREEN'S FUNCTION

This section introduces Green's function for a channel. The input to the channel will be a forcing function $\{f\}$ and the output will be a state variable in the channel $\{\psi\}$. At this point it is difficult to visualize what $\{f\}$ might be. Actually, $\{f\}$ is only being used to derive Green's function.

Throughout this section we use $\{\}$ to denote column vectors and $[]$ to denote matrices.

8.3.1 Eigenfunction Representation of Green's Function

We saw in Chapter 4 that if the eigenfunctions for the infinite well are $1 \times N$ column vectors $\{\phi_i\}$, then we can write an $N \times N$ matrix whose columns are the eigenfunctions:

$$[\Phi] = [\{\phi_1\} \ \{\phi_2\} \ \cdots \ \{\phi_N\}]. \tag{8.9}$$

We know that any state variable can be written as a $1 \times N$ column vector, and this column vector can be written as a superposition of the eigenfunctions

$$\{\psi\} = [\Phi]\{c\}. \tag{8.10}$$

The column vector $\{c\}$ consists of the complex coefficients. Actually, we know we can decompose any function in the channel into a superposition of the eigenstates of the channel, so a forcing function, f, can be written as

$$\{f\} = [\Phi]\{d\}. \tag{8.11}$$

We assume that we know $\{d\}$ because we know the forcing function, and now we want to find $\{c\}$, which determines $\{\psi\}$ in the channel via Equation (8.10). The matrix version of Equation (8.2) is

$$E[I]\{\psi\} - [H]\{\psi\} = \{f\}, \tag{8.12}$$

where I is the identity matrix. Substituting Equation (8.10) and Equation (8.11) into Equation (8.12) gives

$$(E[I] - [H])[\Phi]\{c\} = [\Phi]\{d\}. \tag{8.13}$$

If we multiply this equation from the left by the conjugate transpose of Φ, we get

$$[\Phi^+](E[I] - [H])[\Phi]\{c\} = \{d\}.$$

Then, we can solve for $\{c\}$:

$$\{c\} = [G_e(E)]\{d\}, \tag{8.14}$$

where

$$G_e(E) = \{[\Phi^+](E[I] - [H])[\Phi]\}^{-1}. \tag{8.15}$$

The function $G_e(E)$ is called the *Green's function*. The subscript $_e$ indicates that we are working in the *eigenfunction representation*.

Let us look closer at Equation (8.15):

$$[\Phi^+](E[I]-[H])[\Phi]=E[I]-[\Phi^+][H][\Phi].$$

Using Equation (8.9), we can write

$$[H][\Phi]=[\varepsilon_1\{\phi_1\}\quad \varepsilon_2\{\phi_2\}\quad \ldots \quad \varepsilon_N\{\phi_N\}],$$

and

$$[\Phi^+][H][\Phi]=\begin{bmatrix} \varepsilon_1 & 0 & & \\ 0 & \varepsilon_2 & 0 & \\ & 0 & \ldots & 0 \\ & & 0 & \varepsilon_N \end{bmatrix}=[H_e],$$

which is the eigenfunction representation of the Hamiltonian (see Eq. 4.31). We can then rewrite Green's function as:

$$G_e(E)=(E[I]-[H_e])^{-1}=\begin{bmatrix} \dfrac{1}{E-\varepsilon_1} & 0 & \ldots \\ 0 & \dfrac{1}{E-\varepsilon_2} & \ldots \\ \ldots & \ldots & \ldots & \dfrac{1}{E-\varepsilon_N} \end{bmatrix}.$$

We have already seen that it is mathematically and practically desirable to have a loss term related to the eigenenergies. We will assume that the loss at each eigenenergy n is an imaginary term $i\gamma_n/2$, so we will write Green's function as

$$G_e(E)=\begin{bmatrix} \dfrac{1}{E-\varepsilon_1+i\gamma_1/2} & 0 & \ldots \\ 0 & \dfrac{1}{E-\varepsilon_2+i\gamma_2/2} & 0 & \ldots \\ \ldots & 0 & \ldots & 0 \\ & \ldots & 0 & \dfrac{1}{E-\varepsilon_N+i\gamma_N/2} \end{bmatrix}. \quad (8.16)$$

The spectral matrix that we introduced in the previous section can be written directly from the Green's function and its transpose conjugate:

$$A(E)=i[G_e(E)-G_e^+(E)].$$

Each term on the diagonal becomes

$$\frac{i}{E-\varepsilon_n+i\gamma_n/2}-\frac{i}{E-\varepsilon_n-i\gamma_n/2}=\frac{\gamma_n}{(E-\varepsilon_n)^2+(\gamma_n/2)^2}.$$

If we multiply by 2π we have the eigenspace representation of the spectral function that we defined in Equation (8.8):

$$A_e(E)=i\left([G_e(E)]-[G_e(E)]^+\right)$$

$$=2\pi\begin{bmatrix}\dfrac{\gamma_1/2\pi}{(E-\varepsilon_1)^2+(\gamma_1/2)^2} & 0 & \cdots \\ 0 & \dfrac{\gamma_2/2\pi}{(E-\varepsilon_n)^2+(\gamma_1/2)^2} & \cdots \\ 0 & 0 & \cdots\end{bmatrix}. \qquad (8.17)$$

We will see in the following section that it is easier to just calculate $A(E)$ directly in the real space representation. This section was necessary to show that Green's function is linked to the spectral function via Equation (8.17).

8.3.2 Real Space Representation of Green's Function

Now let us start over in the *real space* representation. Once again, we start with

$$[IE-H]\{\psi\}=\{f\}, \qquad (8.18)$$

and solve for

$$\{\psi\}=[IE-H]^{-1}\{f\}=[G(E)]\{f\}.$$

This $[G(E)]$ is the real space Green's function,

$$[G(E)]=[IE-H]^{-1}. \qquad (8.19)$$

We saw previously that the eigenspace Green's function is

$$G_e(E)=\{[\Phi^+][IE-H][\Phi]\}^{-1}. \qquad (8.15)$$

Recall from Chapter 4 that the matrix Φ is unitary, which means that its transpose conjugate is equal to its inverse: $\Phi^{-1}=\Phi^+$. Utilizing this and the matrix identity $[AC]^{-1}=C^{-1}A^{-1}$, we can write

$$G_e(E) = \{[\Phi^+][IE - H][\Phi]\}^{-1}$$
$$= [\Phi]^{-1}\{[\Phi^+][IE - H]\}^{-1}$$
$$= [\Phi]^{-1}[IE - H]^{-1}[\Phi^+]^{-1}$$
$$= [\Phi^+][IE - H]^{-1}[\Phi] = [\Phi^+][G(E)][\Phi].$$

This means the $G(E)$ of Equation (8.19) is the real space representation of Green's function. The real space spectral matrix is

$$A(x, x'; E) = i[G(E) - G^+(E)]. \tag{8.20}$$

Recall that we said in Section 8.2 that we could get the density matrix from the real space spectral function

$$\rho(x, x') = \int_{-\infty}^{\infty} \frac{dE}{2\pi} f_F(E - \mu) A(x, x'; E). \tag{8.21}$$

The important point is that A is calculated from G in Equation (8.20) and G is determined by Equation (8.19). Equation (8.21) does *not* assume that we know the eigenvalues and eigenfunctions already. It uses the Hamiltonian directly. Once we have the spectral function, we can find the density matrix. The electron density is just the diagonal of the density matrix. Therefore, if we have the matrix version of the Hamiltonian, we can find quantities like the electron density without calculating the eigenfunctions.

Figure 8.4 illustrates the calculation of the electron density directly from the eigenfunctions (Eq. 8.3) and from the diagonal of the density matrix using the spectral matrix (Eq. 8.21).

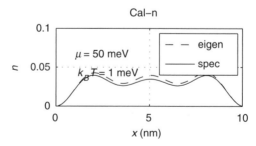

FIGURE 8.4 The electron density calculated from the spectral (spec) function via Equation (8.21) and compared with the eigenfunction (eigen) method of Equation (8.3).

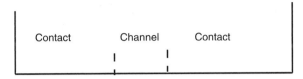

FIGURE 8.5 One large infinite well is divided into three sections that we will think of as a channel with a contact on either side.

FIGURE 8.6 The cell labeling at the left edge of the channel.

8.4 THE SELF-ENERGY MATRIX

The self-energy matrix is a matrix that is added to the Hamiltonian to connect the channel to the outside. Think back to Chapter 2 and our very simple model of the field-effect transistor (FET) where we assumed that a large infinite well was subdivided into three sections, as shown in Figure 8.5. We assumed there were barriers that particles could tunnel through. Now we are going to assume that there is no barrier, and we have arbitrarily chosen a certain number of cells in the middle that we will refer to as the channel.

Let us start by taking a close look at the cells around the left edge of the channel, and number them as shown in Figure 8.6.

Assuming there is zero potential through the well, the time-independent Schrödinger equation is

$$E\psi(x) = -\frac{\hbar^2}{2m}\frac{\partial^2}{\partial x^2}\psi(x).$$

Using the same finite-differencing procedure that we used to develop the simulation programs, this equation becomes

$$E\psi_n \cong -\frac{\hbar^2}{2m}\frac{\psi_{n-1} - 2\psi_n + \psi_{n+1}}{(\Delta x)^2} \tag{8.22}$$
$$= \chi_0(-\psi_{n-1} + 2\psi_n - \psi_{n+1}),$$

where, as before, we define

$$\chi_0 = \frac{\hbar^2}{2m(\Delta x)^2}.$$

We now assume that any waveform propagating in this well can be written as the sum of positive traveling and negative traveling plane waves:

$$\psi_n = Pe^{ikn\cdot\Delta x} + Qe^{-ikn\cdot\Delta x}. \tag{8.23}$$

At cell number 0,

$$\psi_0 = P + Q. \tag{8.24a}$$

At cell number -1:

$$\begin{aligned}
\psi_{-1} &= Pe^{-ik\cdot\Delta x} + Qe^{ik\cdot\Delta x} \\
&= Pe^{-ik\cdot\Delta x} + (\psi_0 - P)e^{ik\cdot\Delta x} \\
&= -2iP\sin(k\cdot\Delta x) + \psi_0 e^{ik\cdot\Delta x}.
\end{aligned} \tag{8.24b}$$

Equation (8.22) becomes

$$\begin{aligned}
E\psi_0 &= \chi_0(-\psi_{-1} + 2\psi_0 - \psi_1) \\
&= \chi_0 2iP\sin(k\cdot\Delta x) - \chi_0\psi_0 e^{ik\cdot\Delta x} + \chi_0 2\psi_0 - \chi_0\psi_1.
\end{aligned} \tag{8.25}$$

The k in Equation (8.23) is dictated by the E on the left side, because if we assume the only energy is kinetic energy, then

$$E = \frac{\hbar^2 k^2}{2m}.$$

We can rewrite Equation (8.25) as:

$$E\psi_0 = \left(-\chi_0 e^{ik\cdot\Delta x} + 2\chi_0\right)\psi_0 - \chi_0\psi_1 + F_{in}, \tag{8.26a}$$

$$F_{in} = \chi_0 2iP\sin(k\cdot\Delta x). \tag{8.26b}$$

Now we can look upon $-\chi_0 e^{ik\cdot\Delta x}$ as a loss term that represents the fact that the waveform can leak out of the channel. The F_{in} is a source that can put waveforms into the channel.

Before continuing, notice that the loss term is $-\chi_0 e^{ik\cdot\Delta x}$. This looks like a plane wave moving to the right, even though it represents the loss going out the left side of the channel. Look back at Equation (8.22) and assume $n = 0$, that is, we are at the far left edge of the channel. We assume we are dealing with plane waves leaving the channel, so they are of the form $e^{-ik\cdot\Delta x}$. We also want to write the term just outside the well at $n = -1$ as a function of the term just inside the well at $n = 0$:

$$E\psi_0 = \chi_0\left(-\left(\frac{\psi_{-1}}{\psi_0}\right)\psi_0 + 2\psi_0 - \psi_1\right).$$

The term Ψ_{-1}/Ψ_0 can be written as:

$$\frac{e^{-ik(-1\cdot\Delta x)}}{e^{-ik(0\cdot\Delta x)}} = e^{ik\cdot\Delta x},$$

which is the loss term of Equation (8.26a).

Recall the matrix version of the Hamiltonian that was developed in Chapter 4:

$$[H] = \begin{bmatrix} 2\chi_0 & -\chi_0 & 0 & \\ -\chi_0 & 2\chi_0 & & 0 \\ 0 & & \cdots & -\chi_0 \\ & 0 & -\chi_0 & 2\chi_0 \end{bmatrix}. \tag{8.27}$$

Obviously, the upper left hand corner has to be modified to include the loss term. Furthermore, we could repeat the process on the other side of the channel and obtain a loss term in the lower right hand corner. Therefore, we can write the Schrödinger equation as:

$$E\{\psi\} = [H + \Sigma]\{\psi\} + \{F_{in}\}, \tag{8.28a}$$

where

$$[\Sigma] = \begin{bmatrix} -\chi_0 e^{ik\cdot\Delta x} & 0 & \cdots & \\ 0 & 0 & & \cdots \\ \cdots & & & 0 \\ & \cdots & 0 & -\chi_0 e^{ik\cdot\Delta x} \end{bmatrix}, \quad \{F_{in}\} = \begin{Bmatrix} f_1 \\ 0 \\ \cdots \\ f_N \end{Bmatrix}. \tag{8.28b}$$

$[\Sigma]$ is referred to as the *self-energy matrix*. We can write Equation (8.28a) as:

$$[IE - H - \Sigma]\{\psi\} = \{F_{in}\}.$$

Now our Green's function is

$$[G] = [IE - H - \Sigma]^{-1}, \tag{8.29}$$

a real space function that already has a loss term.

As before, we can develop a spectral function from this Green's function,

$$[A(E)] = i[G(E) - G^+(E)], \tag{8.30}$$

and from the spectral function we can determine the density matrix

$$\rho(x, x') = \int_{-\infty}^{\infty} \frac{dE}{2\pi} f_F(E - \mu) A(x, x'; E). \tag{8.31}$$

The electron density is calculated from

$$n(x) = diag[\rho(x, x')]. \tag{8.32}$$

We did this in Section 8.3, but now we do not have to add an artificial loss term. Notice that the corners of the spectral matrix will have a term that looks like

$$\gamma = i[\Sigma(1, 1) - \Sigma^+(1, 1)] = i[-\chi_0 e^{ik \cdot \Delta x} + \chi_0 e^{-ik \cdot \Delta x}] = 2\chi_0 \sin(k \cdot \Delta x). \tag{8.33}$$

Let us take a closer look at what γ really is. First, notice that we can write

$$\sin(k \cdot \Delta x) \cong k \cdot \Delta x,$$

because the cell size must be much smaller than any of the wavelengths of interest for the basic finite-difference approximation to be valid. Also, recall that

$$\chi_0 = \frac{\hbar^2}{2m(\Delta x)^2}.$$

So now we write

$$\gamma = 2\chi_0 \sin(k \cdot \Delta x) \cong 2\left(\frac{\hbar^2}{2m(\Delta x)^2}\right) k \cdot \Delta x = \frac{\hbar^2 k}{m \cdot \Delta x}.$$

We also know that the velocity corresponding to this value of k is

$$v = \frac{p}{m} = \frac{\hbar k}{m},$$

so

$$\frac{\gamma}{\hbar} = \frac{v}{\Delta x} = \frac{1}{\tau}. \tag{8.34}$$

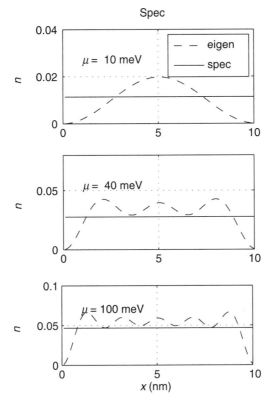

FIGURE 8.7 Calculation of the electron density using the eigenfunction as in Equation (8.34) (dashed lines) and using the spectral matrix as in Equation (8.31) (solid lines). A value of $k_B T = 1$ meV was used.

The value τ is the time for a section of the waveform to leave the cell and is called the *escape rate*.

Figure 8.7 shows the electron density in a 10 nm channel at $k_B T = 1$ meV for three different values of the chemical potential, μ. In each plot, the electron density is calculated by two methods. First, it is calculated directly from the eigenfunctions

$$n(x) = \sum_m |\phi_m(x)|^2 f_F(\varepsilon_m - \mu). \tag{8.35}$$

Then it is calculated by taking the diagonal of the density matrix, Equation (8.31), which is calculated using the spectral matrix of Equation (8.30). A very low $k_B T$ value is used so most eigenstates are "in or out." We see that the eigenfunction calculation of n reflects the shape of the eigenfunctions as the chemical potential μ is increased. Remember that the eigenfunctions of the 10 nm infinite well all go to zero at the edges. The calculation of n from

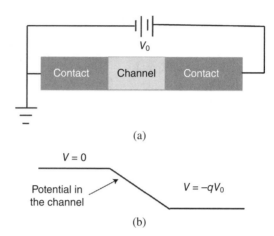

FIGURE 8.8 (a) A voltage V_0 is placed across the contacts that span a channel. (b) The voltage V_0 results in a potential through the channel.

Equation (8.31), however, includes the self-energy, and the electron density is not forced to zero at the walls. Remember, the self-energy tells the channel that it is connected to the contact.

8.4.1 An Electric Field across the Channel

Let us look at the situation shown in Figure 8.8a where a channel has contacts on each side. We assume that the contacts are metal. Therefore, they have available an almost infinite number of closely spaced states. Furthermore, there can be no electric field inside the metal, so each contact can only be at one potential. A voltage source is placed between the two contacts. The voltage source results in a linear potential through the channel, as illustrated in Figure 8.8b. Notice that because the contact on the left side is grounded, we assume the potential on the left side remains at zero.

Let us go back and separate the contacts by writing

$$[\Sigma_1] = \begin{bmatrix} -\chi_0 e^{ik_1 \cdot \Delta x} & 0 & \cdots \\ 0 & 0 & \cdots \\ \cdots & & \cdots & 0 \\ & \cdots & 0 & 0 \end{bmatrix}, \quad [\Sigma_2] = \begin{bmatrix} 0 & 0 & \cdots \\ 0 & 0 & & \cdots \\ \cdots & & \cdots & 0 \\ & \cdots & 0 & -\chi_0 e^{ik_2 \cdot \Delta x} \end{bmatrix}.$$

Previously, we calculated k assuming the energy E was purely kinetic because we assumed there was no potential in the channel. This is no longer the case. Now the k values on the left and right side of the channel are, respectively:

$$k_1 = \frac{\sqrt{2m(E_{\text{total}} - V(0 \text{ nm}))}}{\hbar}, \tag{8.36a}$$

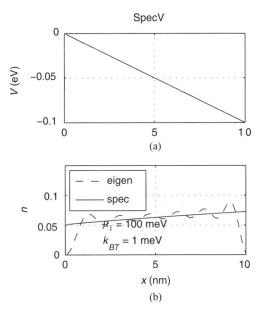

FIGURE 8.9 (a) The resulting change in potential when 0.1 V is applied across a 10 nm channel. (b) The resulting electron density using the eigenfunction method (dashed line) and the spectral method (solid line).

$$k_2 = \frac{\sqrt{2m(E_{\text{total}} - V(10\text{ nm}))}}{\hbar}. \qquad (8.36\text{b})$$

We can then write Green's function as

$$[G] = [IE - H - \Sigma_1 - \Sigma_2]^{-1}.$$

The spectral function is calculated from Equation (8.30), the density matrix from Equation (8.31), and finally the electron density from Equation (8.32).

If we repeat the calculation of Figure 8.7 for $k_B T = 1$ meV and $\mu = 0.1$ eV, but with the potential of Figure 8.9a also in the channel, we get the results shown in Figure 8.9b. The eigenfunction method is different because the eigenfunctions are slightly different due to the potential in the channel. The spectral matrix method gave a different result primarily because the k value at the right side was changed by the potential, as calculated in Equation (8.36b).

8.4.2 A Short Discussion on Contacts

Up until now, we have thought of the contacts on each side of the channel as large spaces. True contacts are metal, or very heavily doped semiconductors that almost behave like a metal. In the semiconductor literature, they are often referred to as "reservoirs." Reservoirs may be thought of as having an infinite number of available states. Because they are connected to the outside, empty states in the source can be replenished and filled states in drain can be emptied.

EXERCISES

8.2 The Density Matrix and the Spectral Matrix

   8.2.1 Write a MATLAB program to calculate the spectral function in the 10 nm infinite well, similar to Equation (8.5).

8.3 The Matrix Version of the Green's Function

   8.3.1 Write a MATLAB program to calculate the electron density, both directly from the eigenfunctions, and by using the spectral matrix from Green's function similar to Figure 8.4. (Note: you will have to add a loss term η, as in Eq. 8.13. I suggest $\eta = 10^{-3}$.)

8.4 The Self-Energy Matrix

   8.4.1 Find the terms equivalent to Equation (8.26) if the analysis is done on the right edge of the channel instead of the left edge as shown in Figure 8.6.

   8.4.2 Write a MATLAB program to duplicate the results of Figure 8.7. Use the spectral matrix method and do not use periodic boundary condition. Your results should most closely resemble the dashed lines in the figure.

   8.4.3 Modify the program from Exercise 8.4.1 to include a potential difference across the channel, as shown in Figure 8.8.

REFERENCES

1. S. Datta, *Quantum Transport—Atom to Transistor*, Cambridge, UK: Cambridge University Press, 2005.
2. S. Datta, "Nanoscale device simulation: The Green's function formalism," *Superlattice. Microst.*, Vol. 28, p. 253, 2000.
3. S. Datta, *Electronic Transport in Mesoscopic Systems*, Cambridge, UK: Cambridge University Press, 1995.
4. http://nanohub.org/topics/ElectonicsFromTheBottomUp.

9

TRANSMISSION

To a large extent, all previous chapters led us up to this chapter where we finally calculate transmission through a channel. Section 9.1 develops the transmission for a channel with only one state. This is a simple example to illustrate the concepts. The next section develops the mathematical connection between the state variable and current flow. In Section 9.3, the full transmission function is developed as well as the calculation of current through a channel once the transmission function is known. Section 9.4 describes quantum conductance. Section 9.5 introduces Büttiker probes, a method of modeling incoherent scattering in the channel. The last section is an example to illustrate the simulation methods described in this chapter.

As mentioned in Chapter 8, the work on transmission in this chapter is based on the pioneering work by Supriyo Datta [1–3] and is part of an overall effort to develop a new approach to electronic devices [4].

9.1 THE SINGLE-ENERGY CHANNEL

Let us return to our simple model of a channel with contacts on each side and a voltage between the contacts, as illustrated in Figure 9.1. This time, we restrict ourselves to a channel that can only contain one state at energy ε_0. The voltage V_{DS} between the drain and source has two effects: (1) It creates a graded

Quantum Mechanics for Electrical Engineers, First Edition. Dennis M. Sullivan.
© 2012 The Institute of Electrical and Electronics Engineers, Inc.
Published 2012 by John Wiley & Sons, Inc.

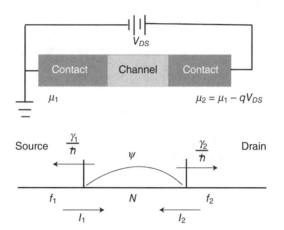

FIGURE 9.1 A channel with only one energy level at ε_0.

potential in the channel, something we will ignore during this example, and (2) it pushes the chemical potential in the right contact qV_{DS} eV lower than the left contact.

The *source* contact, on the left side, will try to set the channel to have

$$f_1 = f_F(\varepsilon_0 - \mu_1) \qquad (9.1a)$$

electrons, while the *drain* contact on the right side would like to see

$$f_2 = f_F(\varepsilon_0 - \mu_2) \qquad (9.1b)$$

electrons. (Remember that both f_1 and f_2 are fractions of 1.) We start by assuming the channel contains an average of N electrons, which is somewhere between f_1 and f_2. There is a net flux I_1 across the left junction proportional to

$$I_1 = \frac{q\gamma_1}{\hbar}(f_1 - N), \qquad (9.2a)$$

where $-q$ is the charge of an electron and $\gamma_1/\hbar$ is the rate at which a particle in the channel will escape into the source, as we showed in Equation (8.34). (Refer back to our discussion of broadening in Section 3.2.) Similarly, there is a net flux on the right side

$$I_2 = \frac{q\gamma_2}{\hbar}(f_2 - N), \qquad (9.2b)$$

with corresponding escape rate $\gamma_2/\hbar$.

If we start with the case of $I_1 + I_2 = 0$, we have

$$I_1 + I_2 = \frac{q\gamma_1}{\hbar}(f_1 - N) + \frac{q\gamma_2}{\hbar}(f_2 - N) = 0.$$

Then we solve for N:

$$N = \frac{\gamma_1 f_1 + \gamma_2 f_2}{(\gamma_1 + \gamma_2)}. \tag{9.3}$$

Go back and look at the current flowing at the left partition:

$$I_1 = \frac{q\gamma_1}{\hbar}(f_1 - N).$$

Obviously, the current flowing into the channel is

$$I_{1in} = \frac{q\gamma_1}{\hbar} f_1, \tag{9.4a}$$

while the part flowing out of the channel is

$$I_{1out} = -\frac{q\gamma_1}{\hbar} N. \tag{9.4b}$$

Note that the current flowing in depends only on f_1 while the current flowing out depends on f_1 and f_2 because N depends on both. We can put Equation (9.3) into Equation (9.2a) and get

$$I_1 = \frac{q\gamma_1}{\hbar}\left(f_1 - \frac{\gamma_1 f_1 + \gamma_2 f_2}{(\gamma_1 + \gamma_2)}\right) = \frac{q}{\hbar}\left(\frac{\gamma_1\gamma_2}{\gamma_1 + \gamma_2}\right)(f_1 - f_2). \tag{9.5}$$

We call the term

$$T = \frac{q}{\hbar}\left(\frac{\gamma_1\gamma_2}{\gamma_1 + \gamma_2}\right) \tag{9.6}$$

the *transmission*. Notice that the transmission gives us the current flow through the channel when a voltage across the drain and source V_{DS} results in a separation of the Fermi distributions, $(f_1 - f_2)$.

9.2   CURRENT FLOW

Up to this point we have talked about calculating the electron density n in a channel. Eventually, we will want to talk about the current. The purpose of this section is to relate the electron density n at a velocity v to the current. The development in this section is primarily due to Griffiths [5].

If a state variable represents an electron, as we have assumed through most of our discussions, and the state variable is confined to some area—one cell, for instance—we would say the charge in that area is

$$Q = -q \int_{x_0}^{x_0 + \Delta x} |\psi(x)|^2 \, dx.$$

The current through this area is the time rate of change of Q.

$$I = \frac{\partial Q(x, t)}{\partial t} = -q \int_{x_0}^{x_0 + \Delta x} \frac{\partial}{\partial t} |\psi(x, t)|^2 \, dx.$$

The first step is to think about the rate of change of the state variable

$$\frac{\partial}{\partial t} |\psi(x, t)|^2 = \frac{\partial}{\partial t} (\psi^* \psi) = \psi^* \frac{\partial \psi}{\partial t} + \frac{\partial \psi^*}{\partial t} \psi. \tag{9.7}$$

The state variable and its complex conjugate both have to solve the time-dependent Schrödinger equation,

$$\frac{\partial \psi}{\partial t} = \frac{i\hbar}{2m} \frac{\partial^2 \psi}{\partial x^2} - \frac{i}{\hbar} V \psi, \tag{9.8a}$$

$$\frac{\partial \psi^*}{\partial t} = -\frac{i\hbar}{2m} \frac{\partial^2 \psi^*}{\partial x^2} + \frac{i}{\hbar} V \psi^*. \tag{9.8b}$$

Inserting Equation (9.8) into Equation (9.7) gives

$$\frac{\partial}{\partial t} |\psi|^2 = \frac{i\hbar}{2m} \left(\psi^* \frac{\partial \psi^2}{\partial x^2} - \frac{\partial^2 \psi^*}{\partial x^2} \psi \right)$$

$$= \frac{\partial}{\partial x} \left[\frac{i\hbar}{2m} \left(\psi \frac{\partial \psi^*}{\partial x} - \psi^* \frac{\partial \psi}{\partial x} \right) \right].$$

If we want to know the time rate of change in one cell of size Δx, we integrate over that distance

$$\frac{d}{dt} \int_{x_0}^{x_0 + \Delta x} |\psi(x, t)|^2 \, dx = \frac{i\hbar}{2m} \left(\psi(x) \frac{\partial \psi^*(x)}{\partial x} - \psi^*(x) \frac{\partial \psi(x)}{\partial x} \right) \Bigg|_{x_0}^{x_0 + \Delta x}.$$

The quantity on the right is the *probability current* [5]:

$$J(x) \equiv \frac{i\hbar}{2m} \left(\psi(x) \frac{\partial \psi^*(x)}{\partial x} - \psi^*(x) \frac{\partial \psi(x)}{\partial x} \right). \tag{9.9}$$

In our discussion, we are assuming that we are dealing with incremental variations in the energy, and therefore, with a specific k. We thus assume a plane wave of the form

$$\psi(x) = Ae^{ikx}.$$

Substituting this into Equation (9.9) we get

$$J(x) \equiv \frac{i\hbar}{2m}\left(\psi(x)(-ik)\psi^*(x) - \psi^*(x)(ik)\psi(x)\right)$$

$$\equiv \frac{i\hbar}{2m}(-2ik)|\psi(x)|^2 = \frac{\hbar k}{m}|\psi(x)|^2 = v|\psi(x)|^2.$$

This is what we were trying to determine: The particle flow is related to $v = \hbar k/m$. The probability current through a cell of width Δx is

$$\frac{J(E; x)}{\Delta x} = \left(\frac{v}{\Delta x}\right)|\psi(x)|^2 = \left(\frac{v}{\Delta x}\right)n(x), \tag{9.10}$$

so the current through the one cell is

$$I_{cell} = -q\frac{J(E; x)}{\Delta x} = -q\left(\frac{v}{\Delta x}\right)n(x). \tag{9.11}$$

9.3 THE TRANSMISSION MATRIX

In Chapter 8 we wrote the real space Green's function as

$$[G] = [IE - H - \Sigma]^{-1}. \tag{9.12}$$

Remember that $[IE - H]$ is composed of only real numbers while Σ is all zeros except for the complex numbers at the corners

$$\Sigma = \begin{bmatrix} -\chi_o e^{ik_1 \cdot \Delta x} & 0 & \\ 0 & \cdots & 0 \\ & 0 & -\chi_o e^{ik_2 \cdot \Delta x} \end{bmatrix}.$$

The values k_1 and k_2 can be different for the following reason. Both these values are related to the energy E in Equation (9.12),

$$E = \frac{(\hbar k)^2}{2m} + V(x).$$

However, they can be different because the potential energy $V(x)$ can be different at the two corners.

The spectral function matrix is

$$[A(E)] = i[G(E) - G^+(E)]. \tag{9.13}$$

We start by multiplying Equation (9.13) from the left by $G(E)^{-1}$:

$$G^{-1}A = i[I - G^{-1}G^+],$$

and then from the right by $(G^+)^{-1}$:

$$G^{-1}A(G^+)^{-1} = i[(G^+)^{-1} - G^{-1}]. \tag{9.14}$$

It can be shown (see the "Exercises" section) that

$$(G^+)^{-1} = (G^{-1})^+.$$

Since $[IE - H]$ is real,

$$i[(G^+)^{-1} - G^{-1}] = i\{[EI - H - \Sigma]^+ - [EI - H - \Sigma]\}$$
$$= i[\Sigma - \Sigma^+].$$

We will call this term

$$\Gamma = i[\Sigma - \Sigma^+]. \tag{9.15}$$

We can rewrite Equation (9.14) as

$$G^{-1}A(G^+)^{-1} = \Gamma,$$

or

$$A = G\Gamma G^+. \tag{9.16}$$

This is the expression for the spectral matrix that we will find most useful.

As in Chapter 8, we split the Σ matrix into two parts:

$$\Sigma_1 = \begin{bmatrix} -\chi_o e^{ik_1 \cdot \Delta x} & 0 & \\ 0 & \dots & 0 \\ & 0 & 0 \end{bmatrix}, \tag{9.17a}$$

$$\Sigma_2 = \begin{bmatrix} 0 & 0 & \\ 0 & \dots & 0 \\ & 0 & -\chi_o e^{ik_2 \cdot \Delta x} \end{bmatrix}, \tag{9.17b}$$

and then define the following related functions:

$$\Gamma_1 = i[\Sigma_1 - \Sigma_1^+], \tag{9.18a}$$

$$\Gamma_2 = i[\Sigma_2 - \Sigma_2^+], \tag{9.18b}$$

$$[G] = [IE - H - \Sigma_1 - \Sigma_2]^{-1}, \tag{9.18c}$$

$$A_1 = G\Gamma_1 G^+, \tag{9.18d}$$

$$A_2 = G\Gamma_2 G^+. \tag{9.18e}$$

Obviously the A in Equation (9.16) is the sum

$$A = A_1 + A_2. \tag{9.18f}$$

Let us take a closer look at Γ_1:

$$\Gamma_1(E) = i[\Sigma_1 - \Sigma_1^+]$$

$$= i\begin{bmatrix} -\chi_o \left(e^{ik_1 \cdot \Delta x} - e^{-ik_1 \cdot \Delta x} \right) & 0 & \\ 0 & \cdots & 0 \\ & 0 & 0 \end{bmatrix} = \begin{bmatrix} 2\chi_o \sin(k_1 \cdot \Delta x) & 0 & \\ 0 & \cdots & 0 \\ & 0 & 0 \end{bmatrix}.$$

We will make the approximation that we have used before:

$$\sin(k \cdot \Delta x) \cong k \cdot \Delta x.$$

Multiplying out the terms in the upper left hand corner:

$$2\chi_0(k \cdot \Delta x) \cong 2\left(\frac{\hbar^2}{2m \cdot \Delta x^2} \right)(k \cdot \Delta x) = \frac{\hbar}{\Delta x}\frac{\hbar k}{m} = \frac{\hbar \cdot v}{\Delta x}.$$

Therefore, we can write Γ_1 as

$$\Gamma_1(E) \cong \begin{bmatrix} \hbar v / \Delta x & 0 & \\ 0 & \cdots & 0 \\ & 0 & 0 \end{bmatrix}.$$

9.3.1 Flow into the Channel

We saw that the density matrix is related to the spectral function by

$$\rho(x, x') = \int_{-\infty}^{\infty} \frac{dE}{2\pi} f_F(E - \mu) A(x, x'; E). \tag{9.19}$$

The electron density can be obtained from the diagonal of the density matrix

$$n(x) = diag[\rho(x, x')]$$
$$= diag\left[\int_{-\infty}^{\infty} \frac{dE}{2\pi} f_F(E - \mu) A(x, x'; E)\right].$$

If we multiply the matrix A by $\Gamma_1(E)$ we get

$$\frac{\hbar v_1}{\Delta x} n(x_1) = Tr\left[\int_{-\infty}^{\infty} \frac{dE}{2\pi} f_F(E - \mu_1) \Gamma_1(E) A(x, x'; E)\right],$$

where $n(x_1)$ is the value of n at the far left edge of the channel and v_1 is the corresponding velocity. Notice that the chemical potential μ_1 is also evaluated at the far left side. We have switched to *trace* (indicated by Tr) because Γ_1 only has a nonzero value in the corner. Equation (9.11) says that n is related to the current at that point. So if we multiply both sides by

$$-\frac{q}{\hbar},$$

we get the current flowing into the channel:

$$I_{1\text{inflow}} = -\frac{qv_1}{\Delta x} n(x_1) = -\frac{q}{\hbar} Tr\left[\int_{-\infty}^{\infty} \frac{dE}{2\pi} f_F(E - \mu_1) \Gamma_1(E) A(x, x'; E)\right]. \quad (9.20)$$

Equation (9.20) calculates the flow going into the channel from the left side. It will be convenient to define the current flow per energy as

$$\tilde{I}_{1\text{lin}} = Tr[\Gamma_1(E) A] f_F(E - \mu_1), \quad (9.21a)$$

and Equation (9.20) can be written as:

$$I_{1\text{inflow}} = -\frac{q}{2\pi\hbar} \int_{-\infty}^{\infty} dE\, \tilde{I}_{1\text{lin}}. \quad (9.21b)$$

9.3.2 Flow out of the Channel

From this point on, we will use

$$f_1 = f_F(E - \mu_1), \quad f_2 = f_F(E - \mu_2)$$

for simplicity. We can write Equation (9.19) as

$$\rho(x, x') = \int_{-\infty}^{\infty} \frac{dE}{2\pi} \{[A_1] f_1 + [A_2] f_2\},$$

from which we can get the current density

$$n(x) = diag \int_{-\infty}^{\infty} \frac{dE}{2\pi} \{[A_1]f_1 + [A_2]f_2\}. \tag{9.22}$$

Analogous to Equation (9.20), we can say that the total current leaking from the left side is

$$I_{1outflow} = \frac{q}{\hbar} Tr \int_{-\infty}^{\infty} \frac{dE}{2\pi} \Gamma_1 \{[A_1]f_1 + [A_2]f_2\},$$

or

$$\tilde{I}_{1out} = Tr\{\Gamma_1[A_1 f_1 + A_2 f_2]\}, \tag{9.23a}$$

$$I_{1outflow} = -\frac{q}{2\pi\hbar} \int_{-\infty}^{\infty} dE \, \tilde{I}_{1out}. \tag{9.23b}$$

Similar to what we found in the single-state channel in Section 9.1, the current flowing into the channel depends on f_1, while the current flowing out of the channel depends on both f_1 and f_2.

9.3.3 Transmission

The total current on the left side of the channel is

$$\begin{aligned}
\tilde{I}_1 = \tilde{I}_{1in} - \tilde{I}_{1out} &= Tr\{\Gamma_1(A_1 + A_2)\}f_1 - Tr\{\Gamma_1 A_1 f_1 + \Gamma_1 A_2 f_2\} \\
&= Tr[\Gamma_1 A_2](f_1 - f_2) \\
&= Tr[\Gamma_1 G \Gamma_2 G^+](f_1 - f_2).
\end{aligned} \tag{9.24}$$

Since the current in the channel must be constant, we can say that the channel current is this input current and write

$$I_{channel} = \frac{q}{2\pi\hbar} \int_{-\infty}^{\infty} dE \, \tilde{I}_1 = \frac{q}{2\pi\hbar} \int_{-\infty}^{\infty} dE \, T(E)(f_1 - f_2), \tag{9.25a}$$

$$T(E) = Tr[\Gamma_1 G \Gamma_2 G^+]. \tag{9.25b}$$

The function in Equation (9.25b) is called the *transmission function*. The transmission function is the probability that an electron will cross a channel as a function of energy. The transmission functions for three different channels are shown in Figure 9.2. Notice that the transmission for the ballistic channel (Fig. 9.2a) is 1. This is not surprising. It means simply that since there is nothing to impede a wave function, there will be current flow for a waveform at any positive energy.

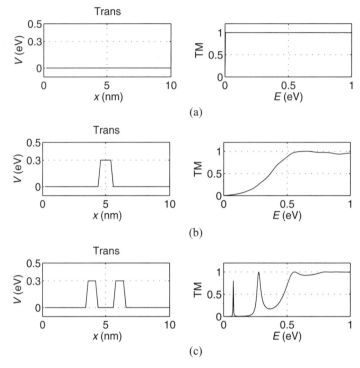

FIGURE 9.2   Three different channels and their corresponding transmissions (TM). (a) A ballistic channel and its TM function. (b) A tunneling barrier and its TM function. (c) A resonant barrier and its TM function.

9.3.4   Determining Current Flow

Once the transmission is determined from Equation (9.25b), the current can be calculated from Equation (9.25a) using the Fermi levels on each side. In Figure 9.3, we start with the tunneling barrier channel and assume the chemical potential on the left side is $\mu_1 = 0.3$ eV. If a voltage $V_{DS} = 0.2$ V is applied across the channel, we see that it produces a slight change in the potential (Fig. 9.3a) and a slight change in the transmission (Fig. 9.3b) compared to Figure 9.2b. The main effect of V_{DS} is to separate the two Fermi distributions, f_1 and f_2, as shown in Figure 9.3c. In Figure 9.3d, the dashed line is the difference in the Fermi functions at each energy and the shaded area is $(f_1 - f_2) \cdot T(E)$. The current of 2.53 µA is the integral of this shaded region times $(q/\hbar)$. The conductance G is the current divided by V_{DS}.

Figure 9.4 shows a similar plot for the ballistic channel containing no barriers. Once again, the chemical potential on the left side is $\mu_1 = 0.3$ eV and $V_{DS} = 0.2$. The conductance is close to 38.7 µS, a theoretical limit called the *quantum conductance*. This will be discussed in Section 9.4.

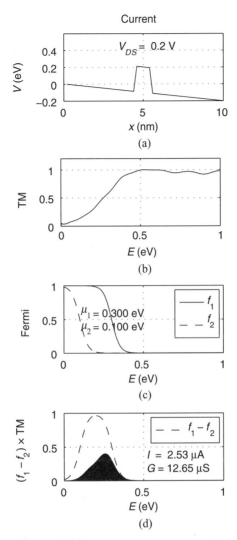

FIGURE 9.3 Calculation of the current flow across a tunneling barrier. (a) The input voltage of $V_{DS} = 0.2$ V causes a slanting in the potential through the channel. (b) The transmission (TM) as a function of energy. (c) The chemical potential of the left edge of the channel, μ_1, is 0.3 eV. The right edge is $\mu_2 = \mu_1 - qV_{DS} = 0.1$ eV. The corresponding Fermi distributions are shown. (d) The dashed line is $f_1 - f_2$. The shaded area is this difference multiplied by the TM. The current is the total shaded area times (q/h) as calculated by Equation (9.25a). The conductance G is the current divided by V_{DS}.

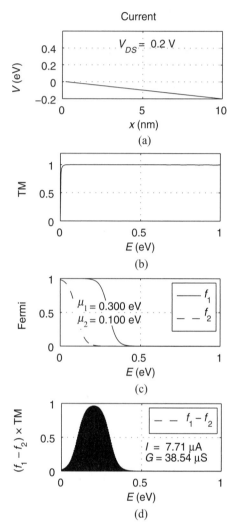

FIGURE 9.4 Calculation of the current flow across a ballistic channel. (a) The input voltage of $V_{DS} = 0.2$ V causes a slanting in the potential through the channel. (b) The transmission is unity as expected for a ballistic channel. (c) The chemical potential of the left edge of the channel, μ_1, was input as 0.3 eV. The right edge is $\mu_2 = \mu_1 - V_{DS}$. The corresponding Fermi distributions are shown. (d) The current is the total shaded area times (q/h) as calculated by Equation (9.21a). The conductance G is the current divided by V_{DS}, which in this case is about equal to the quantum conductance G_0.

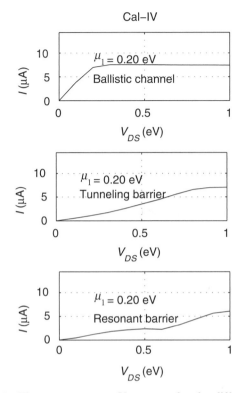

FIGURE 9.5 The current versus V_{DS} curves for the different channels.

It is a simple iterative procedure to plot the current as a function of the voltage across the channel, V_{DS}. This is done for different channels in Figure 9.5. The chemical potential on the left side is $\mu_1 = 0.2$ eV. The chemical potential on the right side varies with V_{DS}.

9.4 CONDUCTANCE

In the previous section we calculated the current flow and then determined the conductance by dividing by the drain-source voltage. Our classical training tells us that for the ballistic channel, which has no barriers, the conductance should approach infinity. Instead, the conductance leveled out at about 38 μS. It turns out that there is a fundamental constant called the *quantum conductance* that limits the conductance of a single mode channel. The development of the quantum conductance is the subject of this section.

In Section 9.3 we developed an expression for the current

$$I_{\text{channel}} = \frac{q}{2\pi\hbar} \int_{-\infty}^{\infty} dE\, T(E)[f_F(E - \mu_1) - f_F(E - \mu_2)]. \tag{9.26}$$

When no external voltage is applied, the channel is in equilibrium and

$$\mu_1 = \mu_2,$$

so no current flows. Now assume a bias current is applied to the drain. This changes the chemical potential on the right side

$$\mu_2 = \mu_1 - qV_{DS}.$$

We will start by assuming that the change in the transmission $T(E)$ due to V_{DS} is negligible, and multiply Equation (9.26) by (qV/qV):

$$
\begin{aligned}
I_{\text{channel}} &= \frac{q^2 V}{2\pi\hbar} \int_{-\infty}^{\infty} dE\, T(E) \left[\frac{f_F(E - \mu_1) - f_F(E - (\mu_1 - qV))}{qV} \right] \\
&\cong \frac{q^2 V}{2\pi\hbar} \int_{-\infty}^{\infty} dE\, T(E) \left[-\frac{\partial f_F(E)}{\partial E} \right].
\end{aligned}
\tag{9.27}
$$

The expression in the brackets is known as the *thermal broadening function*:

$$F_T(E - \mu) = -\frac{\partial f_E(E - \mu)}{\partial E}. \tag{9.28}$$

It can be shown that

$$\int_0^{\infty} F_T(E)\, dE = 1. \tag{9.29}$$

The broadening function corresponding to two different Fermi–Dirac distributions is shown in Figure 9.6.

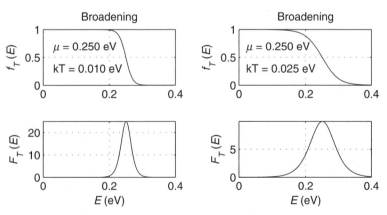

FIGURE 9.6 Two different Fermi–Dirac distributions (top) with their corresponding thermal broadening functions (bottom).

Remember that the transmission $T(E)$ is the probability that a particle of energy E will cross the channel from the left to the right. Obviously, it has a maximum value of 1. Therefore, the quantity

$$T_0 \equiv \int_{-\infty}^{\infty} T(E) F_T(E) dE \tag{9.30}$$

also has a maximum value of 1. Now we can rewrite Equation (9.27) as:

$$I_{\text{channel}} = \frac{q^2 V}{h} T_0. \tag{9.31}$$

Therefore, we conclude that the channel *conductance* is

$$G = \frac{q^2}{h} T_0. \tag{9.32}$$

(Recall that $h = 2\pi\hbar$.) The quantity

$$G_0 = \frac{q^2}{h} = 38.7 \ \mu S = (25.8 \ k\Omega)^{-1} \tag{9.33}$$

is the *quantum conductance*. It is a fundamental constant that limits the conductance for a single mode channel.

9.5 BÜTTIKER PROBES

Until now, we have been concerned with what is termed "ballistic" transport, where a particle is assumed to be capable of crossing a channel without a collision. A collision is the term used for an interaction with an ion or an impurity in the lattice. There are two types of collision. *Coherent* scattering occurs when the particle interacts without transferring energy to the lattice. This is illustrated in Figure 9.7 where the particle hits a potential described by

$$V(x) = -0.2 \ eV. \tag{9.34}$$

A small part of the waveform is reflected and most of it is transmitted through, as expected.

Another type of collision occurs when the particle collides with a defect in the lattice, for instance, and transfers some of its energy to the lattice, causing it to vibrate. This vibration is quantized into packets of energy called *phonons*. This energy can then be returned to the original particle or another particle. This is called *phase-breaking* or *incoherent* scattering. Clearly this will not be adequately modeled by a simple alteration of the potential, as in Equation

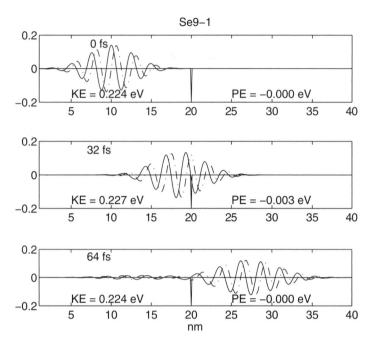

FIGURE 9.7 An impurity in the lattice is modeled by $V(x) = -0.2$ eV. This is an example of coherent scattering. PE, potential energy; KE, kinetic energy.

(9.34). Büttiker [6] suggested modeling such scattering events as fictitious contacts scattered throughout the channel, and so are called *Büttiker probes*.

As an example, we model a Büttiker probe by a fictitious contact potential described by

$$\Sigma_3 = -i0.3 \text{ eV}. \tag{9.35}$$

We assume it is located in the middle of the channel, as illustrated in Figure 9.8a.

The fact that we have another contact complicates the calculation of the transmission. Since we have three contacts we have the three transmissions: T_{12}, T_{13}, and T_{23}. These all must be considered in calculating the total transmission between contacts 1 and 2. Fortunately, the mathematics is analogous to using the sum of conductances to calculate the total conductance of a circuit (Fig. 9.8b). The total conductance between contacts 1 and 2 is

$$G = g_{12} + \frac{g_{13} g_{23}}{g_{13} + g_{23}}.$$

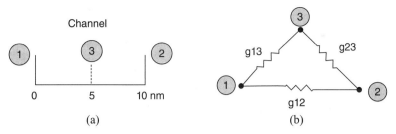

FIGURE 9.8 (a) A channel with contacts on each side, labeled "1" and "2" and a Büttiker probe in the middle modeled as contact number "3." (b) The circuit model of the channel with the Büttiker probe.

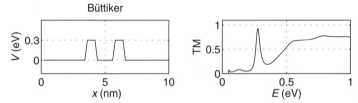

FIGURE 9.9 Calculation of the transmission (TM) through the resonant barrier with a Büttiker probe added at 5 nm. The probe is modeled as $\Sigma_3 = -i0.3$ eV. This is an example of incoherent scattering.

Similarly, the total transmission between 1 and 2 is

$$T_{\text{total}} = T_{12} + \frac{T_{13}T_{23}}{T_{13} + T_{23}}, \tag{9.36a}$$

where

$$T_{12} = Tr\left(\Gamma_1 G \Gamma_2 G^+\right), \tag{9.36b}$$

$$T_{13} = Tr\left(\Gamma_1 G \Gamma_3 G^+\right), \tag{9.36c}$$

and

$$T_{23} = Tr\left(\Gamma_2 G \Gamma_3 G^+\right). \tag{9.36d}$$

An example is shown in Figure 9.9. This is a repeat of the transmission calculation for the resonant channel, but a Büttiker probe has been added in the middle of the channel at 5 nm. Clearly the transmission of the channel with the Büttiker probe has been changed compared to a similar resonant channel without a probe as shown in Figure 9.2.

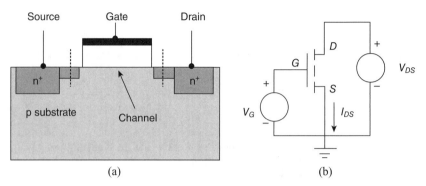

FIGURE 9.10 (a) Diagram of an n-channel enhancement-mode MOSFET. The dashed lines indicate the borders of the area to be simulated. (b) A circuit containing the MOSFET.

9.6 A SIMULATION EXAMPLE

In this section, we examine a metal–oxide–semiconductor field-effect transistor or MOSFET that will help put in perspective the concepts from this chapter. Figure 9.10a is a diagram of a MOSFET. Figure 9.10b is a simple circuit diagram containing a MOSFET. We will use the methods learned in this chapter to simulate a simple, one-dimensional simulation of the channel in the MOSFET.

The first decision to be made for the simulation is to define the problem space in the simulation. Obviously, it must include the channel directly under the gate and will include some of the n-type material of the source and drain. That region is indicated by the dashed lines in Figure 9.10a.

Now we assume that with no voltages applied to the MOSFET, the potential of the channel with the surrounding area has the profile shown in Figure 9.11a. By applying a voltage of 0.3 V to the gate, we reduce the potential as shown in Figure 9.11b. As shown in the circuit of Figure 9.10b the source is grounded, so when a voltage of 0.4 V is added to the drain, the potential is further modified as shown in Figure 9.11c. The transmission for the potential of Figure 9.11c is shown in Figure 9.11d.

We will take the chemical potential of the source terminal to be at $\mu_1 = 0.5$ eV. The drain-source voltage of 0.4 V puts the chemical potential of the drain at $\mu_2 = 0.1$ eV. Assuming room temperature, $k_B T = 0.0259$, we can calculate the Fermi–Dirac distributions shown in Figure 9.11e. The total area of the transmission times the difference in the Fermi functions is used to calculate the current by Equation (9.25a) as shown in Figure 9.11f.

Figure 9.12 illustrates the influence of the gate voltage on current flow. If the gate voltage is reduced to 0.0 V, the transmission is altered so particles with energies less than about 0.55 eV cannot flow through the channel.

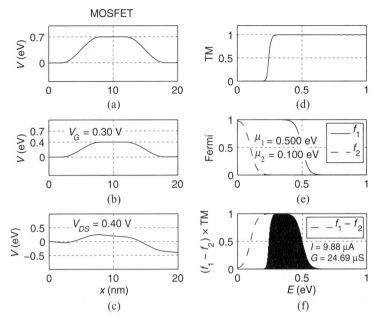

FIGURE 9.11 (a) The potential through the channel with no gate voltage. (b) The potential after a gate voltage of 0.3 V has been applied. (c) The potential after a drain voltage of 0.4 V has been applied. (d) The transmission for the potential of (c). (e) The Fermi distributions for kT = 0.025 eV, assuming the chemical potential on the left side is 0.5 eV. (f) The dark area is the difference between the Fermi distributions in (e) multiplied by the transmission in (d).

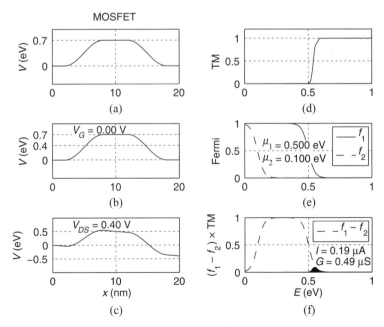

FIGURE 9.12 A simulation similar to the one illustrated in Figure 9.11, except that a gate voltage of 0 V is used. The primary effect is on the transmission which does not allow particles with a minimum energy of 0.55 eV to pass. This substantially reduces the total current flow.

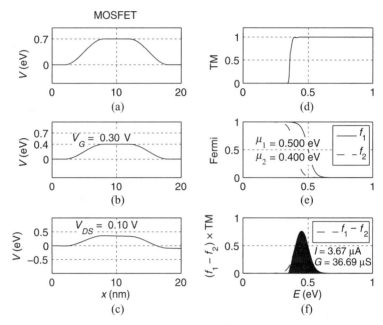

FIGURE 9.13 A simulation similar to the one illustrated in Figure 9.11, except that the drain-source voltage of 0.1 V is used. Although this has some effect on the potential in (c), and, therefore, the transmission in (d), the primary effect is a reduction in the difference between Fermi distributions in (e), which leads to a smaller current flow.

Figure 9.13 illustrates the influence of the drain-source voltage. If the drain-source voltage is reduced, the difference between the Fermi distributions is reduced, which also leads to a smaller total current flow.

It should be emphasized that the example in this section is extremely simplistic and does not correspond to a realistic simulation of a MOSFET. The Green's function method is used in the simulation of nanoscale devices, but the formulations of the problems are far more complex [1, 7].

EXERCISES

9.1 Single Energy Channel

9.1.1 Develop an expression for the transmission in the single energy channel, Equation (9.6), by starting from the right side of the channel.

9.2 Current Flow

9.2.1 Derive Equation (9.8) from Equation (9.7) using the time-dependent Schrödinger equation and its complex conjugate.

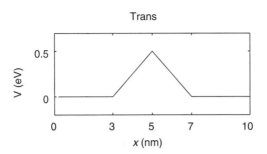

FIGURE 9.14 A triangular potential in a 10 nm well.

9.3 Transmission Matrix

9.3.1 Prove that $(G^+)^{-1} = (G^-)^+$. (Hint: use the identity $[A \cdot B]^+ = [B]^+[A]^+$.)

9.3.2 Prove that $A = A_1 + A_2$.

9.3.3 Calculate the transmission matrix similar to Equation (9.21b) by considering current flow on the right side of the channel.

9.3.4 In Figure 9.2 the transmission of the resonant barrier has a peak at about 0.12 eV. Looking at the resonant structure, how would you have predicted this?

9.3.5 Write a MATLAB program to calculate transmission across a 10 nm channel with the triangle barrier shown in Figure 9.14.

9.3.6 Find a potential that will attenuate particles in the energy range of 0.2–0.4 eV.

9.4 Conductance

9.4.1 Prove Equation (9.29).

9.5 Büttiker Probes

9.5.1 Suppose the channel is modeled with two Büttiker probes. Write the corresponding total transmission.

9.5.2 In Figure 9.8, explain why the transmission peak at 0.12 eV is so greatly diminished with the Büttiker probe added.

9.5.3 Write a MATLAB programs to add a Büttiker probe to the transmission calculation, as in Figure 9.8.

REFERENCES

1. S. Datta, *Quantum Transport—Atom to Transistor*, Cambridge, UK: Cambridge University Press, 2005.

2. S. Datta, *Electronic Transport in Mesoscopic Systems*, Cambridge, UK: Cambridge University Press, 1995.

3. S. Datta, "Nanoscale device simulation: The Green's function formalism," *Superlattice. Microst.*, Vol. 28, p. 253, 2000.

4. http://nanohub.org/topics/ElectonicsFromTheBottomUp.

5. D. J. Griffiths, *Introduction to Quantum Mechanics*, Englewood Cliffs, NJ: Prentice Hall, 1994.

6. M. Buttiker, "Symmetry of electrical conduction," *IBM J. Res. Dev.*, Vol. 32, p. 317, 2001.

7. R. Venugopal, S. Goasguen, S. Datta, and M. S. Lundstrom, "Quantum mechanical analysis of channel access geometry and series resistance in nanoscale transistors," *J. Appl. Phys.*, Vol. 95, pp. 292–305, 2004.

10

APPROXIMATION METHODS

As we have seen, few problems in quantum mechanics exist for which analytic solutions are available. This means that approximation methods play even greater roles in quantum mechanics than other disciplines. The two approximation methods described in this chapter—the variational method and the perturbation method—are widely used in quantum mechanics, and have applications to other disciplines as well. Section 10.1 describes the variational method and gives a simple example of how it is used to estimate eigenenergies and eigenfunctions. Sections 10.2–10.4 explain the perturbation method for increasingly complex examples. Section 10.2 starts with the time-independent perturbation theory for nondegenerate states. Section 10.3 covers the time-independent perturbation for degenerate states, and introduces the idea of "good" states. In Section 10.4 we come to time-dependent perturbation theory. This is a very extensive section with several important examples. It ends with a discussion of Fermi's golden rule.

10.1 THE VARIATIONAL METHOD

For many situations, an analytic solution is difficult or impossible for a certain Hamiltonian H. Sometimes a good starting point for the solution is to find the

Quantum Mechanics for Electrical Engineers, First Edition. Dennis M. Sullivan.
© 2012 The Institute of Electrical and Electronics Engineers, Inc.
Published 2012 by John Wiley & Sons, Inc.

ground state energy ε_0. The variational principal says that if you pick any normalized function ψ, then

$$\varepsilon_0 \le \langle H \rangle = \langle \psi | H | \psi \rangle, \tag{10.1}$$

that is, your guess for ψ will always have a greater or equal expected value of energy than that of the true ground state energy, ε_0.

Proof. Even though the eigenfunctions of H are unknown, they exist:

$$H\phi_n = \varepsilon_n \phi_n.$$

Furthermore, they form a complete set so we can write our selected function ψ as:

$$\psi = \sum_n c_n \phi_n.$$

If the ϕ_n are normalized

$$\sum_n |c_n|^2 = 1,$$

then

$$\langle \psi | H | \psi \rangle = \left\langle \sum_m c_m \phi_m \middle| H \sum_n c_n \phi_n \right\rangle = \left\langle \sum_m c_m \phi_m \middle| \sum_n c_n \varepsilon_n \phi_n \right\rangle$$

$$= \sum_m c_m^* \phi_m^* \sum_n c_n \varepsilon_n \phi_n = \sum_n \varepsilon_n |c_n|.$$

But by definition, $\varepsilon_0 \le \varepsilon_n$, so

$$\varepsilon_0 \le \sum_n \varepsilon_n |c_n| = \langle \psi | H | \psi \rangle.$$

This suggests the following procedure to find the ground state energy:

1. Guess the ground state eigenfunction.
2. Vary one or more parameters until the lowest possible expected value of the Hamiltonian is found.

Example. Suppose we have a 10 nm infinite well, but we put a V-shaped potential in it, as in Figure 10.1. Find the ground state eigenenergy and eigenfunction using the variational method.

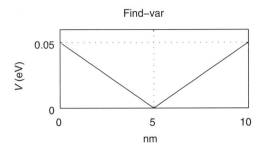

FIGURE 10.1 Quantum well with a V-shaped potential.

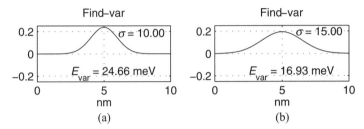

FIGURE 10.2 Approximations to the first eigenstate of the potential in Figure 10.1 using Equation (10.2).

Solution. The ground state eigenfunction will probably resemble the ground state of the one-dimensional infinite well, but "squeezed" toward the middle. So as a first guess, we will take a Gaussian pulse

$$\psi(x) = A \exp\left[-\frac{1}{2}\left(\frac{(x - x_c)}{\sigma}\right)^2\right], \tag{10.2}$$

where x_c is the center position of the well and A is the normalizing coefficient. We will start with $\sigma = 10$ nm, which gives us the waveform in Figure 10.2a. In MATLAB, we can calculate the energy by

$$E = \text{psi} * H * \text{psi}'. \tag{10.3}$$

Next we try $\sigma = 15$ nm and the energy drops a little lower as seen in Figure 10.2b.

After a little trial and error, we can settle on $\sigma = 18$ nm, as shown in Figure 10.3a. This proves to be our best estimate. Figure 10.3b shows the eigenfunction and eigenenergy calculated by the MATLAB program *find_eigen.m*.

So it seems we have a pretty efficient means of calculating the ground state energy. All well and good, but will that be sufficient? Suppose we want to calculate the next highest energy using the variational method. Will the same method work? For instance, as we keep changing the parameters and get lower

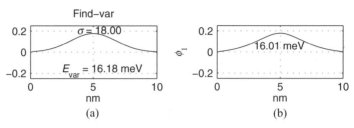

FIGURE 10.3 (a) The lowest ground state found using the variational method; (b) the actual ground state.

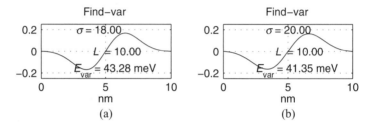

FIGURE 10.4 Estimates of the second eigenfunction of the potential in Figure 10.1 using Equation (10.4). (a) The first guess is $\sigma = 18$, $L = 10$. (b) The second guess is $\sigma = 20$, $L = 10$, and leads to a lower energy.

energies, how will we know that we are not just chasing the ground state eigenfunctions again? We will know that is not true as long as our new test function is orthogonal to the ground state function that we have already found.

For instance, in looking for the second function, suppose we use

$$\psi(x) = A\exp\left[-\frac{1}{2}\left(\frac{(x-x_c)}{\sigma}\right)^2\right]\sin\left(\frac{2\pi(x-x_c)}{L}\right). \tag{10.4}$$

This is a Gaussian envelope around a sinusoid that is asymmetric with the center of the well, so it will be orthogonal to the ground state function.

Note that in Equation (10.4) there are two parameters, σ and L, that can be varied. Examples are shown in Figure 10.4. Figure 10.5 shows the best estimate along with the actual second eigenfunction.

10.2   NONDEGENERATE PERTURBATION THEORY

In this section, we move to a different approximation method called perturbation theory. This method assumes that the eigenenergies and eigenfunctions

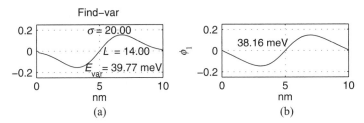

FIGURE 10.5 (a) The best estimation of the second eigenfunction for the potential of Figure 10.7; (b) the second eigenfunction calculated by find_eigen.m.

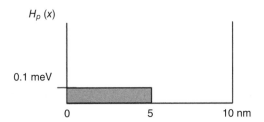

FIGURE 10.6 An infinite well with a perturbation.

of a structure are known, but then a perturbation is added. Suppose we have Hamiltonian H^0 for the one-dimensional infinite well and a corresponding set of eigenfunctions and eigenenergies:

$$H^0\phi_n^0 = \varepsilon_n^0\phi_n^0. \tag{10.5}$$

The eigenfunctions are of course orthonormal. Now we look at the problem of finding eigenenergies and eigenfunctions for the infinite well with a "perturbation," as given in Figure 10.6.

10.2.1 First-Order Corrections

Instead of trying to solve this problem for the new H, let us start by thinking of the Hamiltonian in terms of the original

$$H = H^0 + \lambda H_p, \tag{10.6}$$

where H_p describes the perturbation and λ is a parameter that allows us to vary the perturbation (typically $\lambda \ll 1$). This new Hamiltonian will have a new set of eigenfunctions and eigenenergies that we will write as:

$$\varepsilon_n \cong \varepsilon_n^0 + \lambda\varepsilon_n^1 + \lambda^2\varepsilon_n^2 + \dots, \tag{10.7a}$$

$$\phi_n \cong \phi_n^0 + \lambda\phi_n^1 + \lambda^2\phi_n^2 + \dots, \tag{10.7b}$$

where ε_n^1 is the first-order correction to the nth eigenenergy and ϕ_n^1 is the first-order correction to the nth eigenfunction. We will put these in the time-independent Schrödinger equation

$$\left(H^0 + \lambda H_P\right)\left(\phi_n^0 + \lambda\phi_n^1\right) = \left(\varepsilon_n^0 + \lambda\varepsilon_n^1\right)\left(\phi_n^0 + \lambda\phi_n^1\right),$$

and then rewrite the equation as:

$$H^0\phi_n^0 + \lambda\left(H_P\phi_n^0 + H^0\phi_n^1\right) + \lambda^2 H_P\phi_n^1 = \varepsilon_n^0\phi_n^0 + \lambda\left(\varepsilon_n^1\phi_n^0 + \varepsilon_n^0\phi_n^1\right) + \lambda^2\left(\varepsilon_n^1\phi_n^1\right).$$

By virtue of Equation (10.5) we can eliminate the non-λ terms. The λ^2 will be very small, so we can start by looking at the terms multiplied by a single λ:

$$H_P\phi_n^0 + H^0\phi_n^1 = \varepsilon_n^1\phi_n^0 + \varepsilon_n^0\phi_n^1. \tag{10.8}$$

Now we will multiply each term from the left by ϕ_n^0 and take the inner product:

$$\left\langle\phi_n^0\middle|H_P\phi_n^0\right\rangle + \left\langle\phi_n^0\middle|H^0\phi_n^1\right\rangle = \varepsilon_n^1 + \varepsilon_n^0\left\langle\phi_n^0\middle|\phi_n^1\right\rangle. \tag{10.9}$$

Remember that H^0 is Hermitian, so the second term on the left in Equation (10.9) is

$$\left\langle\phi_n^0\middle|H^0\phi_n^1\right\rangle = \left\langle H^0\phi_n^0\middle|\phi_n^1\right\rangle = \varepsilon_n^0\left\langle\phi_n^0\middle|\phi_n^1\right\rangle,$$

which leaves

$$\varepsilon_n^1 = \left\langle\phi_n^0\middle|H_P\phi_n^0\right\rangle. \tag{10.10}$$

This says that the first-order correction to each eigenenergy is simply the expectation value of the perturbation for the unperturbed wave function.

To look at the first order correction to the wave function, let us return to Equation (10.8) and rewrite it as

$$\left(H_P - \varepsilon_n^1\right)\phi_n^0 = -\left(H^0 - \varepsilon_n^0\right)\phi_n^1. \tag{10.11}$$

Since the zero-order (i.e., unperturbed) wave functions are a complete set, we can write

$$\phi_n^1 = \sum_{n\neq m} c_m^{(n)}\phi_m^0. \tag{10.12}$$

Note that this calculation does not include the nth term. Now insert Equation (10.12) into Equation (10.11):

$$\left(H_P - \varepsilon_n^1\right)\phi_n^0 = -\left(H^0 - \varepsilon_n^0\right)\sum_{n \neq m} c_m^{(n)}\phi_m^0$$

$$= -\left(\varepsilon_m^0 - \varepsilon_n^0\right)\sum_{n \neq m} c_m^{(n)}\phi_m^0,$$

and multiply from the right by ϕ_m^0 and take the inner product

$$\left\langle \phi_m^0 \middle| H_P \phi_n^0 \right\rangle = -\left(\varepsilon_m^0 - \varepsilon_n^0\right)c_m^{(n)}.$$

From this, we solve for $c_m^{(n)}$:

$$c_m^{(n)} = \frac{\left\langle \phi_m^0 \middle| H_P \phi_n^0 \right\rangle}{\varepsilon_n^0 - \varepsilon_m^0}.$$

In summary, the first-order corrections to the eigenenergies and eigenfunctions are:

$$\varepsilon_n^1 = \left\langle \phi_n^0 \middle| H_P \phi_n^0 \right\rangle, \qquad (10.13a)$$

$$\phi_n^1 = \sum_{n \neq m} \frac{\left\langle \phi_m^0 \middle| H_P \phi_n^0 \right\rangle}{\varepsilon_n^0 - \varepsilon_m^0} \phi_m^0. \qquad (10.13b)$$

Be aware that the perturbations to the energies in Equation (10.13a) tend to be accurate; the perturbations to the eigenfunctions in Equation (10.13b) are not always so accurate [1].

Example. In a 10 nm infinite well, suppose the following perturbation is added (Fig. 10.6):

$$H_P(x) = \begin{cases} 0.1\,\text{meV} & 0 \leq x \leq 5\,\text{nm}, \\ 0 & 5\,\text{nm} \leq x \leq 10\,\text{nm}. \end{cases}$$

Find the first-order perturbation to the energies and eigenfunctions.

Solution. The first-order correction to the eigenenergies is given by Equation (10.13a). Each one of these is the integration of eigenfunctions squared over half the well multiplied by the perturbation energy, that is,

$$\varepsilon_n^1 = 0.05\,\text{meV}.$$

Perturbations to the eigenfunctions are given by Equation (10.13b). If we look at just the first two terms we get

$$\phi_1^1 \cong \frac{\left\langle \phi_2^0 \middle| H_P \phi_1^0 \right\rangle}{(3.75 - 15)\,meV}\phi_2^0 + \frac{\left\langle \phi_3^0 \middle| H_P \phi_1^0 \right\rangle}{(3.75 - 33.75)\,meV}\phi_3^0.$$

10.2.2   Second-Order Corrections

Now we look at the first two terms in Equation (10.7),

$$\varepsilon_n \cong \varepsilon_n^0 + \lambda\varepsilon_n^1 + \lambda^2\varepsilon_n^2, \tag{10.14a}$$

$$\phi_n \cong \phi_n^0 + \lambda\phi_n^1 + \lambda^2\phi_n^2. \tag{10.14b}$$

(Remember: ε_n^2 means the second-order correction to the nth eigenenergy.) If we put these into the time-independent Schrödinger equation

$$H\phi_n = \varepsilon_n\phi_n$$

and then equate all of the second-order terms, that is, those with a λ^2 factor, we get

$$H^0\phi_n^2 + H_P\phi_n^1 = \varepsilon_n^0\phi_n^2 + \varepsilon_n^1\phi_n^1 + \varepsilon_n^2\phi_n^0.$$

Now we take the inner product with ϕ_n^0:

$$\langle \phi_n^0 | H^0\phi_n^2 \rangle + \langle \phi_n^0 | H_P\phi_n^1 \rangle = \varepsilon_n^0 \langle \phi_n^0 | \phi_n^2 \rangle + \varepsilon_n^1 \langle \phi_n^0 | \phi_n^1 \rangle + \varepsilon_n^2.$$

Once again, because $\langle \phi_n^0 | H^0\phi_n^2 \rangle = \langle H^0\phi_n^0 | \phi_n^2 \rangle = \varepsilon_n^0 \langle \phi_n^0 | \phi_n^2 \rangle$, we can eliminate the first terms on the left and right and get an expression for

$$\varepsilon_n^2 = \langle \phi_n^0 | H_P\phi_n^1 \rangle - \varepsilon_n^1 \langle \phi_n^0 | \phi_n^1 \rangle.$$

According to Equation (10.11), ϕ_n^1 is composed of elements of each of the original eigenfunctions with the exception of ϕ_n^0, that is,

$$\langle \phi_n^0 | \phi_n^1 \rangle = 0.$$

So from Equation (10.13b):

$$\varepsilon_n^2 = \sum_{n\neq m} \frac{|\langle \phi_m^0 | H_P\phi_n^0 \rangle|^2}{\varepsilon_n^0 - \varepsilon_m^0}. \tag{10.15}$$

We could calculate the second-order corrections to the wave functions, but this is seldom done. Similarly, we could calculate higher order corrections to the eigenenergies, but this also is rarely done [1].

10.3   DEGENERATE PERTURBATION THEORY

In quantum mechanics, the term "degenerate" is used to indicate two different states with the same energy. One example that we have already discussed is

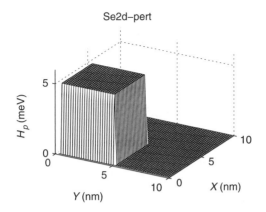

FIGURE 10.7 A perturbation that is added to a 10×10 nm infinite well.

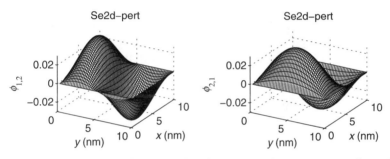

FIGURE 10.8 The $\phi_{1,2}(x, y)$ and $\phi_{2,1}(x, y)$ eigenfunctions of a two-dimensional 10×10 nm infinite well.

the two-dimensional infinite well with identical dimensions. We will assume that the well is 10 nm on each side. It has the following eigenfunctions and eigenenergies:

$$\phi_{l,m}(x,y) = \frac{2}{L}\sin\left(\frac{l\pi x}{L}\right)\sin\left(\frac{m\pi y}{L}\right), \quad L = 10 \text{ nm}, \tag{10.16a}$$

$$\varepsilon_{l,m} = \frac{\pi^2\hbar^2}{2m_e L^2}\left(l^2 + m^2\right) = \left(l^2 + m^2\right)3.75 \text{ meV}. \tag{10.16b}$$

We then assume the perturbation in Figure 10.7 is added.

Clearly the eigenfunctions $\phi_{1,2}(x, y)$ and $\phi_{2,1}(x, y)$ (Fig. 10.8) will have the same energy,

$$\varepsilon_{1,2} = \varepsilon_{2,1} = (3.75 \text{ meV})\left(1^2 + 2^2\right) = 18.75 \text{ meV},$$

but they are distinct states that are orthogonal to each other. Remember that the perturbation method depends on taking an inner product with nonperturbed eigenstates to eliminate all terms except one. Since we have a degenerate state, this poses a problem. We get around this is by constructing a state that is a linear combination of the degenerate states:

$$\phi^0 = \alpha\phi_{1,2}^0 + \beta\phi_{2,1}^0. \tag{10.17}$$

We start as before with, the approximations

$$\varepsilon \cong \varepsilon^0 + \lambda\varepsilon^1, \tag{10.18a}$$

$$\phi \cong \phi^0 + \lambda\phi^1. \tag{10.18b}$$

However, in this case

$$\varepsilon^0 = \alpha\varepsilon_{1,2}^0 + \beta\varepsilon_{2,1}^0. \tag{10.19}$$

Inserting Equations (10.17) and (10.18) into Equation (10.5) gives us

$$\left(H^0 + \lambda H_P\right)\left(\phi^1 + \lambda\phi^1\right) = \left(\varepsilon^0 + \lambda\varepsilon^1\right)\left(\phi^1 + \lambda\phi^1\right).$$

Once again, we retain only λ terms, and take the inner product with $\phi_{1,2}^0$, which results in the following equation:

$$\alpha W_{\alpha\alpha} + \beta W_{\alpha\beta} = \alpha\varepsilon^1,$$

where

$$W_{\alpha\alpha} = \langle\phi_{1,2}|H_P\phi_{1,2}\rangle,$$

$$W_{\alpha\beta} = \langle\psi_{1,2}|H_P\psi_{2,1}\rangle.$$

Obviously, we could start this whole process again, taking the inner product with $\phi_{2,1}^0$ and get a second equation of the form

$$\alpha W_{\beta\alpha} + \beta W_{\beta\beta} = \beta\varepsilon^1.$$

We want to calculate ε^1. The trouble is, we also don't know α and β. We know the value of W, but we do not know α and β. We can write

$$\begin{bmatrix} W_{\alpha\alpha} & W_{\alpha\beta} \\ W_{\beta\alpha} & W_{\beta\beta} \end{bmatrix} \begin{bmatrix} \alpha \\ \beta \end{bmatrix} = \varepsilon^1 \begin{bmatrix} \alpha \\ \beta \end{bmatrix}, \tag{10.20}$$

and solve the determinant to get the eigenvalues, ε^1:

$$\varepsilon_\pm^1 = \frac{1}{2}\left[W_{\alpha\alpha} + W_{\beta\beta} \pm \sqrt{\left(W_{\alpha\alpha} - W_{\beta\beta}\right)^2 - 4W_{\beta\alpha}W_{\alpha\beta}}\right]. \tag{10.21}$$

Once we have these "eigenvalues" for Equation (10.21), we can solve for α and β to get the corresponding functions. However, a caveat goes with this whole process. Notice that the solution to α and β will give us two functions corresponding to the two values of ε^1:

$$\phi^{(1)} = \alpha^{(1)}\phi_{1,2}^0 + \beta^{(1)}\phi_{2,1}^0, \qquad (10.22a)$$

$$\phi^{(2)} = \alpha^{(2)}\phi_{1,2}^0 + \beta^{(2)}\phi_{2,1}^0. \qquad (10.22b)$$

If we were to remove the perturbation, we would expect that this would take us back to the two original degenerate eigenstates given by Equation (10.16). However, this is not necessarily the case. Our selection of α and β might take us to two linear combinations of $\phi_{1,2}(x, y)$ and $\phi_{2,1}(x, y)$. Instead, we would like to be sure that the new perturbed eigenstates selected in Equation (10.22) are *good* states that return to the original eigenstates when the perturbation is removed. One such approach is to diagonalize the matrix W in Equation (10.20) [1].

To demonstrate the difference between good states and bad states, we will start with the two-dimensional infinite well initialized with $\phi_{2,1}$ eigenfunction in Figure 10.8. We will add a perturbation in one corner of the well as shown in Figure 10.7. Using a Hanning window (see Appendix A), the perturbation will be added and withdrawn over a period of 500 femtoseconds (fs). The results are illustrated by the contour diagrams of the state function in Figure 10.9. Clearly, we did not start with a good state because when the perturbation is removed, the wave function does not go back to its original form.

Now we repeat the simulation starting with the superposition state $\psi = \phi_{1,2} - \phi_{2,1}$ shown in Figure 10.10. After the simulation is removed, the waveform returns to its original state, as illustrated in Figure 10.11, demonstrating that the waveform of Figure 10.10 is a good state. In Figure 10.11, the contour at 750 fs does not look exactly like the starting contour at 0 fs. This is the slight error due to the finite-difference simulation.

10.4 TIME-DEPENDENT PERTURBATION THEORY

In considering time-dependent perturbation, we write the state function as a superposition of the eigenstates of an unperturbed Hamiltonian H_0 with time-dependent coefficients [2]

$$\psi(x,t) = \sum_{n=1}^{\infty} c_n(t)\phi_n(x)e^{-i(\varepsilon_n/\hbar)t}, \qquad (10.23)$$

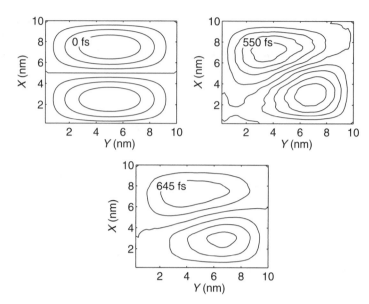

FIGURE 10.9   The contour diagrams of the state function $\phi_{2,1}$ at various times show the effect of the perturbation. Notice that after the perturbation is added and then removed the waveform does not go back to its original state.

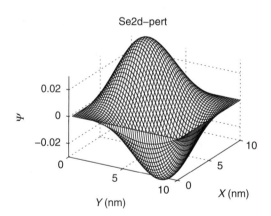

FIGURE 10.10   The superposition state $\psi = \phi_{1,2} - \phi_{2,1}$.

where ϕ_n are the eigenfunctions and ε_n are the corresponding eigenenergies. Note that each state has a time-dependent coefficient $c_n(t)$, as well as the time dependence due to the eigenenergy. Taking $H_P(x,t)$ as the time-dependent perturbation, the time-dependent Schrödinger equation is now

$$i\hbar\frac{\partial \psi(x,t)}{\partial t} = (H_0 + H_P(x,t))\psi(x,t). \tag{10.24}$$

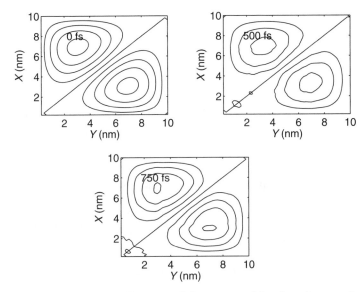

FIGURE 10.11   The contour diagrams of the superposition function $\psi = \phi_{1,2} - \phi_{2,1}$ at various times show the effect of the perturbation. Notice that after the perturbation is added and then removed the waveform returns to its original state.

If we substitute Equation (10.23) into Equation (10.24), we get

$$i\hbar\left(\sum_{n=1}^{\infty}\left(\dot{c}_n(t) - i\frac{\varepsilon_n}{\hbar}c_n(t)\right)\phi_n(x)e^{-i(\varepsilon_n/\hbar)t}\right) = (H_0 + H_P(x,t))\sum_{n=1}^{\infty}c_n(t)\phi_n(x)e^{-i(\varepsilon_n/\hbar)t}.$$

Since

$$H_0\phi_n(x) = \varepsilon_n\phi_n(x),$$

we can eliminate the unperturbed terms on both sides, leaving

$$i\hbar\left(\sum_{n=1}^{\infty}\dot{c}_n(t)\phi_n(x)e^{-i(\varepsilon_n/\hbar)t}\right) = H_P(x,t)\sum_{n=1}^{\infty}c_n(t)\phi_n(x)e^{-i(\varepsilon_n/\hbar)t}.$$

Now we take the inner product with $\phi_m(x)$

$$i\hbar\dot{c}_m(t)e^{-i(\varepsilon_m/\hbar)t} = \sum_{n=1}^{\infty}c_n(t)\langle\phi_m|H_P|\phi_n\rangle\phi_n(x)e^{-i(\varepsilon_n/\hbar)t}.$$

Finally,

$$\dot{c}_m(t) = \frac{1}{i\hbar}\sum_{n=1}^{\infty}c_n(t)\langle\phi_m|H_P|\phi_n\rangle e^{-i\omega_{mn}t}, \tag{10.25}$$

where

$$\omega_{mn} = \frac{\varepsilon_m - \varepsilon_n}{\hbar} = \omega_m - \omega_n. \tag{10.26}$$

Equation (10.25) is the fundamental equation for time-dependent perturbation. It says that we are tracking the effect of the perturbation by the time-dependent coefficients $c_n(t)$ of the eigenfunction expansion of Equation (10.23).

10.4.1 An Electric Field Added to an Infinite Well

Suppose we start with a particle in the ground state of the 10 nm infinite well. At time $t = 0$, an electric field of 10^6 V/m is added across the well. This corresponds to a change in potential of

$$V_{max} = (10^6 \text{ V/m})(10 \text{ nm}) = 10 \text{ meV,}$$

as shown in Figure 10.12.

The time-dependent perturbation can be written as:

$$H_P(x,t) = H_P(x)u(t),$$

where $H_P(x)$ is the spatial part of the perturbation, as shown in Figure 10.12, while the $u(t)$ is the step function that indicates that the electric field is turned on at time zero. Recall that the ground state is at 3.75 meV. The second state is at 15 meV and the third state is way up at 33.75 meV. Therefore, the most likely transition due to the perturbation is to the second eigenstate; so we calculate

$$H_{P12} = \langle \phi_2 | H_P(x) | \phi_1 \rangle = \int_0^L \phi_1(x)\phi_2(x) \cdot H_P(x)\,dx = 1.82 \text{ meV.}$$

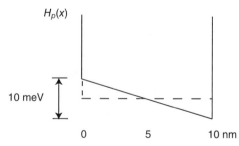

FIGURE 10.12 The change in potential across a 10 nm well shown here simulates an electric field of 10^6 V/m across the channel.

Because we start with a particle in the ground state, we can say that at time $t = 0$, $c_1(0) = 1$ and $c_n(0) = 0$ for all other n.

To find the probability of a transition to the second eigenstate, start with

$$\dot{c}_2(t) = \frac{-i}{\hbar} H_{P21} u(t) e^{i\omega_{21}t},$$

and integrate both sides:

$$c_2(t) = \frac{-i}{\hbar} H_{P21} \int_0^t e^{i\omega_{21}\tau} d\tau$$

$$= \frac{-i}{\hbar} H_{P21} \left[\frac{e^{i\omega_{21}t} - 1}{i\omega_{21}} \right].$$

The probability of a transition is

$$|c_2|^2 = \frac{H_{P21}^2}{\hbar^2} \left[\frac{e^{i\omega_{21}t} - 1}{i\omega_{21}} \right] \left[\frac{e^{-i\omega_{21}t} - 1}{-i\omega_{21}} \right] = \frac{H_{P21}^2}{(\varepsilon_2 - \varepsilon_1)^2} [2 - 2\cos(\omega_{21}t)]. \tag{10.27}$$

The predicted frequency of the oscillation is

$$f_{21} = \frac{\varepsilon_2 - \varepsilon_1}{h} = \frac{(2^2 - 1^1)3.75\,\text{eV} \times 10^{-3}}{4.135 \times 10^{-15}\,\text{eV} \cdot \text{s}} = 2.72 \times 10^{12}\,\text{s},$$

with the corresponding time period of

$$T_0 = \frac{1}{f_{21}} = 0.363 \times 10^{-12} = 363\,\text{fs}.$$

This can be simulated with one of our finite-difference simulations by initializing the state variable in the ground state of a 10 nm infinite well, and adding the potential of Figure 10.12. In Figure 10.13a, the particle starts in the ground state at time $t = 0$. After 205 fs, we can recognize that a portion of the particle is in the second eigenstate. After 375 fs, the particle has returned to the ground state, and by 560 fs a portion is again in the second state. The oscillation of the probabilities is tracked by Figure 10.13b. Since we have restricted ourselves to the first two eigenstates, it must be that

$$|c_2(t)|^2 + |c_1(t)|^2 = 1. \tag{10.28}$$

10.4.2 Sinusoidal Perturbations

Often the time-dependent part of the perturbation is sinusoidal, that is,

$$H_P(x,t) = H_P(x)\cos(\omega_L t)u(t). \tag{10.29}$$

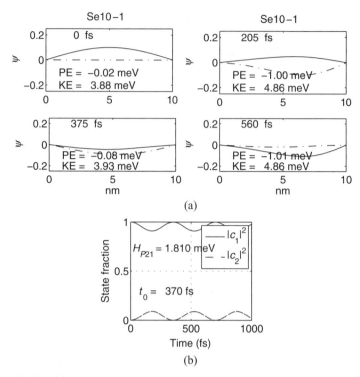

FIGURE 10.13 (a) The oscillation between the first and second eigenstate of the 10 nm infinite well after the perturbation of Figure 10.12 had been added at time $t = 0$; (b) a plot of $|c_1(t)|^2$ and $|c_2(t)|^2$ as functions of time. PE, potential energy; KE, kinetic energy.

The perturbation might be a light from a laser, for instance. We assume that the spatial part of the perturbation is the same as in the previous section. We again start with the particle in the ground state, and look for the transition to the second eigenstate, so Equation (10.25) becomes

$$\dot{c}_2(t) = \frac{-i}{\hbar} H_{P21} \cos(\omega_L t) u(t) e^{i\omega_{21} t}. \tag{10.30}$$

Integrating both sides gives

$$c_2(t) = \frac{-i}{\hbar} H_{P21} \int_0^t \left[\frac{e^{i\omega_L \tau} + e^{-i\omega_L \tau} +}{2} \right] e^{i\omega_{21}\tau} d\tau$$

$$= \frac{-iH_{P21}}{2\hbar} \left[\frac{e^{i(\omega_{21}+\omega_L)t} - 1}{i(\omega_{21} + \omega_L)} + \frac{e^{i(\omega_{21}-\omega_L)t} - 1}{i(\omega_{21} - \omega_L)} \right].$$

Obviously

$$(\omega_{21} + \omega_L) \gg (\omega_{21} - \omega_L),$$

so we can say

$$
\begin{aligned}
c_2(t) &\cong \frac{-iH_{P21}}{2\hbar} \left[\frac{e^{i(\omega_L - \omega_{21})t} - 1}{i(\omega_L - \omega_{21})} \right] \\
&= \frac{-iH_{P21}}{2\hbar} \left[\frac{e^{i(\omega_L - \omega_{21})t/2} - e^{-i(\omega_L - \omega_{ba})t/2}}{i(\omega_L - \omega_{21})} \right] e^{i(\omega_L - \omega_{ba})t/2} \\
&= \frac{-iH_{P21}}{\hbar} \left[\frac{\sin((\omega_L - \omega_{21})t/2)}{(\omega_L - \omega_{21})} \right] e^{i(\omega_L - \omega_{21})t/2}.
\end{aligned}
\tag{10.31}
$$

The probability that state 2 is occupied is

$$|c_2(t)|^2 = \frac{|H_{P21}|^2}{\hbar^2} \frac{\left[\sin^2(0.5(\omega_L - \omega_{21})t) \right]}{(\omega_L - \omega_{21})^2}. \tag{10.32}$$

The time period of the oscillation is determined by the difference between ω_L and ω_{21}:

$$T_0 = \frac{2\pi}{\omega_L - \omega_{21}}. \tag{10.33}$$

We will simulate an oscillating electric field at 3.2 THz with a magnitude of 1.5×10^5 V/m that appears across the 10 nm infinite well with a particle that starts in the ground state. This is simulated by a potential similar to Figure 10.12 except that the magnitude oscillates in time. The time-domain portion is added gradually, as shown in Figure 10.14a. This avoids a transient reaction to a sudden perturbation. (See the section on *windowing* in Appendix A.) Figure 10.14b shows the particle at two different times. Figure 10.14c shows the oscillation between states 1 and 2.

Figure 10.15 is a similar simulation for an oscillating electric field at 3.4 THz. Since this is farther away from $\omega_{21} = (\varepsilon_2 - \varepsilon_1)/\hbar$, the coupling is not as strong.

10.4.3   Absorption, Emission, and Stimulated Emission

We have been discussing the transition from the ground state, state 1, to the next highest state, state 2, resulting from a perturbation at frequency ω_L. This process is referred to as *absorption*, because a particle in state 1 must absorb a photon of energy $\hbar\omega_L = \varepsilon_2 - \varepsilon_1$ to make the transition to state 2.

The process of *emission* is the loss of a photon in the transition from a higher state to a lower state. Unfortunately there is no mechanism in the finite difference simulation for the spontaneous emission of a photon.

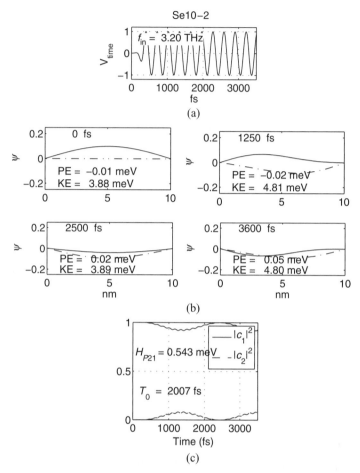

FIGURE 10.14 A particle begins in the ground state of a 10 nm infinite well. (a) Starting at time zero, a 3.20 THz electric field with a magnitude of 1.5×10^5 V/m is gradually added over 0.5 ps. (b) The state function for the particle at two different times. (c) The oscillation between the states. f_{12} is the frequency difference between the first and second eigenstates. T_0 is the time period of the oscillation. PE, potential energy; KE, kinetic energy.

The process of *stimulated emission* is essentially the same as absorption, with the states reversed, as illustrated in Figure 10.16. The particle is initialized in the second eigenstate. Once again, an electric field at 3.2 THz is simulated. Under the influence of the electric field, the particle has a probability of transitioning from the second eigenstate down to the ground state.

10.4.4 Calculation of Sinusoidal Perturbations Using Fourier Theory

Let us consider a sinusoidal perturbation similar to Equation (10.29); however, we will assume that we are going to turn the simulation on and leave it on for

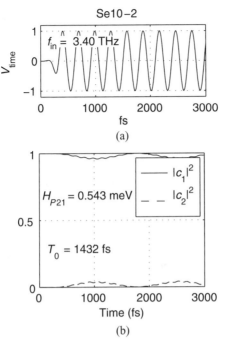

FIGURE 10.15 A particle begins in the ground state of a 10 nm infinite well. (a) Starting at time zero, a 3.4 THz electric field with a magnitude of 1.5×10^5 V/m is gradually added over 0.5 ps. (b) The oscillation between the states. f_{12} is the frequency difference between the first and second eigenstates. T_0 is the time period of the oscillation.

a period of time, t_0. With no loss of generality, we will say that we turn it on at $-t_0/2$ and then turn it off at time $t_0/2$, so we write:

$$H_P(x,t) = H_P(x)\cos(\omega_L t)p_{t_0}(t), \tag{10.34}$$

where the pulse function $p_{t_0}(t)$ is defined by:

$$p_{t_0}(t) = \begin{cases} 1 & -t_0/2 < t < t_0/2 \\ 0 & \text{otherwise} \end{cases}.$$

Equation (10.30) becomes

$$\dot{c}_2(t) = \frac{-i}{\hbar} H_{P21} \cos(\omega_L t) p_{t_0}(t) e^{i\omega_{21}t}. \tag{10.35}$$

We can write

$$\cos(\omega_L t) = \frac{e^{i\omega_L t} + e^{-i\omega_L t}}{2},$$

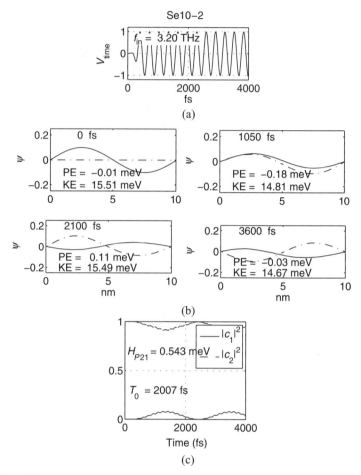

FIGURE 10.16 An example of stimulated emission. A particle begins in the second eigenstate of a 10 nm infinite well. (a) Starting at time zero, a 3.20 THz electric field with a magnitude of 1.5×10^5 V/m is gradually added over 0.5 ps. (b) The state function for the particle at two different times. (c) The oscillation between the states showing that there is a probability that the particle can transition from the second state down to the ground state. f_{12} is the frequency difference between the first and second eigenstates. T_0 is the time period of the oscillation. PE, potential energy; KE, kinetic energy.

but we know from experience that only the negative exponential will be important in our calculations. So we write Equation (10.35) as:

$$\dot{c}_2(t) = \frac{-i}{2\hbar} H_{P21} p_{t_0}(t) e^{i(\omega_{21}-\omega_L)}.$$

Now we integrate

$$c_2(t) = \frac{-i}{2\hbar} H_{P21} \int_{-\infty}^{\infty} p_{t_0}(t) e^{i(\omega_{21}-\omega_L)} dt. \qquad (10.36)$$

Notice that the integral of Equation (10.36) has the form of the Fourier transform, except the exponential is $i(\omega_{21} - \omega_L)$ instead of just $i\omega$. Therefore, we can just use the Fourier transform of $p_{t_0}(t)$ from Table 3.1 and evaluate at $\omega = (\omega_{21} - \omega_L)$:

$$c_2(t) = \frac{-i}{2\hbar} H_{P21} \left[t_0 \frac{\sin(\omega t_0/2)}{(\omega t_0/2)} \right]_{\omega = \omega_{21} - \omega_L}$$

$$= \frac{-i}{\hbar} H_{P21} \left[\frac{\sin((\omega_{21} - \omega_L)t_0/2)}{(\omega_{21} - \omega_L)} \right].$$

This is the same as Equation (10.31) except for the phase term because we started at $t_0/2$ instead of time zero. There is a mathematical difference to this solution. The value c_2 is no longer a function of time because we took the Fourier transform that extends to infinity in time. Time now appears implicitly in the width of the pulse t_0.

The real benefit of this approach is obvious when we are faced with the more complicated time-domain functions.

Example. A more realistic perturbation might be a pulse from a laser. If this pulse is a sinusoid in a Gaussian envelope, then the perturbation is of the form:

$$H_P(x,t) = H_P(x)\cos(\omega_L t)\exp\left(-\frac{1}{2}\left(\frac{t}{\sigma}\right)^2\right), \tag{10.37}$$

where σ is the width of the pulse. If we once again start with the particle in the ground state, what is the probability of a transition to the second state?

Solution. Equation (10.30) becomes

$$\dot{c}_2(t) = \frac{-i}{\hbar} H_{P21} \exp\left[-\frac{1}{2}\left(\frac{t}{\sigma}\right)^2\right]\cos(\omega_L t)e^{i\omega_{21}t}.$$

Solving using the Fourier transforms

$$c_2 = \frac{-i}{2\hbar} H_{P21} \int_{-\infty}^{\infty} \exp\left[-\frac{1}{2}\left(\frac{t}{\sigma}\right)^2\right]e^{i(\omega_{21} - \omega_L)t} dt$$

$$= \frac{-i}{2\hbar} H_{P21} F\left\{\exp\left(-\frac{1}{2}\left(\frac{t}{\sigma}\right)^2\right)\right\}_{\omega = \omega_{mn} - \omega_L}$$

$$= \frac{-i}{2\hbar} H_{P21} \cdot \sqrt{2\pi}\sigma \exp\left(-\frac{\sigma^2(\omega_{21} - \omega_L)^2}{2}\right).$$

So the probability is

$$|c_2|^2 = \frac{\sigma^2 \pi}{2\hbar^2} H_{P21}^2 \exp\left(-\sigma^2(\omega_{21} - \omega_L)^2\right).$$

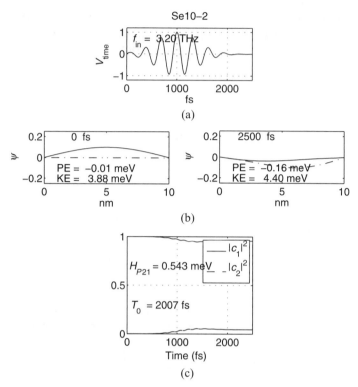

FIGURE 10.17 The simulation of a perturbation consisting of an electric field of 3.2 THz inside a Gaussian pulse of 400 fs. (a) The perturbation. (b) The state variables at $T = 0$ and at $T = 2.5$ ps. (c) The plot of $|c_1(t)|^2$ and $|c_2(t)|^2$ as functions of time. PE, potential energy; KE, kinetic energy.

Not surprisingly, the probability of the transition depends on the width of the pulse, σ.

Example. Repeat the simulation of Figure 10.14 using a perturbation that consists of an electric field at 3.2 THz, but inside a Gaussian envelope of 400 fs.

Solution. See Fig. 10.17.

Example. Suppose the perturbation is of the form

$$H_P(x,t) = H_P(x)e^{-\alpha t}\cos(\omega_L t)u(t).$$

Calculate the probability of a transition from the first to the second eigenstate.

Solution

$$\dot{c}_2(t) = \frac{-i}{\hbar} H_{P21} e^{-\alpha t} \cos(\omega_L t) e^{i\omega_{21} t}$$

$$\cong \frac{-i}{2\hbar} H_{P21} e^{-\alpha t} e^{-i\omega_L t} e^{i\omega_{21} t} u(t),$$

$$c_2 = \frac{-i}{2\hbar} H_{P21} F\{e^{-\alpha t} u(t)\}_{\omega = \omega_{21} - \omega_L}$$

$$= \frac{-i}{2\hbar} H_{P21} \frac{1}{\alpha - i(\omega_{21} - \omega_L)},$$

$$|c_2|^2 = \frac{|H_{P21}|^2}{4\hbar^2} \frac{2\alpha}{\alpha^2 + (\omega_{21} - \omega_L)^2}.$$

The addition of the term $e^{-\alpha t}$ has the effect of broadening in the frequency domain, as discussed in Section 3.2. As a consequence, the strength of the perturbation is not as dependent on how close ω_L is to ω_{21}.

10.4.5 Fermi's Golden Rule

Instead of the infinite well, we will now do a simulation using the harmonic oscillator potential shown in Figure 10.18a. The simulation has the evenly spaced eigenenergies at intervals of 5 meV, as shown in Figure 10.18b. The 5 meV interval corresponds to a frequency of 1.2 THz. As our perturbation, we use an electric field at 1.2 THz inside a Gaussian envelope (Fig. 10.18c). The particle is initialized in the 10th eigenstate. In Figure 10.18d we show the state function and a bar graph displaying $|c_n|^2$. After 4 ps, part of the probability amplitude has scattered into the surrounding states. The key point is this: if we were interested primarily in how much the probability amplitude of the 10th eigenstate decreases, we cannot look only at one other eigenstate. Since the harmonic oscillator has two eigenstates in the same close proximity, any analytical solution to calculate the decay of $|c_{10}(t)|^2$ will have to take into account transitions to both states 9 and 11. In fact, there is some probability of a transition to either state 8 or state 12. This is due in part to the fact that the Gaussian pulse will have frequency components not just at 1.2 THz, but at nearby frequencies as well.

There are cases where we are interested in the decay out of a state that is one of many closely spaced states, as is illustrated in Figure 10.18. This can occur in solids that can have very dense sets of transitions. Suppose in the vicinity of the perturbation frequency ω_L there is a large set of possible transitions. We assume this set is very dense with a density of $g(\hbar\omega)$ per unit energy near the photon energy $\hbar\omega_L$ [3].

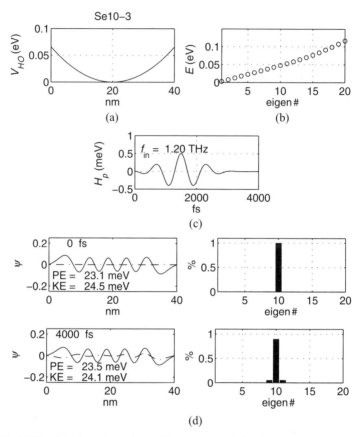

FIGURE 10.18 (a) A harmonic oscillator potential with a reference energy of $E_{ref} = \hbar\omega_0 = 5$ meV. (b) The corresponding eigenenergies. (c) The perturbation is an electric field at 1.2 THz inside a Gaussian envelope. (d) The state function and the corresponding percentage of each eigenstate for times $T = 0$ and $T = 4$ ps. PE, potential energy; KE, kinetic energy.

Equation (10.32) gave us the probability of a transition from state 1 to state 2 assuming a perturbation at frequency ω_L. We will generalize this equation to get the probability of a transition from state m to state n:

$$|c_n(t)|^2 = \frac{|H_{Pnm}|^2 t^2}{4\hbar^2} \operatorname{sinc}^2\left(\frac{(\omega_L - \omega_{nm})t}{2}\right). \tag{10.38}$$

Assume that at time zero, we are at the state m. Now we can think about the probability of any transition out of state m as the summation of the probabilities of transitions to any other state:

$$P_{\text{out of } m} = \sum_{n \neq m} |c_n(t)|^2. \tag{10.39}$$

If we assume that near the mth state, the density of states is given by $g(\omega)$, then we can write

$$P_{\text{out of } m} = \frac{|H_{Pnm}|^2 t^2}{4\hbar^2} \int \text{sinc}^2\left(\frac{(\omega_L - \omega_{nm})t}{2}\right) g(\omega) d\hbar\omega_{nm}. \qquad (10.40)$$

Making the change of variables $x = (\omega_L - \omega_{nm})t/2$ and the fact that

$$\int_{-\infty}^{\infty} \text{sinc}^2(x) dx = \pi,$$

we can write Equation (10.40) as

$$P_{\text{out of } m} = \frac{\pi |H_{Pnm}|^2 t^2}{2\hbar} g(\omega_m).$$

We see that the total probability of a transition out of state m is dependent on t, so we can say that the transition rate is

$$W = \frac{\pi |H_{Pnm}|^2}{2\hbar} g(\omega_{nm}). \qquad (10.41)$$

This is known as *Fermi's golden rule*. (Physics texts write Fermi's golden rule with a factor of 4 greater than Eq. 10.41 because they start with a time-domain function of $\exp(i\omega_L t) + \exp(-i\omega_L t)$ instead of $\cos(\omega_L t)$). Fermi's golden rule is an important result and is used to calculate the optical absorption spectra of solids, among many other things.

EXERCISES

10.1 The Variational Method

10.1.1 A very important canonical structure in quantum mechanics is the harmonic oscillator, which has the potential

$$U_{HO} = \frac{1}{2}m\omega_0^2 x^2 = \frac{1}{2}k_0^2 x^2,$$

$$k_0 = m_0\left(\frac{E_{ref}}{\hbar}\right)^2.$$

Find approximations to the first three eigenfunctions and eigenenergies using the variational method for the case where $E_{ref} = 5$ meV. Use MATLAB to do this, and then use find_eig.m to determine how close you came.

10.2   Nondegenerate Perturbation Theory

10.2.1   The following perturbation is place in a 10 nm infinite well:

$$H'(x) = \begin{cases} 0.1\,\text{meV} & 4.5\,\text{nm} \le x \le 5.5\,\text{nm} \\ 0 & \text{elsewhere} \end{cases}.$$

Calculate the perturbation of the first three eigenenergies. (If you think about the problem, you shouldn't have to do much math at all!)

Repeat this problem using the MATLAB program **find_eig.m**.

10.3   Degenerate Perturbation Theory

10.3.1   In Figure 10.19, the first two eigenstates of the "double V" potential are shown along with the corresponding ground state eigenenergies.

A perturbation is added.

(a)   Estimate the change in energy due to the perturbation. (Do not do a lot of math.)

(b)   The two first eigenstates are close enough in energy that they can be considered degenerate states. Are they *good* states? If not, can you think of a pair of good states?

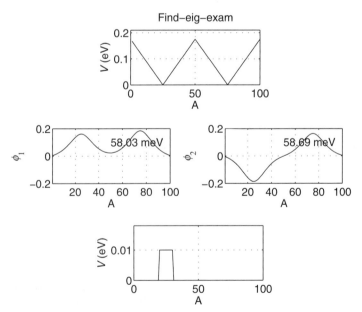

FIGURE 10.19   A "double V" potential and its first two eigenstates. A perturbation of 0.01 eV is added at 25 Å.

10.4 Time-Dependent Perturbation Theory

10.4.1 Assume that a particle is in the third eigenstate of an infinite well. Find a time-dependent perturbation that will cause it to drop to the second eigenstate with a maximum probability of 10%, similar to Figure 10.13. Simulate this with the MATLAB program se10_2.m.

10.4.2 Repeat the simulations of Figures 10.14 and 10.15 using

$$H_P(x,t) = H_P(x)e^{-\alpha t}\cos(\omega_L t)u(t),$$

where $H_P(x)$ is given by Figure 10.12. Choose α so there is less difference in the results between $f_L = 3.2$ THz and $f_L = 3.5$ THz. What else will be different about this simulation?

10.4.3 One of the examples in this chapter was the calculation of the probability of a transition from the first to the second eigenstate of the 10 nm well for a perturbation of the form

$$H_P(x,t) = H_P(x)\cos(\omega_L t)\exp\left(-\frac{1}{2}\left(\frac{t}{\sigma}\right)^2\right).$$

The amplitude was calculated as:

$$|c_2|^2 = \frac{\sigma^2\pi}{2\hbar^2} H_{P21}^2 \exp\left(-\sigma^2(\omega_{mn}-\omega_L)^2\right).$$

Show that this has the correct units.

REFERENCES

1. D. Griffiths, *Introduction to Quantum Mechanics*, Englewood Cliffs, NJ: Prentice Hall, 1995.
2. A. F. J. Levi, *Applied Quantum Mechanics*, 2nd ed., Cambridge, UK: Cambridge University Press, 2003.
3. D. A. B. Miller, *Quantum Mechanics for Scientists and Engineers*, Cambridge, UK: Cambridge University Press, 2009.

11

THE HARMONIC OSCILLATOR

We learned in Chapter 4 that the harmonic oscillator is one of the few potentials for which there is an analytic solution. The harmonic oscillator plays a special role in quantum mechanics for a number of other reasons. For example, many complicated potentials can be approximated with the harmonic oscillator. The fact that the eigenenergies of a harmonic oscillator are evenly spaced gives it interesting properties.

In the first section of this chapter we analyze the one-dimensional harmonic oscillator using creation and annihilation operators. These operators play a significant role in several advanced topics in quantum mechanics. In the second section, the coherent state of the harmonic oscillator is described. This coherent state most closely resembles the behavior of a classical state. The third section describes a two-dimensional harmonic oscillator and explains how it can be used to simulate a quantum dot, a fundamental three-dimensional quantum structure.

11.1 THE HARMONIC OSCILLATOR IN ONE DIMENSION

The potentials that have appeared throughout this book tend to be ideal potentials, for example, the 10 nm infinite well. Although there is no infinite well in nature, it is a good approximation for many practical problems. Look at Figure 11.1. The potential labeled V_{actual} is not likely to be described by any

Quantum Mechanics for Electrical Engineers, First Edition. Dennis M. Sullivan.
© 2012 The Institute of Electrical and Electronics Engineers, Inc.
Published 2012 by John Wiley & Sons, Inc.

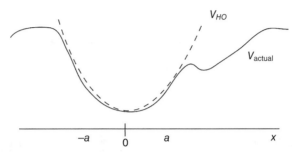

FIGURE 11.1   An irregular potential (solid line) and a harmonic oscillator potential (dashed line) that serves as a good approximation in the interval $-a < x < a$.

simple mathematical expression, much less lead to closed-form solutions of the Schrödinger equation. However, if we were interested only in particles in the area $-a < x < a$, we might decide that the potential could well be approximated by the harmonic oscillator potential

$$V_{\mathrm{HO}}(x) = \frac{1}{2} m \omega_0^2 x^2. \tag{11.1}$$

The ω_0 is a constant that determines the shape of V_{HO}.

The time-independent Schrödinger equation for the harmonic oscillator potential is

$$-\frac{\hbar^2}{2m} \frac{\partial^2 \psi}{\partial x^2} + \frac{1}{2} m \omega_0^2 x^2 \psi = E \psi. \tag{11.2}$$

We will make the following change of variables [1]:

$$\xi = \sqrt{\frac{m \omega_0}{\hbar}} x. \tag{11.3}$$

The Schrödinger equation can now be written as:

$$\frac{1}{2} \left[-\frac{\partial^2}{\partial \xi^2} + \xi^2 \right] \psi = \frac{E}{\hbar \omega_0} \psi. \tag{11.4}$$

The left side can be written using the product of two operators

$$\left[\frac{1}{\sqrt{2}} \left(-\frac{\partial}{\partial \xi} + \xi \right) \cdot \frac{1}{\sqrt{2}} \left(\frac{\partial}{\partial \xi} + \xi \right) + \frac{1}{2} \right] \psi = \frac{E}{\hbar \omega} \psi.$$

We will give names to the individual operators:

$$\hat{a}^\dagger = \frac{1}{\sqrt{2}}\left(-\frac{\partial}{\partial\xi}+\xi\right),$$ (11.5a)

$$\hat{a} = \frac{1}{\sqrt{2}}\left(\frac{\partial}{\partial\xi}+\xi\right).$$ (11.5b)

The first operator, $\hat{a}^\dagger$ (which we call "*a dagger*") is the *creation operator*, while $\hat{a}$ is the *annihilation operator*. Using these definitions, we can rewrite Equation (11.4) as:

$$\left(\hat{a}^\dagger\hat{a}+\frac{1}{2}\right)\psi = \frac{E}{\hbar\omega_0}\psi.$$

The Hamiltonian is

$$H = \hbar\omega_0\left(\hat{a}^\dagger\hat{a}+\frac{1}{2}\right).$$ (11.6)

It will be very desirable to be able to manipulate these operators like algebraic terms. Consider the properties of these operators, all of which can be proved using their definitions. The operators have an *associative* property

$$\hat{a}^\dagger(\hat{a}\psi) = (\hat{a}^\dagger\hat{a})\psi,$$ (11.7a)

and a *distributive* property

$$(\hat{a}+\hat{a}^\dagger)\psi = \hat{a}\psi + \hat{a}^\dagger\psi.$$ (11.7b)

If γ is a constant, then

$$\hat{a}^\dagger(\gamma\psi) = \gamma\hat{a}^\dagger\psi.$$ (11.7c)

However, note that

$$\hat{a}^\dagger\hat{a} \neq \hat{a}\hat{a}^\dagger.$$

An important operator in quantum mechanics is the *commutator*. If $\hat{\alpha}$ and $\hat{\beta}$ are operators, then their commutator is

$$\left[\hat{\alpha},\hat{\beta}\right] = \hat{\alpha}\hat{\beta} - \hat{\beta}\hat{\alpha}.$$

It can be shown that the commutator of the creation and annihilation operators is

$$\left[\hat{a},\hat{a}^\dagger\right] = \hat{a}\hat{a}^\dagger - \hat{a}^\dagger\hat{a} = 1.$$ (11.7d)

(The proof is left as an exercise.) Using the commutator of Equation (11.7d), we can determine an alternative definition of the Hamiltonian of Equation (11.6):

$$H = \hbar\omega_0\left(\hat{a}^\dagger\hat{a} + \frac{1}{2}\right) = \hbar\omega_0\left(\hat{a}\hat{a}^\dagger - 1 + \frac{1}{2}\right) = \hbar\omega_0\left(\hat{a}\hat{a}^\dagger - \frac{1}{2}\right). \qquad (11.8)$$

The use of the properties of Equation (11.7a) through Equation (11.7d) is demonstrated in the following theorem.

Theorem. If ϕ_n is an eigenfunction of the harmonic oscillator Hamiltonian with eigenenergy ε_n, then $\hat{a}^\dagger\phi_n$ is an eigenfunction with eigenenergy $\varepsilon_n + \hbar\omega_0$.

Proof. We are given that

$$\hbar\omega_0\left(\hat{a}^\dagger\hat{a} + \frac{1}{2}\right)\phi_n = \varepsilon_n\phi_n.$$

To prove the theorem, replace ϕ_n with $\hat{a}^\dagger\phi_n$, and then use the properties of Equation (11.7):

$$\hbar\omega_0\left(\hat{a}^\dagger\hat{a} + \frac{1}{2}\right)\hat{a}^\dagger\phi_n = \hbar\omega_0\left(\hat{a}^\dagger\hat{a}\hat{a}^\dagger + \frac{1}{2}\hat{a}^\dagger\right)\phi_n$$

$$= \hbar\omega_0\hat{a}^\dagger\left(\hat{a}\hat{a}^\dagger + \frac{1}{2}\right)\phi_n = \hbar\omega_0\hat{a}^\dagger\left[\left(\hat{a}\hat{a}^\dagger - \frac{1}{2}\right) + 1\right]\phi_n.$$

Using Equation (11.8), the term in the parentheses becomes

$$\hbar\omega_0\hat{a}^\dagger\left[\left(\hat{a}^\dagger\hat{a} + \frac{1}{2}\right) + 1\right]\phi_n = \hat{a}^\dagger\left[\hbar\omega_0\left(\hat{a}^\dagger\hat{a} + \frac{1}{2}\right) + \hbar\omega_0\right]\phi_n$$

$$= \hat{a}^\dagger\hbar\omega_0\left(\hat{a}^\dagger\hat{a} + \frac{1}{2}\right)\phi_n + \hat{a}^\dagger\hbar\omega_0\phi_n.$$

Finally, we use the basic assertion to replace the term in parentheses,

$$\hat{a}^\dagger\varepsilon_n\phi_n + \hat{a}^\dagger\hbar\omega_0\phi_n = (\varepsilon_n + \hbar\omega_0)\hat{a}^\dagger\phi_n,$$

which is what we wanted to prove.

The *creation operator*, $\hat{a}^\dagger$, can also be thought of as a *raising operator*, because it "raises" the state of an eigenfunction from n to $n + 1$. We can also show that $\hat{a}\phi_n$ is an eigenstate with eigenenergy $\varepsilon_n - \hbar\omega_0$. (The proof is left as an exercise.) Therefore, we can think of $\hat{a}$ as a *lowering operator*.

The harmonic oscillator must have a lowest state, which we will call ϕ_0. If ϕ_0 is the lowest state, then it must be that

$$\hat{a}\phi_0 = 0,$$

because we cannot lower the lowest state. Using the definition of the annihilation operator:

$$\hat{a}\phi_0 = \frac{1}{\sqrt{2}}\left(\frac{\partial}{\partial\xi} + \xi\right)\phi_0 = 0,$$

$$\frac{\partial\phi_0}{\partial\xi} = -\xi\phi_0,$$

$$\frac{\partial\phi_0}{\phi_0} = -\xi\,\partial\xi.$$

Integrating both sides we get

$$\ln(\phi_0) = -\frac{\xi^2}{2} + K.$$

Finally, we take the natural logarithm of both sides to get

$$\phi_0 = \exp\left[-\frac{\xi^2}{2} + K\right] = \exp(K)\exp\left(-\frac{\xi^2}{2}\right).$$

This eigenstate must be normalized, which determines the constant e^K, giving us

$$\phi_0 = \frac{1}{\pi^{1/4}}\exp\left(-\frac{\xi^2}{2}\right).$$

What energy does this state have? If we operate on it with the Hamiltonian, then

$$H\phi_0 = \hbar\omega_0\left(\hat{a}^\dagger\hat{a} + \frac{1}{2}\right)\phi_0 = \frac{\hbar\omega_0}{2}\phi_0.$$

So the ground state energy is $\varepsilon_0 = \hbar\omega_0/2$. We can operate on ϕ_0 with the raising operator and get a new function that we will call

$$\hat{a}^\dagger\phi_0 = A_1\phi_1,$$

where A_1 is a constant. We know that this new eigenfunction has eigenenergy

$$\varepsilon_1 = \left(n + \frac{1}{2}\right)\hbar\omega_0 = \frac{3}{2}\hbar\omega_0.$$

By successively using the raising operator, we determine that we have a set of eigenfunctions ϕ_n with corresponding eigenenergies

$$\varepsilon_n = \left(n + \frac{1}{2}\right)\hbar\omega_0. \tag{11.9}$$

If this is true, then by applying the Hamiltonian of Equation (11.6)

$$H\phi_n = \hbar\omega_0\left(\hat{a}^\dagger\hat{a} + \frac{1}{2}\right)\phi_n = \left(n + \frac{1}{2}\right)\hbar\omega_0\phi_n,$$

we conclude that we have a *number operator*,

$$\hat{n} = \hat{a}^\dagger\hat{a}. \tag{11.10}$$

The explicit eigenfunctions are [2]:

$$\phi_n(x) = \left(\frac{m\omega_0}{\pi\hbar}\right)^{1/4}\frac{1}{\sqrt{2^n n!}}H_n(\xi)e^{-\xi^2/2}. \tag{11.11}$$

In Equation (11.11), the eigenfunctions $\phi_n(x)$ are functions of x, but we still use the variable ξ for convenience. The functions $H_n(\xi)$ are the *Hermite* polynomials. The first few are given in Table 11.1. Others can be generated through the recursive formula

$$H_{n+1}(\xi) = 2\xi H_n(\xi) - 2nH_{n-1}(\xi).$$

11.1.1   Illustration of the Harmonic Oscillator Eigenfunctions

In this section we return to usual distance parameter x. It is often convenient to write the harmonic oscillator potential as:

$$V_{HO}(x) = \frac{1}{2}\left(\frac{E_{ref}}{\hbar}\right)^2 x^2.$$

$E_{ref} = \hbar\omega_0$ is a reference energy related to the system. The eigenenergies can then be written as:

$$\varepsilon_n = (n + 1/2)E_{ref}, \quad n = 0,1,2.\ldots$$

TABLE 11.1   The First Few Hermite Polynomials

$H_0 = 1$
$H_1 = 2x$
$H_2 = 4x^2 - 2$
$H_8 = 8x^3 - 12x$

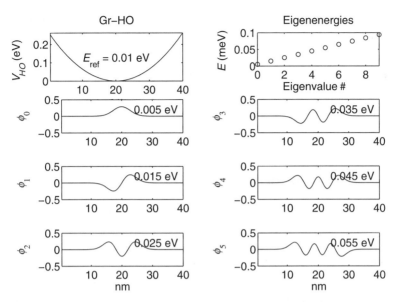

FIGURE 11.2 The harmonic oscillator potential for $E_{ref} = 0.01$ eV (upper left) and the corresponding eigenenergies (upper right). The first six eigenfunctions are also displayed.

We can find the eigenvalues and eigenfunctions using MATLAB command **eig**. The results are shown in Figure 11.2.

11.1.2 Compatible Observables

We briefly mentioned the commutator. If two operators *commute*, we say that they are *compatible observables*. For instance, if $\hat{\alpha}$ and $\hat{\beta}$ are compatible observables, then

$$\left[\hat{\alpha}, \hat{\beta}\right] = \hat{\alpha}\hat{\beta} - \hat{\beta}\hat{\alpha} = 0.$$

We saw that the creation and annihilation operators are *not* compatible observables.

It can be shown that compatible observables share a complete set of eigenfunctions. Sometimes, finding the eigenvectors of an operator $\hat{\alpha}$ is difficult, but if the operator has a compatible observable $\hat{\beta}$, it might be easier to find the eigenfunctions of $\hat{\beta}$ [2].

11.2 THE COHERENT STATE OF THE HARMONIC OSCILLATOR

In this section we consider the coherent state of the harmonic oscillator, an important concept in quantum mechanics. Before doing so, consider two time-domain simulations that lay the groundwork for the significance of the coherent state.

11.2.1   The Superposition of Two Eigentates in an Infinite Well

Suppose we start with a particle that is the superposition of the first two eigenstates of the 10 nm infinite well:

$$\psi(x,t) = c_1\phi_1(x)e^{-i(\varepsilon_1/\hbar)t} + c_1\phi_2(x)e^{-i(\varepsilon_2/\hbar)t}. \tag{11.12}$$

The modulus of the function is

$$|\psi(x,t)|^2 = \psi^*(x,t)\psi(x,t) = |c_1|^2\,\phi_1^2(x) + |c_2|^2\,\phi_2^2(x)$$
$$+ c_1^*\phi_1^*(x)c_2\phi_2(x)e^{-i(\varepsilon_2-\varepsilon_1/\hbar)t} + c_1\phi_1(x)c_2^*\phi^*(x)e^{i(\varepsilon_2-\varepsilon_1/\hbar)t}.$$

If we assume that the c_ns and ϕ_ns are real, then

$$|\psi(x,t)|^2 = |c_1|^2\,\phi_1^2(x) + |c_2|^2\,\phi_2^2(x) + c_1\phi_1(x)c_2\phi_2(x)2\cos\left(\frac{(\varepsilon_2-\varepsilon_1)}{\hbar}t\right). \tag{11.13}$$

This says that the modulus will have an oscillation with a time period determined by

$$T_0 = \frac{h}{\varepsilon_2 - \varepsilon_1}. \tag{11.14}$$

(Do not confuse this with the type of simulation seen in Fig. 2.3 where the modulus remains constant while the real and imaginary parts of an eigenfunction oscillate at the time predicted by the eigenenergy.)

If we start in the equal superposition of the first two eigenstates of the 10 nm infinite well

$$\psi(x,t) = 0.707\phi_1(x)e^{-i(\varepsilon_1/\hbar)t} + 0.707\phi_2(x)e^{-i(\varepsilon_2/\hbar)t},$$

then the time T_0 will depend on the energy difference between the first two eigenenergies of 0.00375 eV and 0.00375 eV $\times 2^2 = 0.015$ eV:

$$T_0 = \frac{h}{\varepsilon_2 - \varepsilon_1} = \frac{4.135\times10^{-15}\text{ s}}{(0.01125\text{ eV})} = 0.367\text{ ps}.$$

The simulation is shown in Figure 11.3. The modulus is the solid line. As the simulation proceeds, the relative amplitudes of the eigenfunctions remain the same but the phases of the eigenfunctions vary. Note that after the predicted time of 0.367 ps, the modulus has returned to the same form. If we had picked two other states—the third and fourth eigenstates, for instance—the time

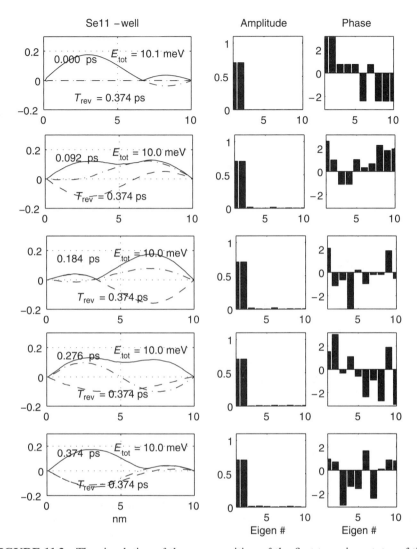

FIGURE 11.3 The simulation of the superposition of the first two eigenstates of the 10 nm infinite well. The modulus is the solid line. After the predicted time, the modulus returns to the original form.

would have been different because the spacing between eigenenergies increases geometrically with n.

11.2.2   The Superposition of Four Eigenstates in a Harmonic Oscillator

Recall that the eigenstates of a harmonic oscillator are given by

$$\varepsilon_n = (n + 1/2)\, E_{\text{ref}}, \quad n = 0, 1, 2 \dots,$$

where $E_{ref} = \hbar\omega_0$. Therefore, in contrast to the infinite well, two neighboring eigenstates of the harmonic oscillator potential—for example, m and $m + 1$—will always have the same revival time because the energy eigenstates are equally spaced:

$$T_0 = \frac{h}{\hbar\omega_0\left[(m+1+1/2)-(m+1/2)\right]} = \frac{h}{\hbar\omega_0} = \frac{2\pi}{\omega_0}. \qquad (11.15)$$

Figure 11.4 is a simulation of the superposition of the first four eigenstates of the harmonic oscillator with a reference energy of $E_{ref} = \hbar\omega_0 = 0.01$ eV. This superposition was created by

$$\psi(x) = \phi_0(x) + \phi_1(x) - \phi_2(x) - \phi_3(x),$$

and then normalized. The negative of the second and third eigenstates were used to give the form shown in the figure at time $T = 0$, that is, it looks as if the particle is located right of center. The predicted revival time is

$$T_0 = \frac{4.135\times10^{-15} \text{ eV} \cdot \text{s}}{0.01 \text{ eV}} = 0.4135 \text{ ps.}$$

Once again, the modulus returns to its original position in the predicted time.

As we look at the modulus in Figure 11.4, there is some sense that the probability amplitude is moving back and forth with respect to the center of the well similar to the way a particle in a bowl might roll back and forth. However, the analogy is limited because the waveform is spread out over a relatively large area and it changes shape during the simulation.

11.2.3   The Coherent State

The coherent state of the harmonic oscillator is a linear superposition of the eigenstates of the harmonic oscillator that corresponds most closely to the expected behavior of a classical particle in a harmonic well [1]. The coherent state for a harmonic oscillator is

$$\psi_N(\xi,t) = \sum_{n=0}^{\infty} c_{Nn}\phi_n(\xi)e^{-i(\varepsilon_n/\hbar)t}, \qquad (11.16)$$

where

$$c_{Nn} = \sqrt{\frac{N^n e^{-N}}{n!}}. \qquad (11.17)$$

It is interesting that

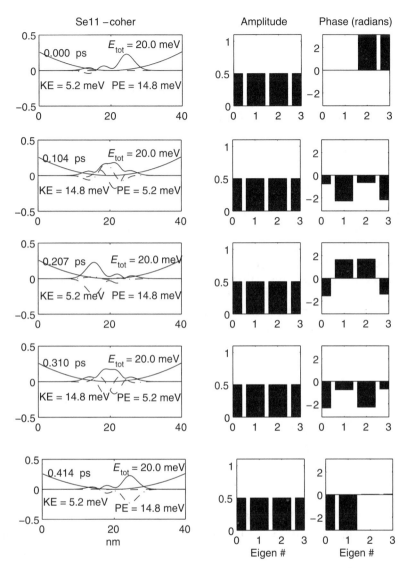

FIGURE 11.4 The simulation of a superposition of the first four eigenstates of a harmonic oscillator potential with $E_{\text{ref}} = 0.01$ eV. After the time predicted by the reference energy $\hbar\omega_0$ of the well, the modulus returns to its original position. PE, potential energy; KE, kinetic energy.

$$|c_{Nn}|^2 = \frac{N^n e^{-N}}{n!} \tag{11.18}$$

is the Poisson distribution with mean N and standard deviation $\sqrt{N}$. One of the main characteristics of the coherent state is that it is a minimum uncertainty packet in the sense of Equation (3.14a) [3].

Figure 11.5 shows a simulation of a coherent state in a harmonic oscillator. This harmonic oscillator also has a reference energy of 0.01 meV, and has the time period given by Equation (11.15). The coherent state was constructed using $N = 10$ in Equation (11.16) through Equation (11.18). Notice that as the particle moves down into the well it exchanges potential energy for kinetic energy, but the total energy remains the same. After the prescribed time period, the particle returns to the original state. Notice also that the distribution of relative amplitudes of the eigenstates is centered around 10, as it should be. One of the key features of this simulation, as opposed to those in Figures 11.3 and 11.4, is that the modulus maintains its Gaussian shape. This is in keeping with the statement that this Gaussian waveform is the minimum uncertainty wave packet, as we saw in Section 3.3.

It can be shown mathematically that a mode of an electromagnetic field is similar to a quantum harmonic oscillator. The coherent state of the quantum harmonic oscillator is the one that most closely resembles the electromagnetic mode [1, 3].

11.3   THE TWO-DIMENSIONAL HARMONIC OSCILLATOR

The two-dimensional harmonic oscillator is described in this section. This section also introduces eigenfunctions that have angular components. The simulation of a quantum dot is described using a two-dimensional harmonic oscillator as a model of the potential.

11.3.1   The Simulation of a Quantum Dot

A quantum dot is a structure that can confine particles in three dimensions (Fig. 11.6). Quantum dots are the subject of much interest in modern semiconductor theory both as possible devices [4] and from a theoretical viewpoint by regarding them as artificial atoms [5].

In actuality, the construction of a true quantum dot is extremely difficult. One approach is to begin by constructing a one-dimensional quantum well using layered media. This is illustrated in Figure 11.7 [5], where confinement occurs in the z-direction. Electrons are considered to be bound laterally in the quantum well by a potential that resembles the two-dimensional harmonic oscillator

$$V(\rho) = \frac{1}{2} m \omega_0^2 \rho^2, \tag{11.19}$$

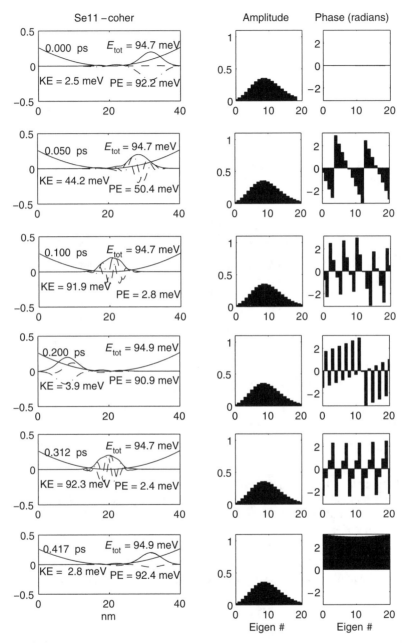

FIGURE 11.5 The simulation of a coherent state in a harmonic oscillator potential. The modulus of the waveform maintains its shape as it moves back and forth in the well. The waveform also exchanges kinetic (KE) and potential energy (PE) as it moves relative to the center of the well.

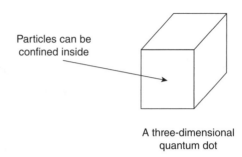

A three-dimensional
quantum dot

FIGURE 11.6 A quantum dot is a structure that can confine particles in three dimensions.

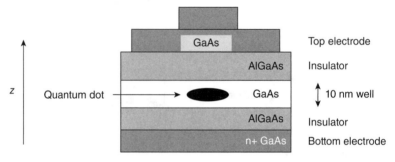

FIGURE 11.7 A quantum dot was constructed by layered media that formed a 10 nm well [5]. The narrow well in the z-direction restricts particles to be in only the ground state.

FIGURE 11.8 Electrons in the z-directed quantum well of Figure 11.7 are also confined by a bowl-shaped potential that can be described by a two-dimensional harmonic oscillator potential.

where ρ is the distance from the center (Fig. 11.8).

The solution to the Schrödinger equation for the potential of Equation (11.19) is usually given in cylindrical coordinates [6]. The eigenenergies are:

$$\varepsilon_{n,l} = \hbar\omega_0 \left(2n + |l| + 1\right), \tag{11.20}$$

where n and l are quantum numbers: n is a positive integer which corresponds to the number of nodes in the wave function moving radially out from the center; l can be positive or negative, but $2|l|$ gives the number of nodes seen moving around in a circle of constant ρ. The first few eigenfunctions are given in Table 11.2. A positive l results in a wave function moving counterclockwise as viewed from above.

TABLE 11.2 The First Six Eigenfunctions of the Two-Dimensional Harmonic Oscillator

$$\varepsilon = \hbar\omega_0 \qquad\qquad n = 0 \qquad\qquad \phi_0(\rho) = \frac{\chi}{\sqrt{\pi}}e^{-\chi^2\rho^2},$$

$$\varepsilon = 2\hbar\omega_0 = \begin{cases} n=0, \quad l=1 \quad \phi_{0,1}(\rho,\theta) = \frac{\chi}{\sqrt{\pi}}\chi\rho e^{-\chi^2\rho^2}e^{i\phi} \\[2mm] n=0, \quad l=-1 \quad \phi_{0,-1}(\rho,\theta) = \frac{\chi}{\sqrt{\pi}}\chi\rho e^{-\chi^2\rho^2}e^{-i\phi} \end{cases},$$

$$\varepsilon = 3\hbar\omega_0 = \begin{cases} n=0, \quad l=2 \quad \phi_{0,2}(\rho,\theta) = \frac{\chi}{\sqrt{\pi}}(\chi\rho)^2 e^{-\chi^2\rho^2}e^{i2\phi} \\[2mm] n=0, \quad l=-2 \quad \phi_{0,2}(\rho,\theta) = \frac{\chi}{\sqrt{\pi}}(\chi\rho)^2 e^{-\chi^2\rho^2}e^{-i2\phi} \\[2mm] n=0, \quad l=-2 \quad \phi_{0,2}(\rho,\theta) = \frac{\chi}{\sqrt{\pi}}(\chi\rho)^2 e^{-\chi^2\rho^2}e^{-i2\phi} \end{cases}.$$

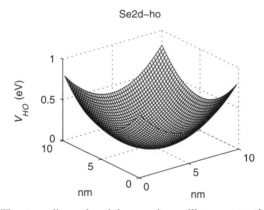

FIGURE 11.9 The two-dimensional harmonic oscillator potential for $E_0 = \hbar\omega_0 = 50$ meV.

As an example, we will simulate two of the eigenfunctions of a two-dimensional harmonic oscillator with $E_0 = \hbar\omega_0 = 50$ meV. The simulation uses 50×50 cells, with cell sizes of 0.2 nm. The potential is shown in Figure 11.9 [7].

Figure 11.10 is a simulation of the ground state energy, that is, $n = 0, l = 0$. Note the oscillation between the real and imaginary parts of the state vectors, similar to the time-domain simulation of the ground state of the one-dimensional infinite well in Chapter 2.

Figure 11.11 is a simulation of the $\phi_{0,1}$ eigenstate, that is, $n = 0, l = 1$. The $l = 1$ indicates that this particle has a component of angular momentum. The energy from Equation (11.20) is 100 meV.

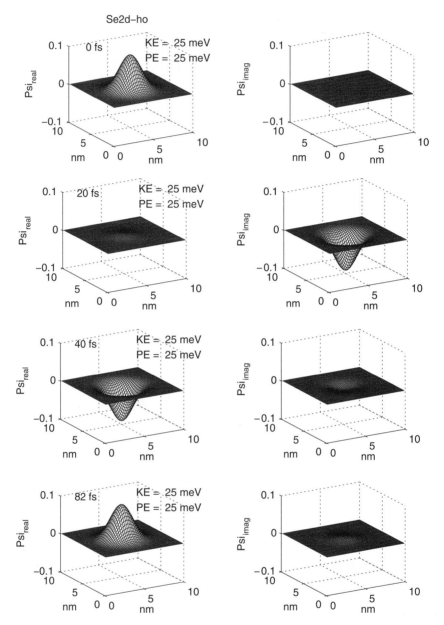

FIGURE 11.10 The ground state eigenfunction of the 50 meV harmonic oscillator (Eq. 11.19). Notice that after the revival time predicted by the energy, the waveform returns to its original state. PE, potential energy; KE, kinetic energy; imag, imaginary.

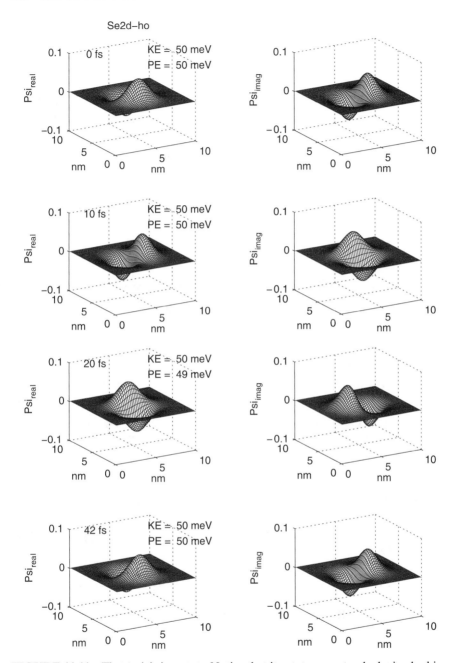

FIGURE 11.11   The $\phi_{0,1}(\rho)$ eigenstate. Notice that it rotates counterclockwise, looking down from the top. After the predicted revival time, the waveform returns to its original position. PE, potential energy; KE, kinetic energy; imag, imaginary.

EXERCISES

11.1 The Harmonic Oscillator in One Dimension

11.1.1 Prove that $\left[\hat{a},\hat{a}^{\dagger}\right] = \hat{a}\hat{a}^{\dagger} - \hat{a}^{\dagger}\hat{a} = 1$ (Eq. 11.7d).

11.1.2 Show that if ϕ_n is an eigenfunction of the harmonic oscillator Hamiltonian with eigenenergy ε_n, then $\hat{a}\phi_n$ is an eigenfunction with eigenenergy $\varepsilon_n - \hbar\omega_0$.

11.1.3 Verify that the normalization constant of the ground state eigenfunction of the harmonic oscillator using the ξ variable is $a = 1/\pi^{1/4}$.

11.1.4 Are the position operator $\hat{x}$ and the momentum operator

$$\hat{p}_x = \frac{\hbar}{i}\frac{\partial}{\partial x}$$

compatible observables? Show why or why not.

11.2 The Coherent State of the Harmonic Oscillator.

11.2.1 In Figure 11.4, after one complete period of 410 fs, the phases of each of the eigenstates has shifted 180°. Explain why.

11.3 The Two-Dimensional Harmonic Oscillator

11.3.1 Simulate a coherent state similar to Figure 11.5, but at 50 meV.

11.3.2 Do a simulation similar to Figure 11.11, but make the particle rotate clockwise when looking down from the top.

REFERENCES

1. D. A. B. Miller, *Quantum Mechanics for Scientists and Engineers*, Cambridge, UK: Cambridge University Press, 2008.

2. D. J. Griffiths, *Introduction to Quantum Mechanics*, Englewood Cliffs, NJ: Prentice Hall, 1995.

3. A. Yariv, *Quantum Electronics*, New York: John Wiley and Sons, 1988.

4. L. Jacak, P. Hawrylak, and A. Wojs, *Quantum Dots*, Berlin, Germany: Springer-Verlag, 1998.

5. R. C. Ashoori, "Electrons in artificial atoms," *Nature*, Vol. 379, p. 413, 1996.

6. C. Cohen-Tannoudji, B. Diu, and F. Laloe, *Quantum Mechanics*, New York: Wiley, 1977.

7. D. M. Sullivan and D. S. Citrin, "Time-domain simulation of two electrons in a quantum dot," *J. Appl. Phys.*, Vol. 89, p. 3841, 2001.

12

FINDING EIGENFUNCTIONS USING TIME-DOMAIN SIMULATION

From the beginning, we have seen the importance of knowing the eigenfunctions of a structure. For the most part, we have worked in one dimension. In one dimension, even if we did not have a closed analytic formula for the eigenfunctions, as we did for the infinite well and the harmonic oscillator, we could use the MATLAB program **eig** to find the eigenenergies and the corresponding eigenfunctions. In this chapter a method is described that can be used to find eigenenergies and eigenfunctions for arbitrary structures in two and three dimensions [1]. For the purpose of illustration, the discussion will start in one dimension.

12.1 FINDING THE EIGENENERGIES AND EIGENFUNCTIONS IN ONE DIMENSION

In this section, we will learn a method to find the eigeneneriges using the finite-difference time-domain (FDTD) simulation. Even though we already know the energies of a 10 nm infinite well, we will proceed as if we did not know them. We begin by putting a test function in the middle of the well, as shown in Figure 12.1. We know that we can write this as a summation of the eigenstates, even if the eigenstates weren't already known,

$$\psi(x,t) = \sum_{n=1}^{N} c_n e^{-i(\varepsilon_n/\hbar)t} \phi_n(x). \tag{12.1}$$

Quantum Mechanics for Electrical Engineers, First Edition. Dennis M. Sullivan.
© 2012 The Institute of Electrical and Electronics Engineers, Inc.
Published 2012 by John Wiley & Sons, Inc.

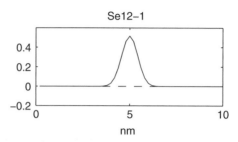

FIGURE 12.1 A pulse initialized in a 10 nm infinite well.

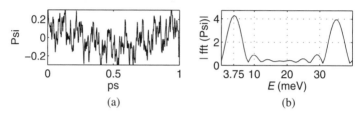

FIGURE 12.2 (a) The time-domain data for the pulse of Figure 12.1 monitored at the 5 nm point. (b) The Fourier transform of the time-domain data in (a).

If we look at the function at any given point, say the $x_0 = 5$ nm point in the middle of the well, Equation (12.1) reduces to a time-dependent function:

$$\psi(x_0, t) = \sum_{n=1}^{N} c_n e^{-i(\varepsilon_n/\hbar)t} \phi_n(x_0). \tag{12.2}$$

Note that this function is composed of sinusoids at frequencies $\omega_n = (\varepsilon_n/\hbar)$. Therefore, a time-domain Fourier transform of Equation (12.2) should reveal the eigenergies.

The FDTD simulation is started, and as the simulation proceeds, we store the time-domain data from the initial starting point at 5 nm. Figure 12.2a shows the time-domain data after 10,000 steps. We then take the Fourier transform of this time-domain data. To increase resolution in the frequency domain, we put the time-domain data in a buffer of $2^{16} = 65,636$. Since our time steps are $\Delta t = 10^{-16}$ s, the resolution of our Fourier transform using the fast Fourier transform (fft) command is:

$$\Delta f = \frac{1}{2^{16} \times 10^{-16} \text{ s}} \cong 0.1526 \text{ MHz}. \tag{12.3}$$

Our course, we prefer to convert this to energy

$$\Delta E = h \cdot \Delta f = 0.63 \text{ meV}.$$

The results are shown in Figure 12.2b. We can see that the first peak is centered at 3.75 meV, as we expect, because that is the ground state eigenenergy of a

10 nm infinite well. The other peak is due to the third eigenfunction at 33.75 meV. Notice there is no peak corresponding to the second eigenfunction, because it has a null at 5 nm.

Between the two peaks of the Fourier transform of Figure 12.2b are many ripples that can be regarded as error. They come from several sources, but one of the main sources is the fact that our simulation begins abruptly at 0 and ends abruptly at $T = 10,000$. The error can be significantly improved by a process called "windowing," that is, multiplying the time-domain data by a function that smoothes the edges. One such window is called the Hanning window [2], given by the function

$$H(t) = \frac{1}{2}\left(1 - \cos\left(\frac{2\pi t}{T_{max}}\right)\right). \tag{12.4}$$

The Hanning window for the 10,000 points is shown in Figure 12.3.

If we go back to Figure 12.2a and add the Hanning window to the time-domain data, we get the data shown in Figure 12.4a. The Fourier transform of this signal is shown in Figure 12.4b. Notice how much smoother it is. The Fourier data of Figure 12.2b certainly could show us the peak of the ground state eigenenergy without windowing. However, it is not as clear. The windowing is a good procedure to determine real peaks from artifacts. Windowing is also discussed in Appendix A.

When we look for the second eigenenergy, we must be sure to select a test function that is not sitting on a null of the second eigenfunction, as what happened when we used the pulse in Figure 12.1. It is also desirable that this test function be orthogonal to the ground state eigenfunction. (Notice the

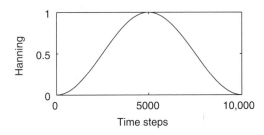

FIGURE 12.3   A Hanning window over 10,000 points.

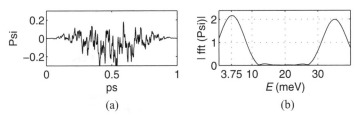

FIGURE 12.4   (a) The same data as Figure 12.2 after the Hanning window of Figure 12.3 has been applied. (b) The Fourier transform of the data in (a).

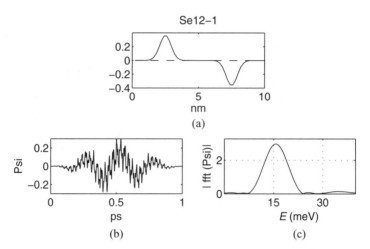

FIGURE 12.5   (a) This is a test function that is orthogonal to the first eigenstate. (b) The time-domain data is collected at the point 2.5 nm. (c) The Fourier transform of the time-domain data reveals the eigenenergy of the second eigenstate.

similarity to the way the variational method looks for progressively higher eigenstates, as described in Section 10.1.) One such choice is given in Figure 12.5a. In this case we monitor the time-domain data at 2.5 nm. The windowed time-domain data is in Figure 12.5b and the corresponding Fourier transform is given in Figure 12.5c. Notice that a peak has been found very close to the theoretical value of 15 meV. If we want better accuracy, we probably have to make the Fourier transform buffer greater than $2^{16} = 65,636$. We might also need to model the infinite well itself with better accuracy than 0.2 nm.

12.1.1   Finding the Eigenfunctions

Now that we know the eigenenergies, the question is how to construct the corresponding eigenfunctions. Once again, we start with the test function in Figure 12.1 and the eigenfunction decompositions of Equation (12.1). The test function $\psi(x,t)$ contains the ground state eigenfunction

$$\phi_1(x,t) = c_1 e^{-i(\varepsilon_1/\hbar)t}\phi_1(x).$$

The trick is to extract this from Equation (12.1). If we multiply Equation (12.1) by $e^{i(\varepsilon_1/\hbar)t}$ and integrate over time, we get

$$\int_{-\infty}^{\infty}\psi(x,t)e^{i(\varepsilon_1/\hbar)t}dt = \int_{-\infty}^{\infty}\sum_{n=1}^{N}c_n e^{-i(\varepsilon_n/\hbar)t}\phi_n(x)e^{i(\varepsilon_1/\hbar)t}dt = 2\pi c_1\phi_1(x).$$

In effect, what we have done is take the Fourier transform at one frequency, $\omega_1 = (\varepsilon_1/\hbar)$. In the simulation programs it is not possible to calculate a true Fourier transform from minus to plus infinity. However, a good approximation can be made by the following summation, provided that M represents a long enough time:

$$\tilde{\phi}_1^M(x,\omega_1) = \sum_{m=1}^{M} \psi(x,t) e^{i\omega_1(\Delta t \cdot m)} \Delta t \cong \int_{-\infty}^{\infty} \psi(x,t) e^{i(\varepsilon_1/\hbar)t} dt. \quad (12.5)$$

This is a discrete Fourier transform at the frequency $\omega_1 = \varepsilon_1/\hbar$. Remember, this has to be calculated at each position x, or in our case, at each of the cells in the problem space. Even Equation (12.5) would be computationally prohibitive except for one more trick. We write Equation (12.5) as:

$$
\begin{aligned}
\tilde{\phi}_1^M(x,\omega_1) &= \sum_{m=1}^{M} \psi(x,t) e^{i\omega_1(\Delta t \cdot m)} \\
&= \sum_{m=1}^{M-1} \psi(x,t) e^{i\omega_1(\Delta t \cdot m)} + \psi(x,t) e^{i\omega_1(\Delta t \cdot M)} \\
&= \tilde{\phi}_1^{M-1}(x,\omega_1) + \psi(x,\Delta t \cdot M)\left(e^{i\omega_1 \Delta t}\right)^M.
\end{aligned}
\quad (12.6)
$$

This means we calculate the new value $\tilde{\phi}_1^M(x)$ from the previous value $\tilde{\phi}_1^{M-1}(x)$ plus the state function times the Fourier term.

In actuality, we use the following:

$$\tilde{\phi}_1^M(x,\omega_1) = \tilde{\phi}_1^{M-1}(x,\omega_1) + w(M \cdot \Delta t) \cdot \psi(x, M \cdot \Delta t)\left(e^{-i\omega_1 \cdot \Delta t}\right)^M,$$

where $w(M)$ is the Hanning window of Equation (12.4). If we just used the time-domain data to calculate Equation (12.6), then the result would depend heavily on when we stopped. Also, the large value of ψ in the middle of the problem space as the simulation starts would cause errors. The disadvantage is that we must anticipate the size of the simulation at the beginning in order to establish the value T_{max} in Equation (12.4).

The construction of the ground state eigenfunction using Equation (12.6) is shown in Figure 12.6.

12.2 FINDING THE EIGENFUNCTIONS OF TWO-DIMENSIONAL STRUCTURES

The process of finding two-dimensional eigenenergies and eigenfunctions is fundamentally the same. Once again, we start with a canonical structure that has an analytic solution so we can evaluate the accuracy of the method. Suppose we have an infinite well that is 10×10 nm. We calculated the eigenenergies for such a structure back in Chapter 6:

$$E_{m,n} = \frac{\hbar^2 \pi^2}{2m}\left(\frac{m^2}{a^2} + \frac{n^2}{b^2}\right). \quad (6.10)$$

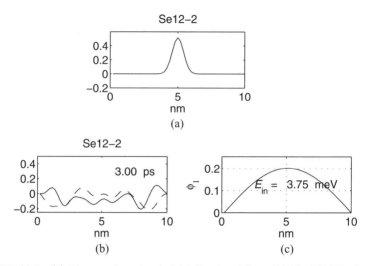

FIGURE 12.6 (a) The test function is initialized at 1.5 ps. (b) The FDTD simulation after 15,000 iterations. (c) The eigenfunction constructed from Equation (12.6).

In this case, $a = b = 10$ nm, so

$$E_{m,n} = (m^2 + n^2)3.75 \, \text{meV}. \qquad (12.7)$$

and the ground state is $E_{1,1} = 7.5$ meV. We start by putting a narrow test function in the middle of the well, as shown in Figure 12.7. The simulation begins and by 20 femtoseconds, the waveform has spread out and interacted with the walls. While this simulation is proceeding, the state variable ψ at the point of origin is stored. After 10,000 iterations, the program is halted and the fast Fourier transform of the stored time-domain data is taken. The results are shown in Figure 12.8. A peak at 7.88 meV was found. This is in reasonable agreement with our analytic result. However, the second peak is at 39.4 meV. The next highest eigenenergy should be $E_{1,2} = E_{2,1} = 18.75$ meV. So what happened? The second eigenfunction has a null at the center. However, $E_{1,3} = E_{3,1} = 37.5$ meV because the third eigenfunction, like the first, has a maximum in the center.

Now that we have the eigenenergy for the ground state, we can find the corresponding eigenfunction. The procedure is almost exactly the same as was used for the one-dimensional eigenfunctions. It still desirable to use the Hanning window during the reconstruction process as well. So Equation (12.6) in the previous section becomes

$$\tilde{\phi}_{1,1}^M (x,y) = \tilde{\phi}_{1,1}^{M-1} (x,y) + w(M \cdot \Delta t) \cdot \psi (x,y,M \cdot \Delta t)(e^{-i\omega_{1,1} \cdot \Delta t})^M, \qquad (12.8)$$

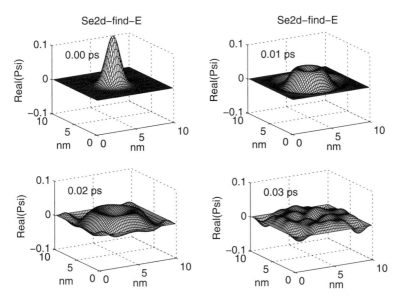

FIGURE 12.7 The test function as it begins to interact with the two-dimensional infinite well.

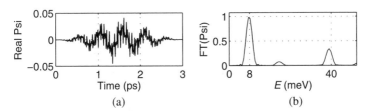

FIGURE 12.8 (a) After 30,000 iterations, the program is stopped. The real part of the state variable at the source point is shown. (b) The magnitude of the Fourier transform (FT) of (a). The first peak occurs at 7.88 meV and the second at 39.4 meV.

where $w(M)$ is the Hanning window of Equation (12.4). The result is shown in Figure 12.9.

The wave function of Figure 12.9 can be used as the initial waveform for another simulation. As the wave function evolves, it must maintain normalization, as always. But it also has to revive in the time prescribed by

$$T_{\text{revival}} = \frac{h}{\varepsilon_{1,1}} = \frac{4.135 \times 10^{-15} \text{ eV} \cdot \text{s}}{0.0079 \text{ eV}} = 0.53 \text{ ps.}$$

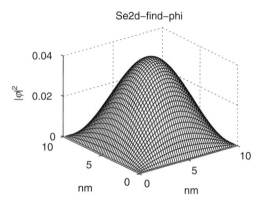

FIGURE 12.9   After 10,000 iterations, the ground state eigenfunction at 7.88 meV was constructed using Equation (12.8).

This is illustrated in Figure 12.10.

Now we want to look for the next highest eigenstate, which in this case is

$$E_{1,2} = (1^2 + 2^2)3.75 \, \text{meV} = 18.75 \, \text{meV}.$$

Unfortunately, the test function that we used in Figure 12.7 will not "see" this state because it has a null at the middle of the problem space. So we will start with the one shown in Figure 12.11a. After 30,000 iterations, we have the windowed time-domain data in Figure 12.11b, and the corresponding Fourier transform in Figure 12.11c. Using the peak energy of 19.5 meV, the corresponding eigenfunction is reconstructed and shown in Figure 12.12.

12.2.1   Finding the Eigenfunctions in an Irregular Structure

Now that we have a certain amount of confidence in the method, we can look for eigenfunctions in structures that do not lend themselves to analysis. Figure 12.13a is a potential that might be typical of those found in semiconductor channels. A test function is initialized in the middle and the time-domain data at that point is displayed in Figure 12.13b. The Fourier transform in Figure 12.13c reveals the ground state energy at 52 meV. The constructed function is shown in Figure 12.14. Once again, this constructed eigenfunction can be used to initialize a simulation that revives in the time prescribed by

$$T_{\text{revival}} = \frac{h}{\varepsilon_1} = \frac{4.135 \times 10^{-15} \, \text{eV} \cdot \text{s}}{0.052 \, \text{eV}} = 0.0875 \, \text{ps},$$

as seen in Figure 12.15.

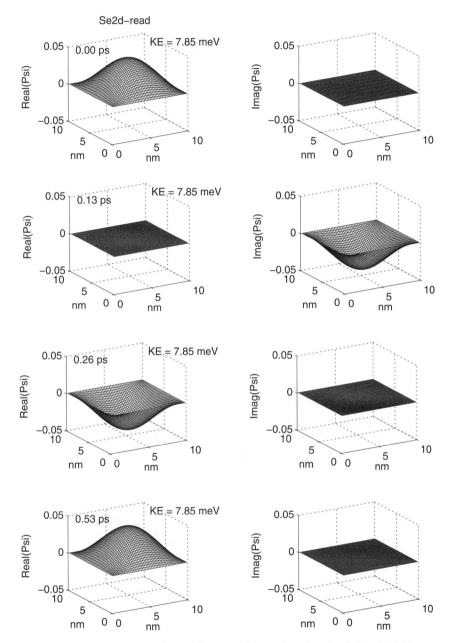

FIGURE 12.10 Time evolution of the ground state function in Figure 12.9. The particle returns to it original state after the revival time of 0.53 ps. KE, kinetic energy; Imag, imaginary.

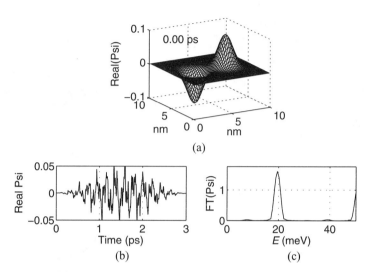

(a)

(b)   (c)

FIGURE 12.11   (a) The test function used to look for the second eigenfunction; (b) the time-domain data collected at one of the peaks; and (c) the Fourier transform (FT) showing the first peak at 19.5 meV.

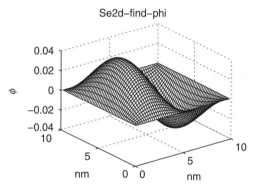

FIGURE 12.12   The eigenstate corresponding to the eigenenergy found in Figure 12.11.

In finding the second eigenstate, we used the test function shown in Figure 12.16a, which resulted in the windowed time-domain data in Figure 12.16b. This identified an eigenenergy at 63.6 meV in Figure 12.16c, and a corresponding eigenfunction is shown in Figure 12.16d. The revival time is calculated to be

$$T_{\text{revival}} = \frac{h}{\varepsilon_2} = \frac{4.135 \times 10^{-15} \text{ eV} \cdot \text{s}}{0.0635 \text{ eV}} = 0.065 \text{ ps.}$$

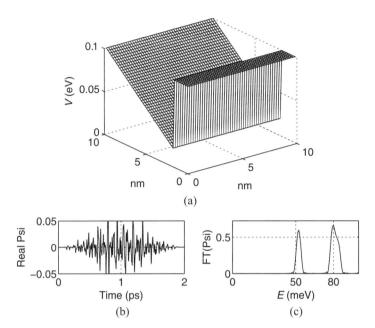

(a)

(b)   (c)

FIGURE 12.13 (a) This is the potential for which we are looking for the eigen-functions. (b) The time-domain data collected at the source of the test function. (c) The Fourier transform (FT) of the time-domain signal in (b). The first peak is at 52 meV.

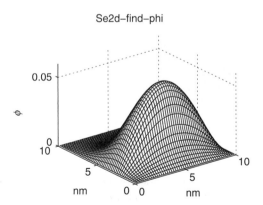

Se2d–find–phi

FIGURE 12.14 The eigenfunction corresponding to the 52 meV peak in Figure 12.13c.

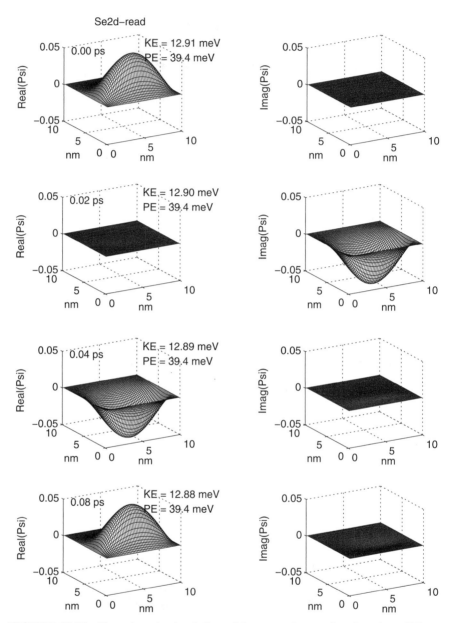

FIGURE 12.15   Time-domain simulation of the ground state eigenfunction of Figure 12.14. KE, kinetic energy; PE, potential energy; Imag, imaginary.

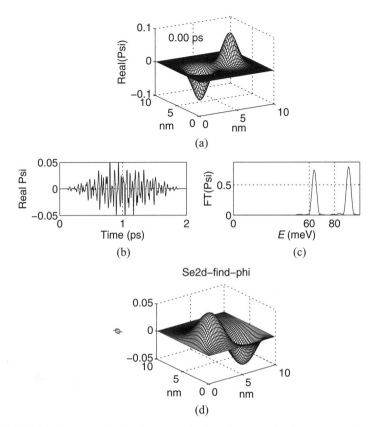

FIGURE 12.16   The search for the second eigenfunction for the potential of Figure 12.13a. (a) The test function, (b) the time-domain data, (c) the Fourier transform showing a peak at 63.5 meV, and (d) the constructed eigenfunction (only the real part is shown).

The revival is shown in Figure 12.17.

The techniques for finding the eigenfunctions for three-dimensional structures follow exactly the same procedures [3].

12.3 FINDING A COMPLETE SET OF EIGENFUNCTIONS

The examples above demonstrate how to find a few eigenfunctions in an arbitrary structure, but a more systematic approach is needed to find a complete set to be used for a meaningful analysis [4]. First, even though the Fourier transform at a point in the structure reveals many of the eigenenergies, as shown in Figure 12.8 for instance, we only look for the lowest energy eigenfunction function. Then the next time, we start with a test function that will

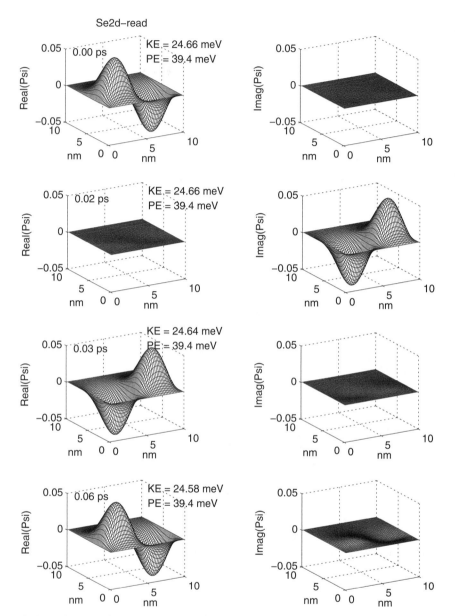

FIGURE 12.17 Time evolution of the eigenfunction found in Figure 12.16. KE, kinetic energy; PE, potential energy; Imag, imaginary.

only give eigenenergies corresponding to eigenfunctions that are orthonormal to the previous one, as shown in Figure 12.11. Notice that this is similar to the procedure used by the variational method described in Chapter 10.

Here is a summary of the procedure to find the eigenfunctions:

1. Initialize a test function at a point (x_0, y_0) in the structure. This test function must be orthogonal to any previously determined eigenfunctions. Start the FDTD simulation, and as the simulation proceeds, store the time-domain data at (x_0, y_0). At the end of the simulation, take the Fourier transform of the stored time-domain data. The first peak will be the eigenenergy.
2. Repeat the simulation with the same test function and calculate the discrete Fourier transform at the eigenenergy at every cell in the problem space. This procedure will construct the eigenfunctions.
3. Test the new eigenfunction. It must maintain normalization and revive in the time prescribed by the eigenenergy. It must also be orthogonal to any previously determined eigenfunctions.

EXERCISES

12.1 Finding the Eigenenergies and Eigenfunctions in One Dimension

12.1.1 Using the methods of this chapter, find the eigenenergies and eigenfunctions of the V potential in a 40 nm well in Figure 2.9.

12.2 Finding the Eigenfunctions of Two-Dimensional Structures

12.2.1 Find the lowest five eigenenergies and corresponding eigenfunctions for the 10×10 nm infinite well with the perturbation of Figure 10.7.

REFERENCES

1. D. M. Sullivan and D. S. Citrin, "Determination of the eigenfunctions of arbitrary nanostructures using time domain simulation," *J. Appl. Phys.*, Vol. 91, p. 3219, 2002.
2. A. V. Oppenheim and R. W. Schafer, *Digital Signal Processing*, Englewood Cliffs, NJ: Prentice Hall, 1975.
3. D. M. Sullivan and D. S. Citrin, "Determining quantum eigenfunctions in three-dimensional nanoscale structures," *J. Appl. Phys.*, Vol. 97, p. 104305, 2005.
4. D. M. Sullivan and D. S. Citrin, "Determining a complete three-dimensional set of eigenfunctions for nanoscale structure analysis," *J. Appl. Phys.*, Vol. 98, p. 084311, 2005.

APPENDIX A

IMPORTANT CONSTANTS AND UNITS

TABLE A.1   Fundamental Constants

$h = 6.625 \times 10^{-34}$ J·s	Planck's constant
$\quad = 4.135 \times 10^{-15}$ eV·s	
$\hbar = \dfrac{h}{2\pi} = 1.054 \times 10^{-34}$ J·s	
$\quad\quad = 6.58 \times 10^{-16}$ eV·s	
$k_B = 1.38 \times 10^{-23}$ J/K	Boltzmann constant
$\quad = 8.62 \times 10^{-5}$ eV/K	
$c_0 = 3 \times 10^8$ m/s	Speed of light in a vacuum
$\varepsilon_0 = 8.85 \times 10^{-12}$ F/m	Permittivity of free space
$\mu_0 = 4\pi \times 10^{-7}$ H/m	Permeability of free space
$m_e = 9.109 \times 10^{-31}$ kg	Free space mass of an electron
$-e = 1.6 \times 10^{-19}$ C	Charge of an electron
$G_0 = 38.7$ μS $= (25.8$ kΩ$)^{-1}$	Quantum conductance

Quantum Mechanics for Electrical Engineers, First Edition. Dennis M. Sullivan.
© 2012 The Institute of Electrical and Electronics Engineers, Inc.
Published 2012 by John Wiley & Sons, Inc.

TABLE A.2 Effective Mass (Density of States) for Electrons in Semiconductors

Semiconductor	m/m_e
Silicon (Si)	1.08
Gallium arsenide (GaAs)	0.067
Germanium (Ge)	0.55

TABLE A.3 Conversion Factors

1 Å (angstrom) = 10^{-10} m
1 μm (micron) = 10^{-6} m
2.54 cm = 1 inch
1 J (joule) = 1.6×10^{-19} eV (electron volt)

TABLE A.4 Units

The basic SI units:

Mass	Kilogram	kg
Length	Meter	m
Time	Second	s
Current	Ampere	A
Temperature	Kelvin	K

Derived units:

Charge	Coulomb	C	$[\text{A·s}]$
Force	Newton	N	$\left[\dfrac{\text{kg} \cdot \text{m}}{\text{s}^2}\right]$
Energy	Joule	J	$[\text{N} \cdot \text{m}] = \left[\dfrac{\text{kg} \cdot \text{m}^2}{\text{s}^2}\right]$
Potential	Volt	V	$\left[\dfrac{\text{J}}{\text{C}}\right] = \left[\dfrac{\text{kg} \cdot \text{m}^2}{\text{C} \cdot \text{s}^2}\right]$
Electric field intensity	Volt per meter	E	$\left[\dfrac{\text{V}}{\text{m}}\right] = \left[\dfrac{\text{kg} \cdot \text{m}}{\text{C} \cdot \text{s}^2}\right]$
Capacitance	Farad	F	$\left[\dfrac{\text{C}}{\text{V}}\right] = \left[\dfrac{\text{C}^2 \cdot \text{s}^2}{\text{kg} \cdot \text{m}^2}\right]$

TABLE A.4   *Continued*

Resistance	Ohm	Ω	$\left[\dfrac{V}{A}\right] = \left[\dfrac{kg \cdot m^2}{C^2 \cdot s}\right]$
Conductance	Siemens	S	$\left[\dfrac{1}{R}\right] = \left[\dfrac{C^2 \cdot s}{kg \cdot m^2}\right]$
Magnetic flux	Weber	Wb	$[V \cdot s] = \left[\dfrac{kg \cdot m^2}{C \cdot s}\right]$
Magnetic flux density	Tesla	T	$\left[\dfrac{Wb}{m^2}\right] = \left[\dfrac{kg}{C \cdot s}\right]$
Inductance	Henry	H	$\left[\dfrac{Wb}{A}\right] = \left[\dfrac{kg \cdot m^2}{C^2}\right]$

TABLE A.5   Multiples and Submultiples of Units

Symbol	Prefix	Multiple
E	exa	10^{18}
P	peta	10^{15}
T	tera	10^{12}
G	giga	10^{9}
M	mega	10^{6}
k	kilo	10^{3}
c	centi	10^{-2}
m	milli	10^{-3}
μ	micro	10^{-6}
n	nano	10^{-9}
p	pico	10^{-12}
f	femto	10^{-12}
a	atto	10^{-18}

APPENDIX B

FOURIER ANALYSIS AND THE FAST FOURIER TRANSFORM (FFT)

This is a brief introduction to the fast Fourier transform (FFT) [1], particularly for the applications in this book.

B.1 THE STRUCTURE OF THE FFT

Figure B.1 shows a cosine function and the Fourier transform of this function using the MATLAB command **fft**. For simplicity, we will assume that the cosine function is a time-domain function and that the units are seconds.

The Fourier transform of a cosine is two delta functions:

$$F\{\cos(2\pi f_o t\} = \pi[\delta(f + f_0) + \delta(f - f_0)].$$

The two delta functions appear in the bottom part of Figure B.1, but not where we expect them. This takes some explanation. The MATLAB command **fft** implements a "fast Fourier transform." (This is the algorithm that made Fourier transforms practical in calculating convolutions.) The **fft** puts the positive frequencies in the first $N/2$ positions, and the negative frequencies in the last $N/2$ positions. This is illustrated in Figure B.2 for a buffer that is eight cells.

The lowest frequency that can be represented is one that takes the entire buffer. So if the buffer is eight cells, the lowest frequency is $f_0 = 1/8$. The highest

Quantum Mechanics for Electrical Engineers, First Edition. Dennis M. Sullivan.
© 2012 The Institute of Electrical and Electronics Engineers, Inc.
Published 2012 by John Wiley & Sons, Inc.

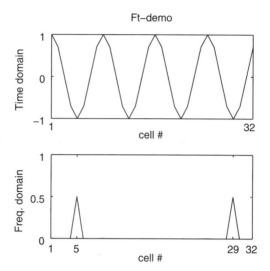

FIGURE B.1   A cosine function in the time domain (top) and its FFT in the frequency domain (bottom).

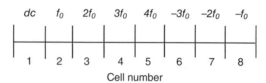

FIGURE B.2   Diagram of where the **fft** command stores the different frequencies.

frequency will be $f_{max} = 4/8 = 1/2$. If N is the total number of cells, then the highest frequency is $N/2$. This frequency occupies only one cell in the Fourier domain, as shown in Figure B.3.

Obviously, we can make a similar representation for any frequency that is an integer division of N.

Let us look at some other waveforms, as shown in Figure B.4. Each of these waveforms has a center frequency of $8/N$. The first one is in a fairly narrow envelope. In the Fourier domain, it is centered at $8 + 1 = 9$, but it is also spread out somewhat. The second pulse has a much broader envelope, but its Fourier transform is very narrow. Is this consistent with what we know about Fourier theory?

Let us look at another example. Figure B.5 shows a sinusoid with a wavelength of 10 cells. Unfortunately, 10 is not an integer division of our buffer size, which is 64. So the Fourier transform is spread out.

If we want more accuracy, a simple solution exists: Do the transforms in a longer buffer. The buffer with 64 cells has an accuracy of $\Delta f = 1/64$. If we go to a buffer with 256 cells, our accuracy is $\Delta f = 1/256$ (Fig. B.6).

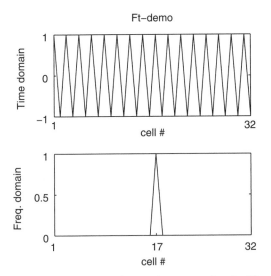

FIGURE B.3   The FFT of the highest frequency for the 32-cell buffer.

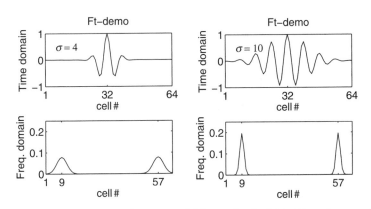

FIGURE B.4   Two pulses at frequency 8/N contained within Gaussian envelopes.

B.2   WINDOWING

The Fourier domain waveform of Figure B.6 still looks a little ragged because when we move the sinusoid to the larger buffer, we abruptly truncate the signal at 64. The result is the same as if we had multiplied the time-domain data by a rectangular function. A sharp transition in the time domain leads to ripples in the frequency domain (Fig. B.7, left side). Instead of this abrupt truncation, we can "window" the data, that is, multiply it by a function that results in a smooth transition [2].

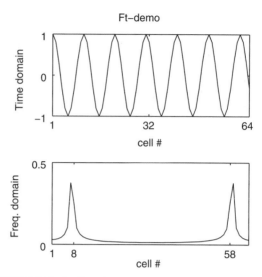

FIGURE B.5   Fourier transform of frequency 1/10.

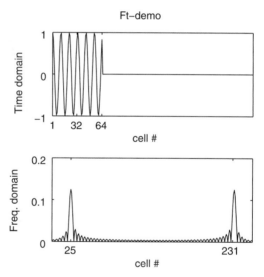

FIGURE B.6   Fourier transform of the same signal shown in Figure B.5, but with a buffer of 256 cells.

One such window, referred to as the Hanning window, is shown on the right side of Figure B.7. Notice how much smoother the **fft** of the Hanning window is compared to the rectangular function. Many different windows are used to optimize different parameters, but they all tend to have a Gaussian-like shape [3].

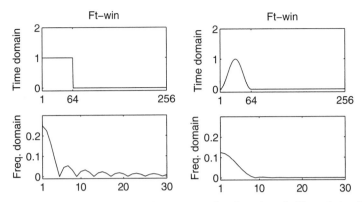

FIGURE B.7   Fourier transform of a rectangular function (left) and the Hanning window (right).

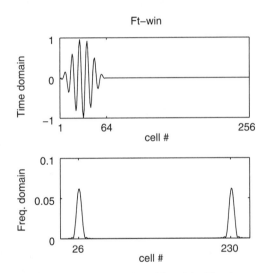

FIGURE B.8   A windowed sinusoid and its Fourier transform.

When we window the sinusoidal data before transforming, we get the smoother result shown in Figure B.8.

Until now, we have assumed that the data represent time domain data. However, we can represent waveforms in the spatial domain as well. Suppose Figure B.1 represents a cosine in the space domain where each cell is 1 μm. Then the Fourier transform would take the signal to the inverse wavelength, or $1/\lambda$. So for the 64-cell buffer in Figure B.1, the lowest frequency would be $1/32$ μm, and the Fourier transform would produce spikes at $\pm 4/32$ μm.

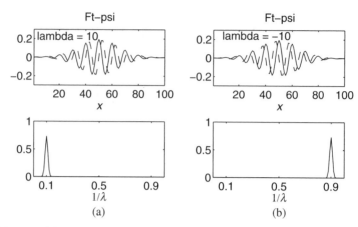

FIGURE B.9   (a) The top diagram shows a spatial Gaussian pulse traveling in the positive direction. The center wavelength is 10; the Fourier transform of the pulse is illustrated on the bottom. Notice that it is centered at 0.1 or 1/10. (b) A similar pulse traveling in the negative direction.

B.3   FFT OF THE STATE VARIABLE

So far, we have been using only real functions in time or space. If Figure B.1 represented the electric field of an electromagnetic pulse we could not tell if it was propagating to the left or right, at least not without seeing the magnetic field as well. In quantum mechanics, we do not have this problem. The state variable ψ is complex, and the phase between the real and imaginary part dictates its direction.

Figure B.9 shows two complex pulses with their respective Fourier transforms. Pulse (a) is traveling from left to right, as indicated by the fact that its Fourier transform only has positive frequencies. Pulse (b) by contrast, is moving right to left, evident from its negative frequencies.

The position in the frequency domain is easy to understand if we once again think about where the different frequency components are stored in the **fft** buffer, similar to Figure B.2. The difference is that we are using spatial frequencies. The longest wavelength in the 100-cell buffer is 100. The wavelength of 10 is going to be the $n = 100/10 = 10$ cell, which is the eleventh cell from the right. The pulse moving in the negative direction will be in the tenth cell from the right.

Figure B.10 shows two similar pulses with different wavelengths.

Physicists prefer plots as functions of $k = 2\pi/\lambda$. However, with functions of k, the axis does not show anything that corresponds directly to the waveform.

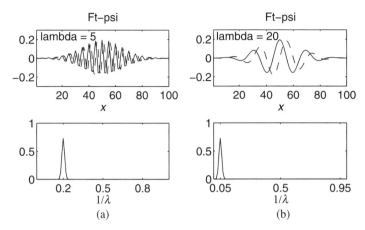

FIGURE B.10 Space domain and wavelength domain plots for wavelengths of (a) five cells and (b) 20 cells.

EXERCISES

B.1 Use the program ft_demo.m to duplicate the results of Figure B.1. Replace the input with the following wavelengths: 64, 32, and 2 cells. Does the **fft** give you the results you expect?

B.2 Duplicate the results of Exercise B.1 with the following differences: Use $N = 100$ instead of 64 for the size of the problem space, but still use the wavelengths of 64, 32, and 2 cells. Assume the cells sizes are 1 nm and label the axis in the space and the $1/\lambda$ domain.

B.3 Repeat Exercise B.2, but assume you are using time and frequency. Assume the time-domain cells are 1 fs, and label your **fft** domain appropriately.

B.4 Using the program ft_psi.m, initialize a complex wave traveling in the positive direction with a center wavelength of 20 cells inside a Gaussian pulse. Add another domain that plots the **fft** as a function of energy, similar to Figure 3.10.

REFERENCES

1. E. O. Brigham, *The Fast Fourier Transform and Its Applications*, Englewood Cliffs, NJ: Prentice Hall, 1988.
2. A. V. Oppenheim and R. W. Schafer, *Digital Signal Processing*, Englewood Cliffs, NJ: Prentice Hall, 1975.
3. E. P. Cunningham, *Digital Filtering: An Introduction*, New York: John Wiley and Sons, 1995.

APPENDIX C

AN INTRODUCTION TO THE GREEN'S FUNCTION METHOD

The material of this appendix is taken primarily from Matthews and Walker [1].

Start with a linear operator L in an inhomogeneous equation with a forcing function f:

$$Lu(x) - \gamma u(x) = f(x). \tag{C.1}$$

The homogeneous version of this equation

$$Lu(x) - \gamma u(x) = 0,$$

has eigenfunctions $u_n(x)$ and corresponding eigenvalues γ_n:

$$Lu_n(x) - \gamma_n u_n(x) = 0.$$

Any function $u(x)$, including the forcing function $f(x)$, can be expanded in terms of the eigenfunctions

$$u(x) = \sum_n c_n u_n(x), \tag{C.2a}$$

$$f(x) = \sum_n d_n u_n(x). \tag{C.2b}$$

Quantum Mechanics for Electrical Engineers, First Edition. Dennis M. Sullivan.
© 2012 The Institute of Electrical and Electronics Engineers, Inc.
Published 2012 by John Wiley & Sons, Inc.

The coefficients d_n are determined by,

$$d_n = u_n \cdot f = \langle u_n | f \rangle = \int_\Omega u_n^*(x') f(x') dx'.$$

The Ω just means over the domain of the integral. Substituting Equation C.2a and C.2b back into Equation C.1 gives

$$L \sum_n c_n u_n(x) - \gamma \sum_n c_n u_n(x) = \sum_n d_n u_n(x),$$

$$\sum_n c_n (\gamma_n - \gamma) u_n(x) = \sum_n d_n u_n(x).$$

Since the $u_n(x)$ are linearly independent, we can write, for each n,

$$c_n = \frac{d_n}{\gamma_n - \gamma}. \tag{C.3}$$

Unlike the γ_n, which are solutions of the homogeneous equation, γ is a parameter determined by the forcing function in the inhomogeneous Equation C.1. Going back to Equation C.2a, we can write

$$u(x) = \sum_n \left\{ \frac{d_n}{\gamma_n - \gamma} \right\} u_n(x)$$

$$= \sum_n \frac{\int_\Omega u_n^*(x') f(x') dx'}{\gamma_n - \gamma} u_n(x). \tag{C.4}$$

This can be written as:

$$u(x) = \int_\Omega \sum_n \frac{u_n(x) u_n^*(x') f(x')}{\gamma_n - \gamma} dx'$$

$$= \int_\Omega G(x, x') f(x') dx'. \tag{C.5}$$

We now have a Green's function

$$G(x, x') = \sum_n \frac{u_n(x) u_n^*(x')}{\gamma_n - \gamma}. \tag{C.6}$$

Note the following: When γ equals an eigenvalue of the homogeneous equation, like γ_n, the Green's function is infinite and there is no solution $u(x)$ unless

$$\int_\Omega u_n^*(x) f(x) dx = 0.$$

Notice that if the γ_n were complex, we would not have to face that problem. In fact, sometimes a very small positive imaginary part is added to each eigenfunction to make it mathematically tractable. In any case, we will simply set this point aside for the purpose of this discussion.

One last note: Equation C.6 is the Green's function, but it is very hard to digest its meaning. It is often much easier to use the following pair of equations:

$$u(x) = \sum_n \left\{ \frac{d_n}{\gamma_n - \gamma} \right\} u_n(x), \qquad (C.7a)$$

$$d_n = \int_\Omega u_n^*(x') f(x') dx'. \qquad (C.7b)$$

C.1   A ONE-DIMENSIONAL ELECTROMAGNETIC CAVITY

As an example, we will apply the Green's function to find the solution of a one-dimensional electric cavity with a dipole antenna source in the middle, as shown in Figure C.1. In one dimension, the electric field can be described by the Helmholtz equation with the initial conditions specified at the boundaries

$$\frac{\partial^2 u(x)}{\partial x^2} + k^2 u(x) = 0, \quad u(0) = u(L) = 0,$$

where L is the length of the cavity. The eigenfunctions are the standing E fields

$$u_n(x) = \sin\left(\frac{\pi n x}{L}\right).$$

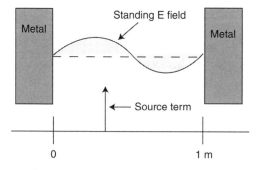

FIGURE C.1   A one-dimensional electric cavity with a dipole antenna in the middle.

The eigenvalues are

$$\gamma_n = k_n^2 = \left(\frac{\pi n}{L}\right)^2.$$

We now assume the forcing function f is an electric dipole source. We say it is a point source located at x_0, oscillating at a frequency ω_{in}. Therefore,

$$f(x, \omega) = \delta(x - x_0)e^{-i\omega_{int}}.$$

In order to formulate a concrete example, let us say that $L = 1$ m , $x_0 = 0.4$ m, and that the source has a frequency of 160 MHz, or $\omega_{in} = 2\pi \times (1.6 \times 10^8)$ rad/s. Look at Equation C.7a. We know the values of γ_n and u_n, but we don't have γ in Equation C.7a. That is determined by the frequency of the forcing function, because for an electromagnetic wave in free space

$$k = \frac{\omega}{c_0},$$

where c_0 is the speed of light in a vacuum. Therefore,

$$\gamma = k^2 = \left(\frac{\omega_{in}}{c_0}\right)^2 = \left(\frac{2.8\pi \times 10^8}{3 \times 10^8}\right)^2 = 1.137\pi^2. \tag{C.8}$$

The values of γ_n are

$$\gamma_n = \left(\frac{\pi n}{L}\right)^2. \tag{C.9}$$

So, γ will lie between the first and second eigenvalues. This is important because of the term $1/(\gamma_n - \gamma)$ in Equation C.7b. The contributions of the third and higher eigenfunctions will diminish as their corresponding values of γ_n move further away from γ in Equation C.8. We define the parameter as:

$$a_n = \frac{1}{|\gamma_n - \gamma|} = \frac{1}{\pi^2 |n^2 - 1.137|}. \tag{C.10}$$

We take the absolute value because it is the distance that matters. The first few values are:

$$a_1 = 0.7354, a_2 = 0.0354, a_3 = 0.0129, \text{ and } a_4 = 0.0068.$$

Now, we turn our attention to the calculation of the d_n. Since the spatial part of the forcing function is a delta function, the inner product terms are particularly easy:

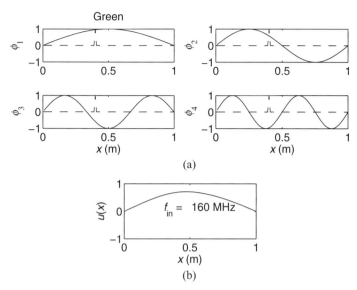

FIGURE C.2   (a) The first four eigenfunctions. (b) The solution calculated by Equation C.11 for a forcing function of 160 MHz located at $x_0 = 0.4$ m.

$$d_1 = \int_0^{1\,m} \sin(\pi x)\delta(x-0.4)\,dx = \sin(\pi 0.4) = 0.95,$$

$$d_2 = \int_0^{1\,m} \sin(2\pi x)\delta(x-0.4)\,dx = \sin(2\pi 0.4) = 0.59,$$

$$d_3 = \int_0^{1\,m} \sin(3\pi x)\delta(x-0.4)\,dx = \sin(3\pi 0.4) = -0.59,$$

$$d_4 = \int_0^{1\,m} \sin(4\pi x)\delta(x-0.4)\,dx = \sin(4\pi 0.4) = -0.95.$$

The first four terms of Equation C.7a are:

$$u(x) \cong \sum_{n=1}^{4} \frac{a_n}{d_n} \sin(n\pi x). \qquad (C.11)$$

This is illustrated in Figure C.2.

If we repeat the above problem for a source at the same location but a frequency of 290 MHz, then the values of d_n remain the same but the a_n values become

$$a_1 = 0.0370,\ a_2 = 0.3864,\ a_3 = 0.0193,\ \text{and}\ a_4 = 0.0083,$$

resulting in the $u(x)$ in Figure C.3.

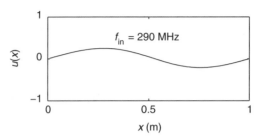

FIGURE C.3 The solution calculated by Equation C.11 for a forcing function of 290 MHz located at $x_0 = 0.4$ m.

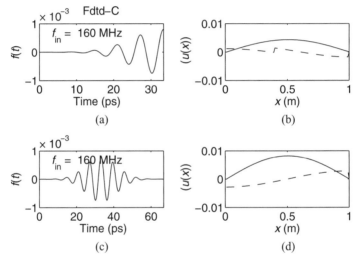

FIGURE C.4 An FDTD simulation of the Green's function analysis of Figure C.2. The forcing function is an electric dipole located at 0.4 m. The dipole oscillates at 160 MHz within a Gaussian envelope. (a) The time-domain forcing function after 35 ps. (b) The E field (solid line) and the H field (dashed line) after 35 ps. (c) The time-domain forcing function after 55 ps. (d) The E field (solid line) and the H field (dashed line) after 55 ps.

An illustration of the above can be obtained by a one-dimensional electromagnetic simulation using the finite-difference time-domain (FDTD) method [2]. This is shown in Figure C.4. The input is assumed to be a sinusoid at 160 MHz inside a Gaussian envelope. The figures in the left columns are the time-domain forcing function. The figures in the right column are the E fields (solid line) and the H fields (dashed lines). Figure C.5 is a similar simulation at 290 MHz.

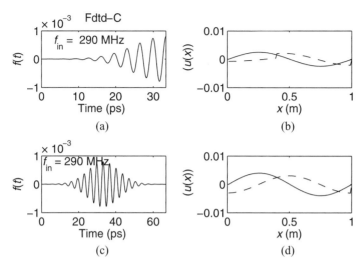

(a) (b)

(c) (d)

FIGURE C.5   An FDTD simulation of the Green's function analysis of Figure C.3. The forcing function is an electric dipole located at 0.4 m. The dipole oscillates at 290 MHz within a Gaussian envelope. (a) The time-domain forcing function after 35 ps. (b) The E field (solid line) and the H field (dashed line) after 35 ps. (c) The time-domain forcing function after 55 ps. (d) The E field (solid line) and the H field (dashed line) after 55 ps.

EXERCISES

C.1   Using the programs Green.m and Fdtd_C.m, repeat the electromagnetic example at 0.4 m but for 340 MHz.

C.2   Using the programs Green.m and Fdtd_C.m, repeat the electromagnetic example at 0.16 m with a frequency of 340 MHz.

REFERENCES

1. J. Mathews and R. L. Walker, *Mathematical Methods of Physics*, 2nd ed., Menlo Park, CA: Benjamin/Cummings Publishing, 1970.
2. D. M. Sullivan, *Electromagnetic Simulation Using the FDTD Method*, New York: IEEE Press, 2000.

APPENDIX D

LISTINGS OF THE PROGRAMS USED IN THIS BOOK

D.1 CHAPTER 1

```
% Se1_1.m.           % This is a 1D FDTD simulation of psi.

clear all

NN = 400;            % Number of points in the problem
space.
hbar = 1.054e-34;    % Plank's constant
m0 = 9.1e-31;        % Free space mass of an electron
meff = 1.0;          % Effective mass: Si is 1.08, Ge is
0.067, GaAs is 0.55
melec = meff*m0;     % Mass of an electron
ecoul = 1.6e-19;     % Charge of an electron
epsz = 8.85e-9;      % Dielectric of free space
eV2J = 1.6e-19;      % Energy conversion factors
J2eV = 1/eV2J;

del_x = .1e-9;       % The cell size
dt = 2e-17;          % Time steps
ra = (0.5*hbar/melec)*(dt/del_x^2);  % ra must be < .15
DX = del_x*1e9;      % Cell size in nm.
XX = (DX:DX:DX*NN);  % Length in nm for plotting
```

Quantum Mechanics for Electrical Engineers, First Edition. Dennis M. Sullivan.
© 2012 The Institute of Electrical and Electronics Engineers, Inc.
Published 2012 by John Wiley & Sons, Inc.

```
% --- Specify the potential --------------------

V=zeros(1,NN);

% Barrier
  for n=NN/2:NN/2+50
%   V(n) = .15*eV2J;
  end
% Semiconductor conduction band
  for n=1:NN/2
%    V(n) = .1*eV2J;
  end
  for n=NN/2+1:NN
%    V(n) = .2*eV2J;
  end
% Electric field
% for n=1:NN
%    V(n) = -(0.2/400)*(n-400)*eV2J;
% end
% ----------------------------------------------------
% Initialize a sine wave in a gaussian envelope

lambda = 50;        % Pulse wavelength
%lambda = 25;       % Pulse wavelength
sigma = 50;         % Pulse width
nc = 150;           % Starting position
prl = zeros(1,NN);  % The real part of the state variable
pim = zeros(1,NN);  % The imaginary part of the state
variable
ptot = 0.;
for n=2:NN-1
prl(n) = exp(-1.*((n-nc)/sigma)^2)*cos(2*pi*(n-nc)/lambda) ;
pim(n) = exp(-1.*((n-nc)/sigma)^2)*sin(2*pi*(n-nc)/lambda) ;
ptot = ptot + prl(n)^2 + pim(n)^2;
end
pnorm = sqrt(ptot);   % Normalization constant

% Normalize and check
ptot = 0.;
for n=1:NN
prl(n) = prl(n)/pnorm;
pim(n) = pim(n)/pnorm;
ptot = ptot + prl(n)^2 + pim(n)^2;
end
ptot                      % This should have the value 1

T = 0;
n_step = 1;
while n_step > 0
```

```
n_step = input('How many time steps  -->');

% -----------This is the core FDTD program -------------
for m=1:n_step
T = T + 1;

for n=2:NN-1
prl(n) = prl(n) - ra*(pim(n-1) -2*pim(n) + pim(n+1))  ...
        + (dt/hbar)*V(n)*pim(n);
end

for n=2:NN-1
pim(n) = pim(n) + ra*(prl(n-1) -2*prl(n) + prl(n+1))  ...
        - (dt/hbar)*V(n)*prl(n);
end

end
% --------------------------------------------------

% Calculate the expected values

PE = 0.;
for n=1:NN
psi(n) = prl(n) + i*pim(n);   % Write as a complex function
PE = PE + psi(n)*psi(n)'*V(n);
end
psi*psi'                 % This checks normalization
PE = PE*J2eV;            % Potential energy

ke = 0. + j*0.;
for n=2:NN-1
lap_p = psi(n+1) - 2*psi(n) + psi(n-1);
ke = ke + lap_p*psi(n)';
end
KE = -J2eV*((hbar/del_x)^2/(2*melec))*real(ke); % Kinetic
energy

subplot(2,1,1)
plot(XX,prl,'k')
hold on
plot(XX,pim,'-.k')
plot(XX,J2eV*V,'--k')
hold off
axis( [ 1 DX*NN -.2 .3 ])
TT = text(5,.15,sprintf('%7.0f fs',T*dt*1e15));
set(TT,'fontsize',12)
TT = text(5,-.15,sprintf('KE = %5.3f eV',KE));
set(TT,'fontsize',12)
TT = text(25,-.15,sprintf('PE = %5.3f eV',PE));
```

```
set(TT,'fontsize',12)
TT = text(25,.13,sprintf('E_t_o_t = %5.3f eV',KE+PE));
set(TT,'fontsize',12)
xlabel('nm')
set(gca,'fontsize',12)
title('Se1-1')
```

D.2   CHAPTER 2

```
% Se2_1.m.  % FDTD simulation with eigenfunction
decomposition

clear all

NN = 800;             % Number of points in the problem space.
hbar = 1.054e-34;     % Plank's constant
melec = 9.1e-31;      % Mass of an electron
eV2J = 1.6e-19;       % Energy conversion factors
J2eV = 1/eV2J;

del_x = .05e-9;       % The cells size
dt = .5e-17;          % Time steps
ra = (0.5*hbar/melec)*(dt/del_x^2)   % ra must be < .1
DX = del_x*1e9;       % Cell size in nm.
XX = (DX:DX:DX*NN);   % Length in nm for plotting

chi0 = hbar^2/(2*melec*del_x^2);  % This is for the eigen
calculation.

V = zeros(1,NN);
% V shaped potential
for n=1:NN
%V(n) = eV2J*(.0005)*(abs(NN/2-n));
end
% E field
for n=1:NN
%V(n) = eV2J*(.002)*(NN/2-n);
end

subplot(3,2,1)
plot(XX,J2eV*V,'k');
set(gca,'fontsize',12)
ylabel('V (eV)')
xlabel('nm')
Umax = max(J2eV*V);
title('Se2-1')
%axis( [ 0 DX*NN 0 Umax ])
```

```
% ----- Eigenvalue calculation -------------------------------

% Specify the Hamiltonian
H = zeros(NN,NN);
H(1,1) = 2*chi0+V(1);
H(1,2) = -1*chi0;
for n=2:NN-1
H(n,n-1)= -1*chi0;
H(n,n)  = 2*chi0+ V(n);
H(n,n+1)= -1*chi0;
end
H(NN,NN-1) = -1*chi0;
H(NN,NN)   = 2*chi0+V(NN);
% Switch to PBC
%H(NN,1) = -1*chi0;
%H(1,NN) = -1*chi0;

[phi,D] = eig(H);   % Put the eigenenergies in D and
functions in phi

%Plot the eigenfunctions
subplot(3,2,3)
plot(XX,phi(:,1),'k')
TT = ylabel('f_1','FontName','Symbol','fontsize',12)
TT = text(5,0.05,sprintf('%7.4f eV',J2eV*D(1,1)));
set(TT,'fontsize',12)
set(gca,'fontsize',12)
title('Se2-1')
subplot(3,2,4)
plot(XX,phi(:,2),'k')
TT = ylabel('f_2','FontName','Symbol','fontsize',12)
TT = text(5,.03,sprintf('%7.4f eV',J2eV*D(2,2)));
set(TT,'fontsize',12)
set(gca,'fontsize',12)
subplot(3,2,5)
plot(XX,phi(:,3),'k')
TT = ylabel('f_3','FontName','Symbol','fontsize',12)
TT = text(5,.03,sprintf('%7.4f eV',J2eV*D(3,3)));
set(TT,'fontsize',12)
set(gca,'fontsize',12)
xlabel('nm')
subplot(3,2,6)
plot(XX,phi(:,4),'k')
TT = ylabel('f_4','FontName','Symbol','fontsize',12)
TT = text(5,.03,sprintf('%7.4f eV',J2eV*D(4,4)));
set(TT,'fontsize',12)
set(gca,'fontsize',12)
xlabel('nm')
```

```
-----------------------------------------------------------------
% ------------------------------------------------

% Initialize the FDTD simulation
sigma = 40;             % Pulse width
lambda = 40;            % Pulse wavelength
nc = NN/2 - 30;         % Starting position of the pulse
prl = zeros(1,NN);
pim = zeros(1,NN);
ptot = 0.;
for n=2:NN-1
%     prl(n) = exp(-1.*((n-nc)/sigma)^2)*cos(2*pi*(n-nc)/
lambda) ;
%     pim(n) = exp(-1.*((n-nc)/sigma)^2)*sin(2*pi*(n-nc)/
lambda) ;
   prl(n) = phi(n,1);            % This initializes an
eigenstate
ptot = ptot + prl(n)^2 + pim(n)^2;
end
pnorm = sqrt(ptot);

% Normalize and check
ptot = 0.;
for n=1:NN
prl(n) = prl(n)/pnorm;
pim(n) = pim(n)/pnorm;
ptot = ptot + prl(n)^2 + pim(n)^2;
end
ptot

T = 0;
n_step = 1;
while n_step > 0
T
n_step = input('How many time steps -->');

% -----------------------
for m=1:n_step
T = T + 1;

for n=2:NN-1
prl(n) = prl(n) - ra*(pim(n-1) -2*pim(n) + pim(n+1)) ...
       + (dt/hbar)*V(n)*pim(n);
end

for n=2:NN-1
pim(n) = pim(n) + ra*(prl(n-1) -2*prl(n) + prl(n+1)) ...
       - (dt/hbar)*V(n)*prl(n);
end
```

```
end

% ------------------------

% Check normalization
ptot1 = 0.;
for n=1:NN
ptot1 = ptot1 + prl(n)^2 + pim(n)^2;
end
ptot1

% Calculate the expected values
PE = 0.;
for n=1:NN
psi(n) = prl(n) + i*pim(n);
PE = PE + psi(n)*psi(n)'*V(n);
end
PE = PE*J2eV;

ke = 0. + j*0.;
for n=2:NN-1
lap_p = psi(n+1) - 2*psi(n) + psi(n-1);
ke = ke + lap_p*psi(n)';
end
KE = -J2eV*((hbar/del_x)^2/(2*melec))*real(ke);

subplot(1,1,1)      % This creates a new window

subplot(3,2,1)
plot(XX,prl,'k')
hold on
plot(XX,pim,'--k')
plot(XX,J2eV*V,'-.k')
hold off
axis( [ 0 DX*NN -.25 .30 ])
TT = text(1,.2,sprintf('%5.3f ps',T*dt*1e12));
set(TT,'fontsize',12)
TT = text(1,-.2,sprintf('KE = %4.0f meV',1e3*KE));
set(TT,'fontsize',12)
TT = text(22,-.2,sprintf('PE = %4.1f meV',1e3*PE));
set(TT,'fontsize',12)
E_tot = KE + PE;
TT = text(22,.2,sprintf('E_t_o_t = %4.1f meV',1e3*(KE+PE)));
set(TT,'fontsize',12)
set(gca,'fontsize',12)
xlabel('nm')
%TT = ylabel('f_1','FontName','Symbol','fontsize',12)
TT = ylabel('y','FontName','Symbol','fontsize',12)
```

```
title('Se2-1')

% The eigenfunction decomposition
angle = zeros(1,40);
for m=1:40
xeig(m) = m;
Crl = 0.;
Cim = 0.;
for n=1:NN
Crl = Crl + prl(n)*phi(n,m);
Cim = Cim + pim(n)*phi(n,m);
end
P_eig(m) = Crl + i*Cim;          % The complex value of cn
end

subplot(3,4,3)
bar(xeig,abs(P_eig))
axis( [ 1 20 0 1.1 ])
%TT = text(5,.85,['% Eigenfunctions']);
%set(TT,'fontsize',10)
set(gca,'fontsize',12)
title('Amplitude')
xlabel('Eigenfunction #')

subplot(3,4,4)
%bar(xeig,angle)
bar(xeig,imag(log(P_eig)))
axis( [ 0 20 -pi pi ])
%TT = text(2,2,['Angle Eigenfunctions']);
%set(TT,'fontsize',12)
set(gca,'fontsize',12)
title('Phase')
xlabel('Eigenfunction #')

end



% Se2_phase.m This program uses the complex coef to
propagate a pules.

MM = 100;              % Size of the problem space
NN = 30;               % Number of eigenfunctions to be used.
phi = zeros(10,MM);
prl = zeros(MM,1);
pim = zeros(MM,1);
phi_rl = zeros(MM,1);
phi_im = zeros(MM,1);
a = zeros(NN,1);
```

```
b = zeros(NN,1);
c = zeros(NN,1);
d = zeros(NN,1);
angle = zeros(NN,1);
omega = zeros(NN,1);

ddx = 0.1
XX = (ddx:ddx:ddx*MM);

%  These are the eigenfunctions
for n=1:NN
for m=1:MM
phi(n,m) = sqrt(2/MM)*sin(m*pi*n/MM);
end
end

%----Plot Psi----------

mc = MM/2-15;
sigma = 10;
lambda = 10;
ptot = 0.;
for m=1:MM
prl(m) = exp(-0.5*( ((m-mc)/sigma))^2)*cos(2*pi*(m-mc)/
lambda);
pim(m) = exp(-0.5*( ((m-mc)/sigma))^2)*sin(2*pi*(m-mc)/
lambda);
ptot = ptot + prl(m)^2 + pim(m)^2;
end
pnorm = sqrt(ptot);

for m=1:MM
prl(m) = prl(m)/pnorm;
pim(m) = pim(m)/pnorm;
end

subplot(4,1,1)

plot(XX,prl,'k')
hold on
plot(XX,pim,'k--')
hold off
TT = text( 9.3, -.2, ['nm']);
axis( [ 0 10 -.3 .3 ])
set(TT,'fontsize',12)
title('Se2-phase')

%----Calculate the Cn----------------------
```

```
for n=1:NN
for m=1:MM
a(n) = a(n) + phi(n,m)*prl(m);
b(n) = b(n) + phi(n,m)*pim(m);
end
c(n) = (a(n) + i*b(n));
angle(n) = atan2(b(n),a(n));
end

X=(1:1:NN);
subplot(4,2,3)
bar(X,abs(c))
ylabel('|C_n|')
axis([ 0 30 0 .5 ])

subplot(4,2,4)
bar(X,angle)
axis([ 0 30 -pi pi ])
ylabel('Angle (rad)')

% --------------------------------------------

% Calculate the phase shift to propagate a
%    pulse forward for a time t0.

t0 = 0.1;

while t0 > 0

t0 = input('t0 (in ps) -->')

% Calculate the dn

for n=1:NN
omega(n) = 5.7*(n^2);        % Eq. (2.2.6)
d(n) = c(n)*exp(-i*omega(n)*t0);
angle(n) = atan2(imag(d(n)),real(d(n)));
end

% Plot the amp and phase of the new complex coef.

subplot(4,2,7)
bar(X,abs(d))
ylabel('|D_n|')
axis([ 0 30 0 .5 ])
xlabel('Eigen #')

subplot(4,2,8)
bar(X,angle)
```

```
axis([ 0 30 -pi pi ])
%ylabel('/_ D_n (radians)')
xlabel('Eigen #')
ylabel('Angle (rad)')

phi_rl = zeros(MM,1);
phi_im = zeros(MM,1);
% Reconstruct the function at a new time
t0( in ps);
for n=1:NN
for m=1:MM
phi_rl(m) = phi_rl(m)+ real(d(n))*phi(n,m);
phi_im(m) = phi_im(m)+ imag(d(n))*phi(n,m);
end
end

% Plot the new pulse

subplot(4,1,3)
plot(XX,phi_rl,'k')
hold on
plot(XX,phi_im,'k--')
hold off
axis( [ 0 10 -.3 .3 ])
TT = text( 9.3, -.2, ['nm']);
set(TT,'fontsize',12)
TT = text( 3, .14, ['Reconstucted pulse ']);
set(TT,'fontsize',12)
TT = text( 3, -.14, ['at T = ',num2str(t0),' ps']);
set(TT,'fontsize',12)

saveas(gcf,'se.png')

end




% se2_pbc.m Eigenfunctions of the periodic boundary
condition.

%--- Initialize ---

MM = 100;
NN = 100;
phi = zeros(10,MM);
phi_re = zeros(MM,1);
prl = zeros(MM,1);
pim = zeros(MM,1);
a = zeros(NN,1);
```

```
b = zeros(NN,1);
c = zeros(NN,1);
angle = zeros(NN,1);

ddx = 0.1;
X=(ddx:ddx:ddx*NN);

% Write the exponential eigenfunctions
for n=1:NN
for m=1:MM
phi(n,m) = (1./sqrt(MM))*exp(-i*(m*2*pi*(n-1)/MM));
end
end

%----Plot Psi ----------

mc = MM/2
sigma = 15;
lambda = 15;
ptot = 0.;
for m=1:MM
prl(m) = exp(-0.5*( ((m-mc)/sigma))^2)*cos(2*pi*(m-mc)/
lambda);
pim(m) = exp(-0.5*( ((m-mc)/sigma))^2)*sin(2*pi*(m-mc)/
lambda);
ptot = ptot + prl(m)^2 + pim(m)^2;
end
pnorm = sqrt(ptot);

for m=1:MM
prl(m) = prl(m)/pnorm;
pim(m) = pim(m)/pnorm;
end

subplot(3,2,1)
plot(X,prl,'k')
hold on
plot(X,pim,'k--')
hold off
axis([ 0 10 -.25 .25 ])
xlabel('nm')
set(gca,'fontsize',12)
title('Se2-pbc')

%----Calculate the Cn--------------

ctot = 0.;
for n=1:NN
for m=1:MM
```

```
c(n) = c(n) + phi(n,m)*(prl(m) + i*pim(m));
end
angle(n) = atan2(real(c(n)),imag(c(n)));
ctot = ctot + abs(c(n))^2;
end
ctot

subplot(3,2,2)
bar(abs(c))
ylabel('|C_n|')
axis([ 0 100 0 1. ])
xlabel('Eigenfunction #')
set(gca,'fontsize',12)
```

```
% Se2_wells.m.;        % 3 wells to form a quasi-FET

clear all

NN = 250;              % Number of points in the problem
space.
hbar = 1.054e-34;      % Plank's constant
melec = 9.1e-31;       % Mass of an electron
eV2J = 1.6e-19;        % Energy conversion factors
J2eV = 1/eV2J;

del_x = .2e-9;         % The cells size
dt = 1e-16;            % Time steps
ra = (0.5*hbar/melec)*(dt/del_x^2) % ra must be < .1
DX = del_x*1e9;        % Cell size in nm.
XX = (DX:DX:DX*NN);    % Length in nm for plotting

% --- Specify the coupled wells ----------------------------

V = zeros(1,NN);
for n=99:101
V(n) = .2*eV2J;
end
for n=149:151
V(n) = .2*eV2J;
end
% Middle well
for n=102:148
V(n) = -0.003*eV2J;    % Mimics a gate voltage
end

% ----------- Initialized a waveform ------------------------
```

```
lambda = 400;            % Pulse wavelength
prl = zeros(1,NN);
pim = zeros(1,NN);
ptot = 0.;
for n=2:99
prl(n) = sin(4*pi*n/lambda) ;
ptot = ptot + prl(n)^2 + pim(n)^2;
end
pnorm = sqrt(ptot);

% Normalize and check
ptot = 0.;
for n=1:NN
prl(n) = prl(n)/pnorm;
pim(n) = pim(n)/pnorm;
ptot = ptot + prl(n)^2 + pim(n)^2;
end
ptot
T = 0;
n_step = 1;
while n_step > 0
n_step = input('How many time steps -->');

% ------------ This is the main FDTD loop------------
for m=1:n_step
T = T + 1;

for n=2:NN-1
prl(n) = prl(n) - ra*(pim(n-1) -2*pim(n) + pim(n+1)) ...
      + (dt/hbar)*V(n)*pim(n);
end

for n=2:NN-1
pim(n) = pim(n) + ra*(prl(n-1) -2*prl(n) + prl(n+1)) ...
      - (dt/hbar)*V(n)*prl(n);
end


end
% -----------------------

% Check normalization
ptot = 0.;
for n=1:NN
ptot = ptot + prl(n)^2 + pim(n)^2;
end
ptot

% Calculate the expected values
```

```
PE = 0.;
for n=1:NN
psi(n) = prl(n) + i*pim(n);
PE = PE + psi(n)*psi(n)'*V(n);
end
PE = PE*J2eV;

ke = 0. + j*0.;
for n=2:NN-1
lap_p = psi(n+1) - 2*psi(n) + psi(n-1);
ke = ke + lap_p*psi(n)';
end
KE = -J2eV*((hbar/del_x)^2/(2*melec))*real(ke);

subplot(3,1,1)
plot(XX,prl,'k')
hold on
plot(XX,pim,'-.k')
plot(XX,J2eV*V,'--k')
hold off
axis( [ 0 DX*NN -.2 .25 ])
TT = text(5,.15,sprintf('%7.2f ps',T*dt*1e12))
set(TT,'fontsize',12)
TT = text(35,.15,sprintf('KE = %5.2f meV',1e3*KE))
set(TT,'fontsize',12)
xlabel('nm')
TT = ylabel('y','FontName','Symbol','fontsize',12)
set(gca,'fontsize',12)
title('Se2-wells')
T

end
```

D.3   CHAPTER 3

```
% DOS.m    This program illustrates the locations of the
available
%          states in a 10 nm infinite well.  It also calculates
%          the broadened density of states.

clear all

E0 = 3.75; % in meV

NN = 400
del_E = .25;
```

```
N = 1e4;
Epoints = zeros(1,N);
EE = (0:del_E:del_E*(N-1));

Epoints(15)  = 1.;
Epoints(60)  = 1.;
Epoints(135) = 1.;
Epoints(2% ---- Calculate the broadened DOS ---

Beig= zeros(N,NN);

N = 1e4;
del_E = .1

Epoints = zeros(1,N);
EE = (0:del_E:del_E*(N-1));

% The loss is proportional to the energy.
for m=1:NN
E(m) = 3.75*m^2;
%eta(m) = .05*E(m);
eta(m) = .5 + .05*E(m);
end

total = 0.;
eig_tot = zeros(1,N);
for m=1:NN
for k=1:N
Beig(k,m) = Beig(k,m) + 2*del_E*eta(m)/( (EE(k) - E(m))^2 +
eta(m)^2);
eig_tot(k) = eig_tot(k) + Beig(k,m);
total = total + Beig(k,m);
end
end
total = total/(2*pi)

subplot(212)
plot(EE,eig_tot,'k')
hold on
%plot(EE,Beig(:,1))
%plot(EE,Beig(:,2),'r')
%plot(EE,Beig(:,3),'g')
%plot(EE,Beig(:,4),'m')
hold off
set(gca,'fontsize',12)
axis( [ 0 Emax 0 .3 ])
xlabel('E (meV)')
title('DOS')45) = 1.;
Epoints(375) = 1.;
```

```
Emax = 100;
subplot(3,1,1)
bar(EE,Epoints)
axis([ 0 Emax 0 1.1 ])
set(gca,'fontsize',12)
xlabel('E (meV)')
title('DOS')

% Uncer.m  This program calculates the total uncertainty of
a
%          waveform in the time and frequency domains.

clear all

NN = 100;
nc = NN/2;

prl = zeros(1,NN);
pim = zeros(1,NN);

sigma = 5.0;
lambda = 10;
pnorm = 0.;
% --- Create the pulse and normalize----
% The Gaussian pulse
for n=1:NN
 prl(n) = exp(-( .5*((n-nc)/sigma))^2)*cos(2*pi*(n-nc)/
lambda);
 pim(n) = exp(-( .5*((n-nc)/sigma))^2)*sin(2*pi*(n-nc)/
lambda);
 psi(n) = prl(n) + i*pim(n);
end

% Two-sided Decaying exponential
alpha = .05;
for n=1:NN
% prl(n) = exp( -alpha*(abs(n-nc)))*cos(2*pi*(n-nc)/lambda);
% pim(n) = exp( -alpha*(abs(n-nc)))*sin(2*pi*(n-nc)/lambda);
% psi(n) = prl(n) + i*pim(n);
end
for n = nc:NN
% prl(n) = exp(-( .5*((n-nc)/sigma)))*cos(2*pi*(n-nc)/
lambda);
% pim(n) = exp(-( .5*((n-nc)/sigma)))*sin(2*pi*(n-nc)/
lambda);
psi(n) = prl(n) + i*pim(n);
end
```

```
pnorm = sqrt(psi*psi')
for n=1:NN
psi(n) = psi(n)/pnorm;    % Normalizesubplot(2,2,1)
plot(real(psi),'k')
hold on
plot(imag(psi),'k--')
plot(abs(psi),'k')
hold off
set(gca,'fontsize',12)
set(gca,'XTick', [ 30 50 70])
%set(gca,'YTick', [ .029 .058])
title('Uncer')
axis( [ 0 NN -.3 .3 ])
xlabel('t (ps)')
grid on

% Calculate expected position
POS = 0.
for n=1:NN
POS = POS + n*(psi(n)*psi(n)');
end

subplot(2,2,2)
plot(K,real(Y),'k')
hold on
%plot(K,imag(Y),'k-.')
plot(K,abs_Y,'k')
hold off
axis( [ 0 .2 -.5 .8 ])
set(gca,'fontsize',12)
set(gca,'XTick', [ 0 .1 .2 .3 ])
set(gca,'YTick', [ -.5 .5])
grid on
xlabel('f (THz)')

% Calculate the expected frequency
FAV = 0.;
for n=1:NN
FAV = FAV + K(n)*abs(Y(n))^2;
end
FAV

% Calculate the uncertainty of frequency
sig_P = 0.;
for n=1:NN
sig_P = sig_P + (K(n)-FAV)^2*abs(Y(n))^2;
end
sig_P = sqrt(sig_P);
```

```
TT = text( .12,.36,'D','FontName','Symbol')
set(TT,'fontsize',14)
TT = text( .135,.30,sprintf('_f = %5.3f',sig_P))
set(TT,'fontsize',12)

TT = text( .050,-.35,'D','FontName','Symbol')
set(TT,'fontsize',14)
TT = text( .0615,-.41,sprintf('_t_o_t_a_l =
%6.4f',sig_X*sig_P))
set(TT,'fontsize',12)% Calculate uncertainty of position
sig_X = 0.;
for n=1:NN
sig_X = sig_X + ((n-POS)^2)*psi(n)*psi(n)';
end
sig_X = sqrt(sig_X)

TT = text( 4,.17,'D','FontName','Symbol')
set(TT,'fontsize',14)
TT = text( 9,.14,sprintf('_t = %3.1f',sig_X))
set(TT,'fontsize',12)

% Fourier Domain

% Take the FFT
Y = (1./sqrt(NN))*fft(psi);
abs_Y = abs(Y);
Y*Y'              % Check normalization

% Define the "frequencies."
tot = 0.;
K(n) = 0.;
for n=2:NN
K(n) = (n-1)/NN;
end
tot

%Se3_1.m % FT to k and E space

clear all

NN = 500;        % Size of the problem space
NC = NN/2;       % Center of the problem space
eV2J = 1.6e-19;
J2eV = 1/eV2J;
hbar = 1.06e-34;
hbar_eV = hbar*J2eV;
melec = (9.11e-31);

% FDTD paramters
```

```
del_x = .2e-9;
dt = .5e-16;   % Time steps are .01 fs.
ra = (0.5*hbar/melec)*(dt/del_x^2)
DX = del_x*1e9;  % The cells are in nm
XX = (DX:DX:DX*NN);

% FT parameters
del_K = 1/(del_x*NN);
KK = (0:del_K:del_K*(NN-1));
EE(1) = 0;
for m=2:NN
EE(m) = (J2eV/(2*melec))*(hbar^2)*(2*pi*(KK(m)))^2 ;
end


% --- Initialize the waveform

% Convert Ein to Lambda
Ein = 0.2;
lambda = (hbar*2*pi)/sqrt(2*melec*(Ein*eV2J));
lambda = 5e-9
lambda = lambda/del_x
% Initialized a sine wave in a gaussian envelope
sigma = 20;            % Pulse width
nc = NN/2-15;     % Starting position of the pulse
prl = zeros(1,NN);
pim = zeros(1,NN);
ptot = 0.;
for n=1:NN
prl(n) = exp(-1.*((n-nc)/sigma)^2)*cos(2*pi*(n-nc)/lambda) ;
pim(n) = exp(-1.*((n-nc)/sigma)^2)*sin(2*pi*(n-nc)/lambda) ;
ptot = ptot + prl(n)^2 + pim(n)^2;
end
pnorm = sqrt(ptot);
%pnorm = 1;

% Normalize and check
ptot = 0.;
for n=1:NN
prl(n) = prl(n)/pnorm;
pim(n) = pim(n)/pnorm;
psi(n) = prl(n) + i*pim(n);
ptot = ptot + prl(n)^2 + pim(n)^2;
end
ptot

% FT the input waveform
PFIN = (1/sqrt(NN))*abs(fft(psi));

T = 0;
```

```
n_step = 1;
while n_step > 0
n_step = input('How many time steps -->');

% --------- This is the main FDTD loop ---------------
for m=1:n_step
T = T + 1;

for n=2:NN-1
prl(n) = prl(n) - ra*(pim(n-1) -2*pim(n) + pim(n+1))  ;
end

for n=2:NN-1
pim(n) = pim(n) + ra*(prl(n-1) -2*prl(n) + prl(n+1))  ;
end

end
% -----------------------

for n=1:NN
psi(n) = prl(n) + i*pim(n);
end
psi*psi'

ke = 0. + j*0.;
for n=2:NN-1
lap_p = psi(n+1) - 2*psi(n) + psi(n-1);
ke = ke + lap_p*psi(n)';
end
KE = -J2eV*((hbar/del_x)^2/(2*melec))*real(ke);

subplot(3,2,1)
plot(XX,prl,'k')
hold on
plot(XX,pim,'--k')
hold off
axis( [ 40 60 -.25 .35 ])
xlabel('nm')
TT = text(50,.25,sprintf('KE = %4.2f eV',KE));
set(TT,'fontsize',12)
set(gca,'fontsize',12)
title('Se3-1')

subplot(3,2,3)
plot(1e-9*KK,PFIN,'k')
axis( [ 0 .5 0 .4 ])
%set(gca,'XTick',[ .05 .1 .15 .2])
set(gca,'fontsize',12)
grid on
```

```
xlabel('1/lambda')
ylabel('Psi(k)')

subplot(3,2,5)
plot(EE,PFIN,'k')
axis( [ 0 .3 0 .5 ])
set(gca,'XTick',[ .06 .2 .3 ])
set(gca,'fontsize',12)
grid on
xlabel('eV')
ylabel('Psi(E)')

end

%Se3_2.m  % Calculate transmission

clear all

NN = 500;          % Size of the problem space
NC = NN/2;         % Center of the problem space
eV2J = 1.6e-19;
J2eV = 1/eV2J;
hbar = 1.06e-34;
hbar_eV = hbar*J2eV;
melec = 9.11e-31;

% FDTD paramters
del_x = .2e-9;     % The cell size
dt = .5e-16;       % Time steps are .01 fs.
ra = (0.5*hbar/melec)*(dt/del_x^2)
DX = del_x*1e9;    % The cell size in nm
XX = (DX:DX:DX*NN);

%Specify the potential

V=zeros(1,NN);
Vback = 0.;

% ---- One barrier
 for n=NC-3:NC+2
 V(n) = .2*eV2J;
 end

% ---- Two barriers
%for n=NC-7:NC-5
%for n=NC-12:NC-10
%   V(n)=0.15*eV2J;
%end
%for n=NC+5:NC+7
```

```
% for n=NC+10:NC+12
%   V(n)=0.15*eV2J;
% end

% FT parameters
del_K = 1/(del_x*NN);
KK = (0:del_K:del_K*(NN-1));
EE(1) = 0;
for m=2:NN
%EE(m) = (J2eV/(2*melec))*(hbar^2)*(2*pi*(KK(m)/del_x))^2 +
Vback;
EE(m) = (J2eV/(2*melec))*(hbar^2)*(2*pi*(KK(m)))^2 + Vback;
end

% --- Initialize the waveform

% Convert Ein to Lambda
Ein = 0.1;
lambda = (hbar*2*pi)/sqrt(2*melec*(Ein*eV2J));
lambda = lambda/del_x
% Initialized a sine wave in a gaussian envelope
sigma = 15;            % Pulse width
nc = NN/2-125;      % Starting position of the pulse
prl = zeros(1,NN);
pim = zeros(1,NN);
ptot = 0.;
for n=1:NN
prl(n) = exp(-1.*((n-nc)/sigma)^2)*cos(2*pi*(n-nc)/lambda) ;
pim(n) = exp(-1.*((n-nc)/sigma)^2)*sin(2*pi*(n-nc)/lambda) ;
ptot = ptot + prl(n)^2 + pim(n)^2;
end
pnorm = sqrt(ptot);
%pnorm = 1;

% Normalize and check
ptot = 0.;
for n=1:NN
prl(n) = prl(n)/pnorm;
pim(n) = pim(n)/pnorm;
psi(n) = prl(n) + i*pim(n);
ptot = ptot + prl(n)^2 + pim(n)^2;
end
ptot

% FT the input waveform
PFIN = (1/sqrt(NN))*abs(fft(psi));

T = 0;
n_step = 1;
```

```
while n_step > 0
n_step = input('How many time steps  -->');

% -----------------------
for m=1:n_step
T = T + 1;

for n=2:NN-1
prl(n) = prl(n) - ra*(pim(n-1) -2*pim(n) + pim(n+1))  ...
        + (dt/hbar)*V(n)*pim(n);
end

for n=2:NN-1
pim(n) = pim(n) + ra*(prl(n-1) -2*prl(n) + prl(n+1))  ...
        - (dt/hbar)*V(n)*prl(n);
end

end
% -----------------------

% Check normalization
ptot1 = 0.;
for n=1:NC
ptot1 = ptot1 +  prl(n)^2 + pim(n)^2;
end
ptot1

ptot2 = 0.;
for n=NC+1:NN
ptot2 = ptot2 +  prl(n)^2 + pim(n)^2;
end
ptot2
if ptot2 < 0.001
    ptot2 = 0.;
end

ptot1 + ptot2

PE = 0.;
for n=1:NN
psi(n) = prl(n) + i*pim(n);
PE = PE + psi(n)*psi(n)'*V(n);
end
PE = PE*J2eV;
if PE < 0.001
    PE = 0.;
end

ke = 0. + j*0.;
```

```
for n=2:NN-1
lap_p = psi(n+1) - 2*psi(n) + psi(n-1);
ke = ke + lap_p*psi(n)';
end
KE = -J2eV*((hbar/del_x)^2/(2*melec))*real(ke);

subplot(3,1,1)
plot(XX,prl,'k')
hold on
plot(XX,pim,'--k')
plot(XX,J2eV*V,'k')
hold off
axis( [ 0 DX*NN -.2 .25 ])
%axis( [ 0 DX*NN -1 1 ])
xlabel('nm')
TT = text(3,.15,sprintf( ' T = %5.2f ps',T*dt*1e12));
set(TT,'fontsize',12)
%TT = text(.7*DX*NN,.15,['KE = ',num2str(KE),' eV']);
TT = text(.7*DX*NN,.15,sprintf('KE = %5.3f eV',KE));
set(TT,'fontsize',12)
 TT = text(.7*DX*NN,-.15,sprintf('Trans = %5.3f',ptot2));
 set(TT,'fontsize',12)
 TT = text(3,-.15,sprintf('Ref = %5.3f',ptot1));
 set(TT,'fontsize',12)
set(gca,'fontsize',12)
title('Se3-2')

% ----- Calculate the FT of the transmitted wave -----
psi_out = zeros(1,NN);
for n=NC+20:NN
psi_out(n) = prl(n) + i*pim(n);
end

PFOUT = (1/sqrt(NN))*abs(fft(psi_out));

subplot(3,2,3)
plot(EE,PFIN,'k--')
hold on
plot(EE,PFOUT,'k');
hold off
axis( [ 0 .4 0 .6 ])
set(gca,'XTick',[ .05 .2 .3 .4])
legend('In','Trans',1)
set(gca,'fontsize',12)
xlabel('eV')
ylabel('Psi(E)')

% -------- Calculate transmission ----------------
trans = zeros(1,NN);
```

```
for n=1:NN
if PFIN(n) > 0.1
   trans(n) = (PFOUT(n)/PFIN(n))^2;
end
end

subplot(3,2,4)
plot(EE,trans,'k')
axis( [ 0 .4 0 1 ])
set(gca,'fontsize',12)
TT = text(0.1,.8,'Transmission')
set(TT,'fontsize',12)
%ylabel('Transmission')
xlabel('eV')


T


end

% Se3_3.m.

clear all

NN = 1000;              % Number of points in the problem
space.
hbar = 1.054e-34;       % Plank's constant
melec = 9.1e-31;        % Mass of an electron
eV2J = 1.6e-19;         % Energy conversion factors
J2eV = 1/eV2J;

del_x = .2e-9;          % The cells size
dt = .5e-16;            % Time steps
ra = (0.5*hbar/melec)*(dt/del_x^2)  % ra must be < .1
DX = del_x*1e9;         % Cell size in nm.
XX = (DX:DX:DX*NN);     % Length in nm for plotting

MM = 2e6;               % Buffer for stored time steps
Time = zeros(1,MM);
del_t = 1e12*dt;        % Time in ps
Tt = (del_t:del_t:MM*del_t);

t0 = hbar^2/(2*melec*del_x^2);
nc = NN/2;

%Add the potential
V = zeros(1,NN);
for n=nc-26:nc-24
V(n) = .1*eV2J;
end
```

```
for n=nc+24:nc+26
V(n)  =  .1*eV2J;
end

% Initialized a sine wave in a gaussian envelope
sigma = 1000;             % Pulse width
lambda = 100;             % Pulse wavelength
prl = zeros(1,NN);
pim = zeros(1,NN);
ptot = 0.;
for n=nc-25:nc+25
pim(n)  =  exp(-1.*((n-nc)/sigma)^2)*cos(2*pi*(n-nc)/lambda) ;
%prl(n)  =  exp(-1.*((n-nc)/sigma)^2)*sin(2*pi*(n-nc)/lambda) ;
ptot = ptot + prl(n)^2 + pim(n)^2;
end
pnorm = sqrt(ptot);

% Normalize and check
ptot = 0.;
for n=1:NN
prl(n)  =  prl(n)/pnorm;
pim(n)  =  pim(n)/pnorm;
ptot = ptot +  prl(nc)^2 + pim(nc)^2;
end
ptot

T = 0;
n_step = 1;
while n_step > 0
n_step = input('How many time steps  -->');


% -----------------------
for m=1:n_step
T = T + 1;

for n=2:NN-1
prl(n)  =  prl(n)  -  ra*(pim(n-1) -2*pim(n) + pim(n+1))  ...
        + (dt/hbar)*V(n)*pim(n);
end

for n=2:NN-1
pim(n)  =  pim(n)  + ra*(prl(n-1) -2*prl(n) + prl(n+1))  ...
        - (dt/hbar)*V(n)*prl(n);
end

Time(T)  =  prl(nc) + i*pim(nc);

end
```

```
% -----------------------

% Check normalization
ptot = 0.;
for n=1:NN
ptot = ptot +  prl(n)^2 + pim(n)^2;
end
ptot

% Calculate the expected values
PE = 0.;
for n=1:NN
psi(n) = prl(n) + i*pim(n);
PE = PE + psi(n)*psi(n)'*V(n);
end
PE = PE*J2eV;

ke = 0. + j*0.;
for n=2:NN-1
lap_p = psi(n+1) - 2*psi(n) + psi(n-1);
ke = ke + lap_p*psi(n)';
end
KE = -J2eV*((hbar/del_x)^2/(2*melec))*real(ke);

subplot(2,1,1)
plot(XX,prl,'k')
hold on
plot(XX,pim,'--k')
plot(XX,J2eV*V,'k')
hold off
%axis( [ 0 DX*NN -.2 .25 ])
axis( [ 90 110 -.2 .25 ])
TT = text(5,.15,sprintf('%7.2f ps',T*dt*1e12))
set(TT,'fontsize',12)
TT = text(35,.15,sprintf('KE = %5.2f meV',1e3*KE))
set(TT,'fontsize',12)
YL = ylabel('Psi(x)')
xlabel('nm')
set(gca,'fontsize',12)
legend('real','imag')
title('Se3-3')

subplot(2,1,2)
plot(Tt,real(Time),'k')
hold on
plot(Tt,imag(Time),'k--')
hold off
YL = ylabel('Psi(t)')
axis([ 0 T*del_t  -.3 .3 ])
```

```
grid on
set(gca,'fontsize',12)
xlabel('ps')

T
end
```

D.4   CHAPTER 4

```
% Gr_eigen.m  Finds the eigenfunctions and
%             eigenenergies of a 1D potential.

clear all

% Use this for the 8-cell potentials
del_x = 1e-9;
NN = 8;

% Use this for the Harmonic oscillator
%del_x = .2e-9;
%NN = 50;

% Convert to nm for plotting
DX = del_x*1e9;
XX = (DX:DX:DX*NN);
mid = DX*NN/2;

hbar = 1.054e-34;
m0 = 9.11e-31;
ecoul = 1.6e-19;
eV2J = 1.6e-19;
J2eV = 1./eV2J;
hbar_eV = hbar*J2eV;

% Energies are J
chi0 = hbar^2/(2*m0*del_x^2)

% ---- Potential -----

V = zeros(1,NN);

V(1) = .1*eV2J;
V(2) = .05*eV2J;
V(NN-1) = .05*eV2J;
V(NN) = .1*eV2J;

% V-shapted potential
Eref = .3*eV2J;
```

```
for n=1:NN
%V(n) = (Eref/(NN/2))*abs(n-NN/2);
end

% Harmonic oscillator
Eref = 0.1*eV2J;
for n=1:NN
%V(n) = 0.5*m0*(Eref/hbar)^2*del_x^2*(n-NN/2)^2;
end

subplot(4,2,1)
plot(XX,J2eV*V,'k');
%TT = text( 4, 0.06, ['Potential' ])
ylabel('V (eV)')
xlabel('nm')
title('Gr-eigen')
Vmax = max(J2eV*V);
%Vmax = .1;
axis( [ DX DX*NN -.01 1.2*Vmax ])
grid on
set(gca,'fontsize',12)

% ---- Create the H matrix ---

H = zeros(NN,NN);

H(1,1) = 2*chi0+V(1);
H(1,2) = -1*chi0;
for n=2:NN-1
H(n,n-1)= -1*chi0;
H(n,n) = 2*chi0+ V(n);
H(n,n+1)= -1*chi0;
end
H(NN,NN-1) = -1*chi0;
H(NN,NN) = 2*chi0+V(NN);
% These two lines add the PBC
% H(1,NN) = -1*chi0
% H(NN,1) = -1*chi0

% This calculates the eigenfunction and
% eigenenergies
[phi,D] = eig(H);

% Write the eigevalues in meV.
for m = 1:NN
E(m) = 1e3*J2eV*D(m,m);
end

%Plot the eigen energies
```

```
subplot(4,2,2)
plot(E,'ok')
axis( [ 1 8 0 1000 ])
title('Eigenenergies')
ylabel('E (meV)')
xlabel('Eigenvalue #')
grid on
set(gca,'fontsize',12)

%Plot the eigenfunctions
amax = .6
subplot(4,2,3)
plot(XX,phi(:,1),'k')
axis( [ DX DX*NN -amax amax ])
TT = text( mid,.2,sprintf('%4.1f meV',E(1)));
set(TT,'fontsize',12)
TT = ylabel('f_1','FontName','Symbol','fontsize',12);
set(gca,'fontsize',12)

subplot(4,2,5)
plot(XX,phi(:,2),'k')
axis( [ DX DX*NN -amax amax ])
TT = text( mid,.2,sprintf('%4.1f meV',E(2)));
set(TT,'fontsize',12)
TT = ylabel('f_2','FontName','Symbol','fontsize',12);
xlabel('nm')
set(gca,'fontsize',12)

subplot(4,2,4)
plot(XX,phi(:,4),'k')
axis( [ DX DX*NN -amax amax ])
TT = text( mid,.2,sprintf('%4.1f meV',E(3)));
set(TT,'fontsize',12)
TT = ylabel('f_3','FontName','Symbol','fontsize',12);
set(gca,'fontsize',12)

subplot(4,2,6)
plot(XX,phi(:,5),'k')
axis( [ DX DX*NN -amax amax ])
TT = text( mid,.2,sprintf('%4.1f meV',E(4)));
set(TT,'fontsize',12)
TT = ylabel('f_4','FontName','Symbol','fontsize',12);
xlabel('nm')
set(gca,'fontsize',12)
```

D.5   CHAPTER 5

```
% Gr_n.m  Electron density

clear all

NN = 100
hbar = 1.054e-34;
m0 = 9.11e-31;
melec = m0;
ecoul = 1.6e-19;
eV2J = 1.6e-19;
J2eV = 1./eV2J;
hbar_eV = hbar*J2eV

del_x = .1e-9;
DX = del_x*1e9;
XX = (DX:DX:DX*NN)

% Energies are J
chi0 = hbar^2/(2*melec*del_x^2)

% ---- Potential -----
V = zeros(1,NN);

% V potential
Eref = .3*eV2J;
for n=1:NN
%V(n) = (Eref/(NN/2))*abs(n-NN/2);
end

subplot(4,2,1)
plot(XX,J2eV*V);
%TT = text( 4, 0.06, ['Potential' ])
ylabel('V (eV)')
xlabel('nm')
title('Gr-eigen')
Vmax = max(J2eV*V);
Vmax = .1;
axis( [ DX DX*NN 0 Vmax ])

% --- Calculate the eigenfunctions

% Create the H matrix

H = zeros(NN,NN);

H(1,1) = 2*chi0+V(1);
H(1,2) = -1*chi0;
```

```
for n=2:NN-1
H(n,n-1)= -1*chi0;
H(n,n)  = 2*chi0+ V(n);
H(n,n+1)= -1*chi0;
end
H(NN,NN-1) = -1*chi0;
H(NN,NN)   = 2*chi0+V(NN);
% These two lines add the PBC
%H(1,NN) = -1*chi0
%H(NN,1) = -1*chi0

% This calculates the eigenfunction and
%   eigenenergies
[phi,D] = eig(H);

% Write the eigenenergies in meV.
for m = 1:NN
E(m) = 1e3*J2eV*D(m,m);
end

%Plot the eigen energies
subplot(4,2,2)
plot(E,'o')
axis( [ 1 8 0 200 ])
title('Eigenenergies')
ylabel('E (meV)')
xlabel('Eigenvalue #')

% -------- Calculate the particle density ----------------

kB = 8.62e-5; % Boltzmann in eV
T = 00;       % Temperature in K

EF = 0.01;    % Fermi energy in eV

F = zeros(1,NN);
rho = zeros(1,NN);
ndens = zeros(1,NN);
n_elec = 0.;
for n=1:8
F(n) = 1./(1. + exp( (1e-3*E(n) - EF)/(kB*T)));
n_elec = n_elec + 2*F(n);
for k=1:NN
rho(k) = rho(k) + 2*F(n)*phi(k,n)*phi(k,n)';
ndens(k) = ndens(k) + 2*F(n)*phi(k,n)*phi(k,n)';
end
end
n_elec
```

```
subplot(3,2,5)
plot(1e-3*E,F,'k--o')
axis( [ 0 .1 0 1.1 ])
TT = text( 0.05,0.8,sprintf('E_f = %6.3f',EF));
set(TT,'fontsize',12)
TT = text( 0.05,0.6,sprintf('T = %3d K',T));
set(TT,'fontsize',12)
TT = text( 0.05,0.3,sprintf('n_e = %5.2f',n_elec));
set(TT,'fontsize',12)
ylabel('Prob. of occup.')
xlabel('E (eV)')
set(gca,'fontsize',12)
title('Gr-n')

subplot(3,2,6)
plot(XX,ndens,'k')
TT = ylabel('n(x)')
%set(TT,'FontName','symbol','Fontsize',12)
set(gca,'fontsize',12)
xlabel('nm')
```

D.6   CHAPTER 6

```
%Se6_1.m; Simulation in a lattice
% Periodic Boundary conditions are imposed.

clear all

NN = 500;
hbar = 1.054e-34;
hplank = 6.625e-34;
melec = 9.1e-31;
eV2J = 1.6e-19;
J2eV = 1/eV2J;

del_x = 0.1e-9;        % The cells are 1 A
dt = 0.02e-15;         % Time steps are 1 fs.
ra = (0.5*hbar/melec)*(dt/del_x^2);
DX = 1e10*del_x;       % Cell size in Angstroms
XX = (DX:DX:NN*DX);    % Length in Angstroms for plotting.

% --- Specify a periodic potential
V=zeros(1,NN);

nspace = 20;           % Spacing of the spikes for a periodic
lattice.
V0 = -.1*eV2J;         % Magnitude of the spikes
```

```
for n=10/nspace:NN/nspace;
V(nspace*n ) = V0;
end

FTK =fft(J2eV*V);              % FFT of the lattice

del_K = (1/NN);               % Smallest wavelengh
KK =(0:del_K:del_K*(NN-1)); % Spacial frequency

% ----- Initialized a sine wave in a gaussian envelope ----
sigma = 35;            % Pulse width (15 or 50)
lambda = 20;           % Wavelenth
nc = 100           % Starting position of the pulse
prl = zeros(1,NN);
pim = zeros(1,NN);
ptot = 0.;
for n=2:NN-1
prl(n) = exp(-1.*((n-nc)/sigma)^2)*cos(2*pi*(n-nc)/lambda) ;
pim(n) = exp(-1.*((n-nc)/sigma)^2)*sin(2*pi*(n-nc)/lambda) ;
ptot = ptot + prl(n)^2 + pim(n)^2;
end
pnorm = sqrt(ptot);

% Normalize and check
ptot = 0.;
for n=1:NN
prl(n) = prl(n)/pnorm;
pim(n) = pim(n)/pnorm;
ptot = ptot + prl(n)^2 + pim(n)^2;
end
ptot

T = 0;
n_step = 1;
while n_step > 0
n_step = input('How many time steps -->');

% --------This is the core FDTD code ----------------
for m=1:n_step
T = T + 1;

% Part of the PBC
prl(1) = prl(1) - ra*(pim(NN-1) -2*pim(1) + pim(2)) ...
      + (dt/hbar)*V(1)*pim(1);
prl(NN) = prl(1);

for n=2:NN-1
prl(n) = prl(n) - ra*(pim(n-1) -2*pim(n) + pim(n+1)) ...
      + (dt/hbar)*V(n)*pim(n);
end
```

```
% Part of the PBC
pim(1) = pim(1) + ra*(prl(NN-1) -2*prl(1) + prl(2)) ...
       - (dt/hbar)*V(1)*prl(1);
pim(NN) = pim(1);

for n=2:NN-1
pim(n) = pim(n) + ra*(prl(n-1) -2*prl(n) + prl(n+1)) ...
       - (dt/hbar)*V(n)*prl(n);
end

end

% -----------------------

% Calculate the expected values

PE = 0.;
for n=1:NN
psi(n) = prl(n) + i*pim(n);
PE = PE + psi(n)*psi(n)'*V(n);
end
psi*psi'             % Check normalization
PE = PE*J2eV

ke = 0. + j*0.;
for n=2:NN-1
lap_p = psi(n+1) - 2*psi(n) + psi(n-1);
ke = ke + lap_p*psi(n)';
end
KE = -((hbar/del_x)^2/(2*melec))*real(ke);
KE = J2eV*KE;

% Calculate the effective mass
m_eff = (.5/KE)*(hplank/(lambda*del_x))^2/melec;

% Display psi on the lattice

subplot(2,1,1)
plot(XX,prl,'k')
hold on
plot(XX,J2eV*V,'k')
%plot(XX,pim,'-.k')
hold off
axis( [ 0 DX*NN -.2 .2 ])
TT = text(50,.15,sprintf('%4.0f fs',T*dt*1e15));
set(TT,'fontsize',14)
TT = text(300,.15,sprintf('KE = %5.3f eV',KE));
set(TT,'fontsize',14)
```

```
TT = text(100,-.15,sprintf('lambda = %3.0f A',lambda));
set(TT,'fontsize',14)
TT = text(300,-.15,sprintf('spacing = %3.0f A',nspace));
set(TT,'fontsize',14)
set(gca,'fontsize',14)
xlabel('A')
TT = ylabel('y');
set(TT,'FontName','symbol')
title('Se6-1')

% Plot the spatial Fourier transform of Psi.

subplot(4,2,7)
plot(KK,(1/sqrt(NN))*abs(fft(psi)),'k');
hold on
plot(.5*KK,.05*abs(FTK),'k')
hold off
axis( [ 0 .15 0 .5 ])
set(gca,'XTick',[0 .05 .1 .15 ])
title('Positive wavelengths')
TT = ylabel('Y','FontName','symbol');
set(gca,'fontsize',14)
TT = text(.05,-.27,'1/1','FontName','symbol','fontsize',14);
TT = text(.07,-.25,'(A^-^1)','fontsize',14);
grid on

subplot(4,2,8)
plot(KK,(1/sqrt(NN))*abs(fft(psi)),'k');
hold on
hold off
set(gca,'fontsize',14)
axis( [ .85 1 0 .5 ])
title('Negative wavelengths')
set(gca,'XTick',[ .85 .9 .95 1])
TT = text(.9,-.27,'1/1','FontName','symbol','fontsize',14);
TT = text(.92,-.25,'(A^-^1)','fontsize',14);
grid on

end


%   This program was developed to illustrate the
modes.  Although it is
%      an FDTD program, running this program with low
dimensions like
%      NN = 25 will lead to errors.

% Se2d_modes.m.
```

```
clear all

NN = 25;                % Number of points in the problem
space.
MM = 120;
hbar = 1.054e-34;       % Plank's constant
melec = 9.11e-31;        % Mass of an electron
eV2J = 1.6e-19;         % Energy conversion factors
J2eV = 1/eV2J;

del_x = 4e-10;          % The cells size
dt = .25e-15;           % Time steps
ra = (0.5*hbar/melec)*(dt/del_x^2) % ra must be < .1
DX = del_x*1e9;         % Cell size in nm.
XX = (DX:DX:DX*NN);     % Length in nm for plotting
YY = (DX:DX:DX*MM);     % Length in nm for plotting

NC = NN/2;              % Starting position of the pulse
MC = MM/2;

V = zeros(NN,MM)

%---- Input -----

nmode = input('X mode -->')
mmode = input('y mode -->')

prl = zeros(NN,MM);
pim = zeros(NN,MM);
ptot = 0.;
for n=2:NN-1
for m=2:MM-1
  prl(n,m) = sin(nmode*pi*n/NN)*sin(mmode*pi*m/MM);
  pim(n,m) = 0.;
  ptot = ptot + pim(n,m)^2 + prl(n,m)^2;
end
end
ptot

ptot1 = 0.;
for n=1:NN
for m=1:MM
prl(n,m) = prl(n,m)/sqrt(ptot);
pim(n,m) = pim(n,m)/sqrt(ptot);
  ptot1 = ptot1 + prl(n,m)^2 + pim(n,m)^2;
end
end
ptot1
```

```
subplot(2,2,1)
mesh(YY,XX,prl)
view(-120,30)
axis( [ 0 DX*MM 0 DX*NN -.03 .03 ])

subplot(2,2,2)
mesh(YY,XX,pim)
view(-120,30)
axis( [ 0 DX*MM 0 DX*NN -.03 .03 ])

T = 0;
n_step = 1;
while n_step > 0
n_step = input('How many time steps  -->');

% -----------This is the core FDTD program ------------
for iT=1:n_step

T = T + 1;

for m=2:MM-1
for n=2:NN-1
prl(n,m) = prl(n,m) - ra*(-4*pim(n,m) + pim(n,m-1) +
pim(n,m+1)  ...
                                    + pim(n-1,m) +
pim(n+1,m) ) ...
        + (dt/hbar)*V(n,m)*pim(n,m);
end
end

for m=2:MM-1
for n=2:NN-1
pim(n,m) = pim(n,m) + ra*(-4*prl(n,m) + prl(n,m-1) +
prl(n,m+1)  ...
                                    + prl(n-1,m) +
prl(n+1,m) )...
        - (dt/hbar)*V(n,m)*prl(n,m);
end
end

end
% ------------------------

% Check normalization
ptot = 0,;
for m=1:MM
for n=1:NN
   psi(n,m) = prl(n,m) + i*pim(n,m);
   ptot = ptot + psi(n,m)*psi(n,m)';
```

```
end
end
ptot

% Calculate the expected values

ke = 0. + j*0.;
for m=3:MM-2
for n=3:NN-2
lap_p = psi(n+1,m) - 4*psi(n,m) + psi(n-1,m) ...
      + psi(n,m-1) + psi(n,m+1);
ke = ke + lap_p*psi(n,m)';
end
end
KE = -J2eV*((hbar/del_x)^2/(2*melec))*real(ke);

subplot(2,2,1)
mesh(YY,XX,prl)
view(-120,30)
axis( [ 0 DX*MM 0 DX*NN -.03 .03 ])
TT = text(DX*MM,DX*NN,.035,sprintf('   KE = %5.2f
meV',1e3*KE));
set(TT,'fontsize',14)
set(gca,'fontsize',14)
xlabel('Y (nm)')
ylabel('X (nm)')
title('Se2d-modes')
%grid on

subplot(2,2,2)
mesh(YY,XX,pim)
view(-120,30)
axis( [ 0 DX*MM 0 DX*NN -.03 .03 ])

end


% DOS_2d.m; Display the density of states in 2d infinite
well

clear all

E0 = 3.75; % in meV

del_E = .01;

NN = 1000;
N = 1e4;
Epoints = zeros(1,N);
```

```
EE = (del_E:del_E:del_E*(N));

for m=1:3              % X direction
for mm=1:5             % Y direction (5 or 15)
i = 375*m^2 + 15*mm^2 ; % Location of allowed states
Epoints(i) = 1.;
end
end

Emax = 40;
subplot(3,1,1)
bar(EE,Epoints,'k')
axis([ 0 Emax 0 1.1 ])
set(gca,'fontsize',12)
xlabel('E (meV)')
title('DOS-2d')


% Se2d_prop.m.   Two dimensional propagationg using different
modes.

clear all

NN = 100;              % Number of points in the problem space.
MM = 100;
hbar = 1.054e-34;      % Plank's constant
melec = 9.11e-31;       % Mass of an electron
eV2J = 1.6e-19;        % Energy conversion factors
J2eV = 1/eV2J;

del_x = .4e-9;         % The cells size
dt = .25e-15;           % Time steps
ra = (0.5*hbar/melec)*(dt/del_x^2) % ra must be < .1
DX = del_x*1e9;        % Cell size in nm.
XX = (DX:DX:DX*NN);    % Length in nm for plotting
YY = (DX:DX:DX*MM);    % Length in nm for plotting

NC = NN/2;             % Starting position of the pulse
MC = NN/2;

V = zeros(NN,MM);

%---- Initialize a pulse propagating in the X direction -----

lambda = input('lambda X -->')
sigma = input('sigma X -->')
mmode = input('y mode -->')

pr1 = zeros(NN,MM);
```

```
pim = zeros(NN,MM);
ptot = 0.;
nc = NN/2;
for n=2:NN-1
for m=2:MM-1
   prl(n,m)  =  exp(-1.*((n-nc)/sigma)^2)  ...
               *cos(2*pi*(n-nc)/lambda)*sin(mmode*pi*m/MM);
   pim(n,m)  =  exp(-1.*((n-nc)/sigma)^2)  ...
               *sin(2*pi*(n-nc)/lambda)*sin(mmode*pi*m/MM);
   ptot = ptot + pim(n,m)^2 + prl(n,m)^2;
end
end
ptot

ptot1 = 0.;
for n=1:NN
for m=1:MM
prl(n,m) = prl(n,m)/sqrt(ptot);
pim(n,m) = pim(n,m)/sqrt(ptot);
   ptot1 = ptot1 + prl(n,m)^2 + pim(n,m)^2;
end
end
ptot1

T = 0;
n_step = 1;
while n_step > 0
n_step = input('How many time steps -->');

% -----------This is the core FDTD program -------------
for iT=1:n_step

T = T + 1;

for m=2:MM-1
for n=2:NN-1
prl(n,m) = prl(n,m) - ra*(-4*pim(n,m)+pim(n,m-1)+
pim(n,m+1)  ...
         +                      pim(n-1,m) + pim(n+1,m) ) ...
         + (dt/hbar)*V(n,m)*pim(n,m);
end
end

for m=2:MM-1
for n=2:NN-1
pim(n,m) = pim(n,m)  +
ra*(-4*prl(n,m)+prl(n,m-1)+prl(n,m+1)  ...
         +                      prl(n-1,m)  + prl(n+1,m) )...
         - (dt/hbar)*V(n,m)*prl(n,m);
```

```
end
end

end
% -----------------------

% Check normalization
ptot = 0.;
xpos = 0.
for m=1:MM
for n=1:NN
  psi(n,m) = prl(n,m) + i*pim(n,m);
  ptot = ptot + prl(n,m)^2 + pim(n,m)^2;
  xpos = xpos + n*(prl(n,m)^2 + pim(n,m)^2);
end
end
ptot
xpos

% Calculate the expected values

ke = 0. + j*0.;
for m=3:MM-2
for n=3:NN-2
lap_p = psi(n+1,m) - 4*psi(n,m) + psi(n-1,m) ...
      + psi(n,m-1) + psi(n,m+1);
ke = ke + lap_p*psi(n,m)';
end
end
KE = -J2eV*((hbar/del_x)^2/(2*melec))*real(ke);

subplot(2,1,1)
mesh(YY,XX,prl)
view(70,30)
axis( [ 0 DX*MM 0 DX*NN -.05 .05 ])
TT = text(0,0,.05,sprintf('   KE = %5.2f meV',1e3*KE));
set(TT,'fontsize',14)
set(gca,'fontsize',14)
xlabel('Y (nm)')
ylabel('X (nm)')
title('Se2d-prop')

end
```

D.7 CHAPTER 7

```
% Se7_1.m. Simulation of spin
% Simulate all three B fields
```

```
clear all

NN = 200;               % Number of points in the problem
space.
hbar = 1.054e-34;       % Plank's constant
melec = 9.1e-31;        % Mass of an electron
eV2J = 1.6e-19;         % Energy conversion factors
J2eV = 1/eV2J;

del_x = .05e-9;         % The cells size
dt = .5e-17;            % Time steps
ra = (0.5*hbar/melec)*(dt/del_x^2) % ra must be < .1
DX = del_x*1e9;         % Cell size in nm.
XX = (DX:DX:DX*NN);     % Length in nm for plotting
gamma = 1.6e-19/melec;% The gyromagnetic ratio

V = zeros(1,NN);        % Initialize the potential to zero

% Read in the spin orientation

theta = input('theta -->');
theta = (pi/180)*theta
phi = input('phi -->');
phi = (pi/180)*phi

% ---- Initialize the Particle --------------

% Initialized a sine wave in a gaussian envelope
lambda = 200;
sigma = 20;         % Pulse width
prl1 = zeros(1,NN);
pim1 = zeros(1,NN);
prl2 = zeros(1,NN);
pimd = zeros(1,NN);
ptot = 0;
for n=1:NN
prl1(n) = cos(theta/2)*sin(pi*n/NN) ;
pim1(n) = 0;
prl2(n) = sin(theta/2)*cos(phi)*sin(pi*n/NN) ;
pim2(n) = sin(theta/2)*sin(phi)*sin(pi*n/NN) ;
ptot = ptot + prl1(n)^2 + pim1(n)^2 + prl2(n)^2 + pim2(n)^2;
end
pnorm = sqrt(ptot);

% Normalize and check
ptot1 = 0.;
ptot2 = 0.;
for n=1:NN
```

```
prl1(n) = prl1(n)/pnorm;
pim1(n) = pim1(n)/pnorm;
prl2(n) = prl2(n)/pnorm;
pim2(n) = pim2(n)/pnorm;
ptot1 = ptot1 + prl1(n)^2 + pim1(n)^2;
ptot2 = ptot2 + prl2(n)^2 + pim2(n)^2;
end
ptot = ptot1 + ptot2

% ---- Set the mangentic fields ----
Bx = 0;
By = 0;
Bz = 50.;

Time_z = (2*pi)/(gamma*Bz)   % Pecession time period.
Time_z/dt                    % Number of dt for one time period

T = 0;
n_step = 1;
while n_step > 0
n_step = input('How many time steps -->');

% -----------This is the core FDTD program -------------
for m=1:n_step
T = T + 1;

% Spin up ----

for n=2:NN-1
prl1(n) = prl1(n) - ra*(pim1(n-1) -2*pim1(n) + pim1(n+1)) ...
          + (dt/hbar)*V(n)*pim1(n) ...
   + 0.5*gamma*dt*(-Bx*pim2(n) + By*prl2(n) - Bz*pim1(n) );
end

for n=2:NN-1
pim1(n) = pim1(n) + ra*(prl1(n-1) -2*prl1(n) + prl1(n+1)) ...
          - (dt/hbar)*V(n)*prl1(n) ...
   + 0.5*gamma*dt*(Bx*prl2(n) + By*pim2(n) + Bz*prl1(n) );
end

% Spin down ----

for n=2:NN-1
prl2(n) = prl2(n) - ra*(pim2(n-1) -2*pim2(n) + pim2(n+1)) ...
          + (dt/hbar)*V(n)*pim2(n) ...
   + 0.5*gamma*dt*(-Bx*pim1(n) - By*prl1(n) + Bz*pim2(n) );
end

for n=2:NN-1
```

```
pim2(n) = pim2(n) + ra*(prl2(n-1) -2*prl2(n) + prl2(n+1)) ...
        - (dt/hbar)*V(n)*prl2(n) ...
    + 0.5*gamma*dt*(Bx*prl1(n)  - By*pim1(n)  - Bz*prl2(n) );
end

end
% ----------------------

% --- Check normalization and calculate S ---
psi1 = prl1 + i*pim1;
psi2 = prl2 + i*pim2;
psi1*psi1' + psi2*psi2'
Sx = psi2*psi1' + psi1*psi2';
Sy = i*(-psi2*psi1' + psi1*psi2');
Sz = ptot1 - ptot2;
Sz = psi1*psi1' - psi2*psi2';

% Calculate the expected values

psi = psi1;
ke = 0. + j*0.;
for n=2:NN-1
lap_p = psi(n+1)  - 2*psi(n) + psi(n-1);
ke = ke + lap_p*psi(n)';
end
KE1 = -J2eV*((hbar/del_x)^2/(2*melec))*real(ke);

psi = psi2;
ke = 0. + j*0.;
for n=2:NN-1
lap_p = psi(n+1)  - 2*psi(n) + psi(n-1);
ke = ke + lap_p*psi(n)';
end
KE2 = -J2eV*((hbar/del_x)^2/(2*melec))*real(ke);

subplot(3,2,1)
plot(XX,prl1,'k')
hold on
plot(XX,pim1,'-.k')
plot(XX,J2eV*V,'--k')
hold off
axis( [ 0 DX*NN -.15 .15 ])
TT = text(.5,.11,sprintf('%7.0f fs',T*dt*1e15));
set(TT,'fontsize',12)
set(gca,'fontsize',12)
 TT = text(1,-.12,sprintf('B: %4.0f %4.0f %4.0f',Bx,By,Bz));
 set(TT,'fontsize',12)
ylabel('Spin Up')
title('Se7-1')
```

```
subplot(3,2,3)
plot(XX,prl2,'k')
hold on
plot(XX,pim2,'-.k')
plot(XX,J2eV*V,'--k')
hold off
axis( [ 0 DX*NN -.15 .15 ])
TT = text(1,-.12,'<s >','FontName','Symbol')
set(TT,'fontsize',12)
TT = text(2.5,-.12,sprintf(':    %4.2f    %4.2f
set(TT,'fontsize',12)
ylabel('Spin Down')
xlabel('nm')
set(gca,'fontsize',12)

T

end

% Se2d_B.m.    2D simulation of a particle in a magnetic
field.

clear all

NN = 120;               % Number of points in the problem space.
MM = 120;
hbar = 1.054e-34;       % Plank's constant
melec = 9.1e-31;        % Mass of an electron
ecoul = 1.6e-19;        % Charge of an electron
eV2J = 1.6e-19;         % Energy conversion factors
J2eV = 1/eV2J;

del_x = 10e-10;         % The cells size
dt = 2e-15;             % Time steps
ra = (0.5*hbar/melec)*(dt/del_x^2)  % ra must be < .1
DX = del_x*1e9;         % Cell size in nm.
XX = (DX:DX:DX*NN);     % Length in nm for plotting
YY = (DX:DX:DX*MM);     % Length in nm for plotting

NC = NN/2;              % Center of the problem space.
MC = NN/2;

% ---- Potential from the magnetic field -------

VB = zeros(NN,MM);

Bz = input('Bz (Tesla) -->')
```

```
V0 = (1/8)*ecoul*(del_x^2/melec)*Bz^2;
% Note: one 'ecoul' term appears here and the other in the
SE.

for n=1:NN
for m=1:MM
radius = (n-NC)^2 + (m-MC)^2;
VB(n,m) = radius*V0;
end
end

Vmax = max(max(VB));
subplot(1,1,1)
mesh(YY,XX,VB)
TT = text(0,30.,1.0*Vmax,sprintf(' B_z = %3.0f T',Bz));
set(TT,'fontsize',16)
set(gca,'fontsize',16)
%xlabel('Y (nm)')
TT = text(80,-60.,0.0, ['Y (nm)']);
set(TT,'fontsize',16)
ylabel('X (nm)')
axis( [ 0 120 0 120 0 Vmax])
set(gca,'XTick',[0,60,120])
set(gca,'YTick',[0,60,120])
view(60,20)
zlabel('V_B (eV)')
title('Se2d-B')

%---- Input -----

ptot = 0.;
lambda = 15;
sigma = 10;
Mcent = MC + 20;            % Starting position of psi.
Ncent = NC;
for n=1:NN
for m=1:MM
aaa = exp(-0.5*((n-Ncent)/sigma)^2) *exp(-0.5*((m-Mcent)/
sigma)^2);
   prl(n,m) = aaa*cos(2*pi*(n-Ncent)/lambda);
   pim(n,m) = aaa*sin(2*pi*(n-Ncent)/lambda);
   ptot = ptot + pim(n,m)^2 + prl(n,m)^2;
end
end
ptot

ptot1 = 0.;
for n=1:NN
for m=1:MM
```

```
prl(n,m) = prl(n,m)/sqrt(ptot);
pim(n,m) = pim(n,m)/sqrt(ptot);
   ptot1 = ptot1 + prl(n,m)^2 + pim(n,m)^2;
end
end
ptot1

T = 0;
n_step = 1;
while n_step > 0
n_step = input('How many time steps -->');

% -----------This is the core FDTD program -------------
for iT=1:n_step

T = T + 1;

for m=2:MM-1
for n=2:NN-1
prl(n,m) = prl(n,m) - ra*(-4*pim(n,m) + pim(n,m-1)+
pim(n,m+1)  ...
          +                            pim(n-1,m) +
pim(n+1,m) ) ...
        + (dt/hbar)*ecoul*VB(n,m)*pim(n,m);
end
end

for m=2:MM-1
for n=2:NN-1
pim(n,m) = pim(n,m) + ra*(-4*prl(n,m) + prl(n,m-1) +
prl(n,m+1)  ...
          +                            prl(n-1,m) +
prl(n+1,m) )...
        - (dt/hbar)*ecoul*VB(n,m)*prl(n,m);
end
end

end
% -----------------------

% Check normalization
ptot = 0.;
for m=1:MM
for n=1:NN
   psi(n,m) = prl(n,m) + i*pim(n,m);
   ptot = ptot + prl(n,m)^2 + pim(n,m)^2;
end
end
ptot
```

T

```
% Calculate the expected values

ke = 0. + j*0.;
pe = 0.;
for m=2:MM-1
for n=2:NN-1
lap_p = psi(n+1,m) - 4*psi(n,m) + psi(n-1,m) ...
      + psi(n,m-1) + psi(n,m+1);
ke = ke + lap_p*psi(n,m)';
pe = pe + VB(n,m)*psi(n,m)*psi(n,m)';
end
end
KE = -J2eV*((hbar/del_x)^2/(2*melec))*real(ke);

subplot(1,1,1)
mesh(YY,XX,prl)
view(50,30)
axis( [ 0 DX*MM 0 DX*NN -.03 .06 ])
 TT = text(100,80,.035,sprintf('T = %7.2f ps',T*dt*1e12));
 set(TT,'fontsize',16)
TT = text(30,0,.05,sprintf('   KE = %5.2f meV',1e3*KE));
set(TT,'fontsize',16)
TT = text(30,0,.035,sprintf('   PE = %5.2f meV',1e3*pe));
set(TT,'fontsize',16)
set(gca,'fontsize',16)
set(gca,'XTick',[0,60,120])
set(gca,'YTick',[0,60,120])
xlabel('Y (nm)')
ylabel('X (nm)')
%grid on

end


% Se7_2.m. Two particles in an infinite well.
% The Coulomb interaction is gradually added.

clear all

NN = 50;            % Number of points in the problem space.
hbar = 1.054e-34;   % Plank's constant
melec = 9.1e-31;    % Mass of an electron
epsz = (1/36*pi)*1e-9;
ecoul = 1.6e-19;     %
eV2J = 1.6e-19;     % Energy conversion factors
J2eV = 1/eV2J;
```

```
del_x = .2e-9;          % The cells size
dt = 8.e-17;            % Time steps
ra = (0.5*hbar/melec)*(dt/del_x^2) % ra must be < .1
DX = del_x*1e9;         % Cell size in nm.
XX = (DX:DX:DX*NN);     % Length in nm for plotting
X2 = (DX:DX:DX*2*NN);    % Length in nm for plotting

ps1 = zeros(1,NN);
YF = zeros(1,NN);
y = zeros(1,NN);

% Potential
V=zeros(1,NN);

% Create the Coulomb potential
nc = 1;
Kcoul = (ecoul^2/(4*pi*epsz*del_x))
Vcoul = zeros(1,2*NN)
Vcoul(1) = Kcoul
for n=2:NN+1
   Vcoul(n) = Kcoul*(1./abs(n-nc));
end
for n=NN+2:2*NN
   Vcoul(n) = Kcoul*(1./abs(2*NN+1-n));
end

% Create the FFT of Vcoul
VF = fft(Vcoul);

% ---- Initialize Particle one --------------

lambda = 100;
sigma = 500000;          % Pulse width
nc = NN/2 - 30;          % Starting position of the pulse
nc = 0.;
prl1 = zeros(1,NN);
pim1 = zeros(1,NN);
ptot = 0.;
for n=2:NN
prl1(n) = exp(-1.*((n-nc)/sigma)^2)*sin(2*pi*(n)/lambda) ;
ptot = ptot + prl1(n)^2 + pim1(n)^2;
end
pnorm = sqrt(ptot);

% Normalize and check
ptot = 0.;
for n=1:NN
prl1(n) = prl1(n)/pnorm;
pim1(n) = pim1(n)/pnorm;
```

```
end
ptot

% ---- Initialize Particle two ----

lambda = 50;
sigma = 50000          % Pulse width
nc = NN/2 + 30;            % Starting position of the pulse
nc = 0;
prl2 = zeros(1,NN);
pim2 = zeros(1,NN);
ptot = 0.;
for n=2:NN-1
prl2(n) = exp(-1.*((n-nc)/sigma)^2)*sin(2*pi*(n)/lambda) ;
ptot = ptot + prl2(n)^2 + pim2(n)^2;
end
pnorm = sqrt(ptot);

% Normalize and check
ptot = 0.;
for n=1:NN
prl2(n) = prl2(n)/pnorm;
pim2(n) = pim2(n)/pnorm;
ptot = ptot + prl2(n)^2 + pim2(n)^2;
end
ptot

% ---- End Initialize ----

T = 0;
n_step = 1;
while n_step > 0
n_step = input('How many time steps -->');

% -----------This is the core FDTD program -------------

for m=1:n_step
T = T + 1;

for n=2:NN-1
prl1(n) = prl1(n) - ra*(pim1(n-1) -2*pim1(n) + pim1(n+1))
...
        + (dt/hbar)*(V(n) + imag(y(n)))*pim1(n);
prl2(n) = prl2(n) - ra*(pim2(n-1) -2*pim2(n) + pim2(n+1))
...
        + (dt/hbar)*(V(n) + real(y(n)))*pim2(n);
end

for n=2:NN-1
```

```
pim1(n) = pim1(n) + ra*(prl1(n-1) -2*prl1(n) + prl1(n+1))
...
          - (dt/hbar)*(V(n) + imag(y(n)))*prl1(n);
pim2(n) = pim2(n) + ra*(prl2(n-1) -2*prl2(n) + prl2(n+1))
...
          - (dt/hbar)*(V(n) + real(y(n)))*prl2(n);
end

for n=1:NN
psi1(n) = prl1(n) + i*pim1(n);
psi2(n) = prl2(n) + i*pim2(n);
end

% psi1 goes in the real part and psi2 in the imag. part.
ps = zeros(1,2*NN);
for n=1:NN
ps(n) = prl1(n)^2 + pim1(n)^2 + i*(prl2(n)^2 + pim2(n)^2);
end
PF = fft(ps);

for n=1:2*NN
YF(n) = PF(n)*VF(n);
end

% This is the Hartee term.
rmax = 1e4;
  ramp = 0.5*(1 - cos(pi*T/rmax));
if T >= rmax
  ramp = 1;
end
y = ramp*ifft(YF);

end

% --- End FDTD -----

% Parameters for particle 1

PE = 0.;
nc = NN/2;
for n=1:NN
psi1(n) = prl1(n) + i*pim1(n);
end

ke = 0. + j*0.;
for n=2:NN-1
lap_p = psi1(n+1) - 2*psi1(n) + psi1(n-1);
ke = ke + lap_p*psi1(n)';
end
```

```
KE = -J2eV*((hbar/del_x)^2/(2*melec))*real(ke);
KE1 = KE

% Plot Particle 1
subplot(3,2,1)
plot(XX,prl1,'k')
hold on
plot(XX,pim1,'-.k')
plot(XX,J2eV*V,'--k')
hold off
axis( [ 0 DX*NN -.3 .3 ])
TT = text(1,.2,sprintf('KE = %5.2f meV',1e3*KE));
set(TT,'fontsize',12)
%xlabel('nm')
ylabel('Partilce 1')
set(gca,'fontsize',12)
title('Se7-2')
%grid on

% Parameters for particle 2

PE = 0.;
for n=1:NN
psi2(n) = prl2(n) + i*pim2(n);
PE = PE + psi2(n)*psi2(n)'*V(n);
end
PE = PE*J2eV;

ke = 0. + j*0.;
for n=2:NN-1
lap_p = psi2(n+1) - 2*psi2(n) + psi2(n-1);
ke = ke + lap_p*psi2(n)';
end
KE = -J2eV*((hbar/del_x)^2/(2*melec))*real(ke);
KE2 = KE;

% FFT of psi1
%PS2 = fft(psi1)

subplot(3,2,3)
plot(XX,prl2,'k')
hold on
plot(XX,pim2,'-.k')
plot(XX,J2eV*V,'--k')
hold off
axis( [ 0 DX*NN -.3 .3 ])
  TT = text(1,.2,sprintf('KE = %5.2f meV',1e3*KE));
  set(TT,'fontsize',12)
%xlabel('nm')
```

```
ylabel('Partilce 2')
set(gca,'fontsize',12)

% Calculate the Coulomb influences

Ecoul12 = 0.;
Ecoul21 = 0.;
for n=1:NN
Ecoul12 = Ecoul12 + real(y(n))*psi2(n)*psi2(n)';
Ecoul21 = Ecoul21 + imag(y(n))*psi1(n)*psi1(n)';
end
J2eV*Ecoul12
J2eV*Ecoul21


subplot(3,2,5)
plot(X2,J2eV*real(y),'k')
hold on
plot(X2,J2eV*imag(y),'k--')
hold off
ymax = J2eV*max(abs(y));
ymax = .5;
TT = text(.1,.85*ymax,sprintf('%7.0f fs',T*dt*1e15));
 set(TT,'fontsize',12)
  TT = text(2,.6*ymax,sprintf('Ec12 = %6.2f
meV',1e3*J2eV*Ecoul12));
  set(TT,'fontsize',12)
% TT = text(1,.2*ymax,sprintf('Ec21 = %5.2f
meV',1e3*J2eV*Ecoul12));
% set(TT,'fontsize',12)
axis( [ 0 10 0 ymax ])
set(gca,'fontsize',12)
ylabel('V_c_o_u_l')
xlabel('nm')

J2eV*real(Ecoul12)

% Analytic calculation
Kcoul = J2eV*(ecoul^2/(4*pi*epsz*del_x));
Eana = Kcoul/100
%   TT = text(2,.4*ymax,sprintf('Eana = %6.2f meV',1e3*Eana));
%   set(TT,'fontsize',12)
ptot1 = psi1*psi1'
ptot2 = psi2*psi2'
Etot = KE1 + KE2 + J2eV*Ecoul12
   TT = text(2,.2*ymax,sprintf('Etot = %6.2f meV',1e3*Etot));
   set(TT,'fontsize',12)
abs(psi1*psi2')
ramp
```

T

```
psi1*psi2'
```

```
end
```

D.8   CHAPTER 8

```
% Cal_rho.m Calculate the electron density
%              and the current density.

clear all

hbar = 1.054e-34;
m0 = 9.11e-31;
ecoul = 1.6e-19;
eV2J = 1.6e-19;
J2eV = 1./eV2J;
hbar_eV = J2eV*hbar

del_x = 1.e-10;
NN = 100;
N_center = NN/2;

% Energies are meV

chi0 = J2eV*hbar^2/(2*m0*del_x^2);

% Convert to nm for plotting
DX = del_x*1e9;
X = (DX:DX:NN*DX);

% This adds the H.O. Potential
U = zeros(1,NN);
Eref = .1;                % Ground state energy in eV
k = m0*(Eref/hbar_eV)^2;
for n=1:NN
%U(n) = J2eV*.5*k*((N_center-n)*a)^2;
end

% ---- Construct the Hamiltonian ---

H = zeros(NN,NN);

H(1,1) = 2*chi0+U(1);
H(1,2) = -1*chi0;

for n=2:NN-1
```

```matlab
H(n,n-1)=  -1*chi0;
H(n,n)   =  2*chi0+ U(n);
H(n,n+1)=  -1*chi0;
end

H(NN,NN-1) =  -1*chi0;
H(NN,NN)   =  2*chi0+U(NN);

% ----------------------------------

%  This MATLAB command calculated the eigenenergies
%    and corresponding eigenfunctions
[phi,D] = eig(H);
E = diag(D);

% ------- Calculate the Fermi function ------

f = zeros(1,NN);

mu = 0.05;               % Chemical potential
kT = 0.001;              %
for m=1:NN
f(m) = 1./(1+exp((E(m)-mu)/kT));
end

subplot(4,2,1)
plot(E,f,'o--k')
axis( [ 0 .1 0 1.2])
xlabel(' E (eV)')
ylabel('f(E)')
grid on
title('Cal-rho')
TT = text( .05,  .8,  'm ','FontName','Symbol' );
set(TT,'fontsize',12)
TT = text( .053,  .8,  [' = ',num2str(mu) ]);
set(TT,'fontsize',12)
TT = text( .05,  .3,  ['k_BT = ',num2str(kT) ]);
set(TT,'fontsize',12)
set(gca,'fontsize',12)

% ------- Calculate rho ----------

rho = zeros(1,NN);
rho2 = zeros(NN,NN);
for m=1:20
for n=1:NN
rho(n) = rho(n) + f(m)*phi(n,m)^2;
for l=1:NN
rho2(l,n) = rho2(l,n) + f(m)*phi(n,m)*phi(l,m);
```

```
end
end
end

subplot(4,2,3)
plot(X,rho,'k')
xlabel(' x  (nm)')
ylabel('n')
set(gca,'fontsize',12)

subplot(2,2,3)
mesh(X,X,rho2)
view(-15,30)
axis( [ 0 10 0 10 -0.005 0.04 ])
zlabel('r','FontName','Symbol')
xlabel('x  (nm)')
ylabel('x^1(nm)')
set(gca,'fontsize',12)


% Spectral.m   This program calculates the spectral function

clear all

hbar = 1.054e-34;
m0 = 9.11e-31;
ecoul = 1.6e-19;
eV2J = 1.6e-19;
J2eV = 1./eV2J;

del_x = 1.e-10;
NN = 100;

V = zeros(1,NN);
% Energies are meV

%  Energies are in eV
chi0 = J2eV*hbar^2/(2*m0*del_x^2);

% Convert to nm for plotting
DX = del_x*1e9;
XX = (DX:DX:NN*DX);

% ------- Construct the Hamiltonian ----------

H = zeros(NN,NN);

H(1,1) = 2*chi0+V(1);
H(1,2) = -1*chi0;
```

```
for n=2:NN-1
H(n,n-1)= -1*chi0;
H(n,n)   =  2*chi0+ V(n);
H(n,n+1)= -1*chi0;
end

H(NN,NN-1)  = -1*chi0;
H(NN,NN)    = 2*chi0+V(NN);

% ----------------------------------------------

%  This MATLAB command calculates the eigenenergies
%    and corresponding eigenfunctions
[phi,D] = eig(H);
E = diag(D);

% ------- Calculate the broadened DOS ---------

Ein = 0.03375;
gamma = 0.0001;

wtot = 0.;
del_E = 0.0005;
EE = (del_E:del_E:del_E*100);
func = zeros(1,100);
for m=1:6
for l=1:100
func(l) = func(l) +  del_E*(gamma/(2*pi))/( ( gamma/2)^2 +
(E(m) - EE(l))^2);
wtot = wtot + func(l);
end
end
wtot

% -------- Calculate the spectral function ----------------

AA = zeros(NN,NN);
for m=1:5
wide =  del_E*(gamma/(2*pi))/( ( gamma/2)^2 +
(Ein - E(m))^2);
for n=1:NN
for l=1:NN
AA(n,l)  = AA(n,l) + wide*phi(n,m)*phi(l,m);
end
end
end

subplot(2,2,1)
```

```
mesh(XX,XX,AA)
ylabel('x^1 (nm)')
xlabel('x (nm)')
zlabel('A(E)')
view(-20,30)
TT = text(5,8,.0006,sprintf('E = %6.4f eV',Ein))
set(TT,'fontsize',12)
set(gca,'fontsize',12)
title('Spectral')


% Spec.m.  Calculate the electron density in the 10 nm
infinie well
%           first by the eigenfunction, and then with the
%           spectral matrix.

hbar = 1.06e-34;
m0 = 9.1e-31;
elec = 1.6e-19;
eV2J = 1.6e-19;
J2eV = 1./eV2J;

mu = 0.01;
kT = 0.001;
%kT = 0.0259;              % Room temp

% del_x is the cell size (m)
del_x = 1.e-10;
NN = 100;
N_center = NN/2;

% Energies are eV
chi0 = (1/elec)*hbar^2/(2*m0*del_x^2);

% Convert to nm for plotting
DX = del_x*1e9;
XX = (DX:DX:NN*DX);
mid = DX*N_center;

% -------- Specify the potential---------
V = zeros(1,NN);
for n=1:NN
%V(n) = .01*((N_center-n)/NN);
end

% ------ Construct the Hamiltonian ------

H = zeros(NN,NN);
```

```
H(1,1) = 2*chi0+V(1);
H(1,2) = -1*chi0;

for n=2:NN-1
H(n,n-1)= -1*chi0;
H(n,n)  = 2*chi0+ V(n);
H(n,n+1)= -1*chi0;
end

H(NN,NN-1) = -1*chi0;
H(NN,NN)   = 2*chi0+V(NN);

% These make it a PBC
%H(1,NN) = -chi0;
%H(NN,1) = -chi0;

%  This MATLAB command calculated the eigenenergies
%   and corresponding eigenfunctions
[phi,D] = eig(H);
E = diag(D);

% --- Calculate the density matrix from the eigenfunctions

rho = zeros(1,NN);
ff = zeros(NN,1);
for m=1:100
ff(m)   = 1/(1 + exp((E(m) - mu)/kT)); % Fermi at each
eigenenergy.
for n=1:NN
rho(n) = rho(n) + ff(m)*phi(n,m)^2;
end
end

% -- Calculate the Fermi-Dirac function ----------

Emax = .4;
Emin = 0.00001;
NE = 250;
EE = zeros(1,NE);
del_E = (Emax-Emin)/NE;
EE = (del_E:del_E:del_E*NE);

fermi =zeros(1,NE);
for m=1:NE
fermi(m) = 1/(1 + exp((EE(m) - mu)/kT));
end

% --- Calculate the electron density via the spectral
function
```

```
sigma = zeros(NN,NN);
eta = 2e-3;              % This is an artificial loss term
n = zeros(NN,1);
for m=1:100
k = sqrt(2*m0*EE(m)*eV2J)/hbar;
sig1 = exp(i*k*del_x);
%sig1 = 0.;              % sig1 = 0. means no contact
potential
 sigma(1,1) = -chi0*sig1;
 sigma(NN,NN) = -chi0*sig1;
G = inv( (EE(m) + i*eta)*eye(NN) - H - sigma);
n = n + fermi(m)*del_E*real(diag(i*(G-G'))/(2*pi));
end

amax = 0.1;
subplot(3,2,1)
plot(XX,rho,'--k')
hold on
plot(XX,n,'k')
hold off
axis( [ 0 10 0 amax ])
set(gca,'fontsize',12)
xlabel('x (nm)')
ylabel('n ')
TT = text(1,.7*amax,'m','FontName','Symbol');
set(TT,'fontsize',12)
TT = text(1.2,.7*amax,sprintf(' = %3.0f meV',1e3*mu));
set(TT,'fontsize',12)
legend('eigen','spec')
title('Spec')


% SpecV.m.   This program calculates the electron density
%            in an infinite well with a potential by
%            the eigenfunctions and by the spectral method.

clear all

hbar = 1.06e-34;
m0 = 0.25*9.1e-31;
ecoul = 1.6e-19;
eV2J = 1.6e-19;
J2eV = 1./eV2J;

mu = 0.25;
kT = 0.025;

del_x = 2.e-10;
```

```matlab
NN = 50;
N_center = NN/2;

% Energies are meV
chi0 = J2eV*hbar^2/(2*m0*del_x^2);

% Convert to nm for plotting
DX = del_x*1e9;
XX = (DX:DX:NN*DX);
mid = DX*N_center;

%  This adds the Potential
V = zeros(1,NN);
for n=1:NN
%V(n) = .004*((N_center-n));
V(n) = .002*(-n);
end

subplot(3,2,1)
plot(XX,V,'k')
set(gca,'fontsize',12)
grid on
ylabel('V (eV)')
title('SpecV')

% --------- Construct the Hamiltonian ------------

H = zeros(NN,NN);

H(1,1) = 2*chi0+V(1);
H(1,2) = -1*chi0;

for n=2:NN-1
H(n,n-1)= -1*chi0;
H(n,n)  = 2*chi0+ V(n);
H(n,n+1)= -1*chi0;
end

H(NN,NN-1) = -1*chi0;
H(NN,NN)   = 2*chi0+V(NN);

% These make it a PBC
%H(1,NN) = -chi0;
%H(NN,1) = -chi0;

% -------------------------------------------------

% This MATLAB command calculated the eigenenergies
%   and corresponding eigenfunctions
```

```
[phi,D] = eig(H);
E = diag(D);

% --- Calculate the density matrix from the eigenfunctions

rho = zeros(1,NN);
ff = zeros(NN,1);
for m=1:NN
ff(m)   = 1/(1 + exp((E(m) - mu)/kT));  % Fermi at each
eigenenergy.
for n=1:NN
rho(n) = rho(n) + ff(m)*phi(n,m)^2;
end
end

% ------- Calculate the Fermi-Dirac function ----------

Emax = .4;
Emin = -.1;
NE = 250;
EE = zeros(1,NE);
del_E = (Emax-Emin)/NE;
EE = (Emin:del_E:Emax);

fermi =zeros(1,NE);
for m=1:NE
fermi(m) = 1/(1 + exp((EE(m) - mu)/kT));
end

% --- Calculate the current density via the spectral
function

sigma1 = zeros(NN,NN);
sigma2 = zeros(NN,NN);
sig1 = 0.;
sig2 = 0.;
eta = 0;
n = zeros(NN,1);
for m=1:NE
k = sqrt(2*m0*(EE(m)-V(1))*eV2J)/hbar;
sig1 = exp(i*k*del_x);
k = sqrt(2*m0*(EE(m)-V(NN))*eV2J)/hbar;
sig2 = exp(i*k*del_x);
 sigma1(1,1) = -chi0*sig1;
 sigma2(NN,NN) = -chi0*sig2;
G = inv( (EE(m) + i*eta)*eye(NN) - H - sigma1 - sigma2);
n = n + fermi(m)*del_E*real(diag(i*(G-G'))/(2*pi));
end
```

```
subplot(2,2,3)
plot(XX,rho,'--k')
hold on
plot(XX,n,'k')
hold off
axis( [ 0 10 0 0.15 ])
set(gca,'fontsize',12)
xlabel('x (nm)')
ylabel('n ')
TT = text(2,.05,'m_1');
set(TT,'FontName','Symbol')
set(TT,'fontsize',12)
TT = text(2.3,.05,sprintf(' = %4.0f meV',1e3*mu));
set(TT,'fontsize',12)
TT = text(2,.02,sprintf('k_BT = %2.0f meV',1e3*kT));
set(TT,'fontsize',12)
grid on
legend('eigen','spec',2)
```

D.9 CHAPTER 9

```
% Trans.m.  Calculate the transmision function

clear all

NN = 50;
hbar = 1.06e-34;
m0 = 9.11e-31;
melec = 0.25*m0
eccoul = 1.6e-19;
eV2J = 1.6e-19;
J2eV = 1./eV2J;


del_x = 2.e-10;
DX = del_x*1e9;
X = (DX:DX:NN*DX);
N_center = NN/2;

% Energies are eV
chi0 = J2eV*hbar^2/(2*melec*del_x^2)

%  ---- Specify the potential ----

V = zeros(1,NN);

% Tunneling barrier
for n=22:27
 V(n) = 0.4;
```

```
end

%Resonant barrier
for n=14:19
%  V(n) = 0.4;
end
for n=31:36
%   V(n) = 0.4;
end

% Triangle potential
for n=15:25
%V(n) = 0.05*(n-15);
end
for n=25:35
%V(n) = 0.5*(1-.1*(n-25));
end

% Impurity
%V(N_center) = -0.15;

subplot(3,2,1)
plot(X,V,'k')
title('Trans')
axis( [ 0 10 -.1 .6 ])
grid on
xlabel(' x  (nm)')
ylabel('V (eV)')
set(gca,'fontsize',12)

% ------ Construct the Hamiltonian ---

H = zeros(NN,NN);

H(1,1) = 2*chi0+V(1);
H(1,2) = -1*chi0;

for n=2:NN-1
H(n,n-1)= -1*chi0;
H(n,n)  = 2*chi0+ V(n);
H(n,n+1)= -1*chi0;
end

H(NN,NN-1) = -1*chi0;
H(NN,NN)   = 2*chi0+V(NN);

% -- Specify the energy range ---

Emax = 1;
```

```
Emin = 0.;
NE = 250;
EE = zeros(1,NE);
del_E = (Emax-Emin)/NE;
EE = (0:del_E:del_E*(NE-1));

% --- Calculate the transmission function

sigma1 = zeros(NN,NN);
sigma2 = zeros(NN,NN);
gamma1 = zeros(NN,NN);
gamma2 = zeros(NN,NN);
sig1 = 0.;
sig2 = 0.;
eta = 1e-12;
n = zeros(NN,1);
for m=1:NE
k = sqrt(2*melec*(EE(m)-V(1))*eV2J)/hbar;
sig1 = exp(i*k*del_x);
k = sqrt(2*melec*(EE(m)-V(NN))*eV2J)/hbar;
sig2 = exp(i*k*del_x);
 sigma1(1,1) = -chi0*sig1;
 sigma2(NN,NN) = -chi0*sig2;
 gamma1 = i*(sigma1-sigma1');
 gamma2 = i*(sigma2-sigma2');
G = inv( (EE(m) + i*eta)*eye(NN) - H - sigma1 - sigma2);
TM(m) = real(trace(gamma1*G*gamma2*G'));
end

subplot(3,2,2)
plot(EE,TM,'k')
grid on
axis( [ 0 1 0 1.2 ])
xlabel('E (eV)')
ylabel('TM')
set(gca,'fontsize',12)




% Current.m.  Calculate current for a given voltage

clear all

NN = 50;
N_center = NN/2;
hbar = 1.054e-34;
m0 = 9.1e-31;
melec = 0.25*m0;
```

```
ecoul = 1.6e-19;
eV2J = 1.6e-19;
J2eV = 1./eV2J;

del_x = 2.e-10;          % The cell size
DX = del_x*1e9;
XX = (DX:DX:NN*DX);

G0= ecoul^2/(2*pi*hbar);   % Quantum conductance

% Energies are eV
chi0 = J2eV*hbar^2/(2*melec*del_x^2)

mu = input('mu (left side) -->')
kT = 0.025;

% --- Channel potential --------------

V = zeros(1,NN);

% Tunneling barrier
for n=22:27
% V(n) = 0.4;
end

%Resonant barrier
for n=14:19
%    V(n) = 0.4;
end
for n=31:36
%    V(n) = 0.4;
end

% Triangle potential
for n=15:25
%V(n) = 0.05*(n-15);
end
for n=25:35
%V(n) = 0.5*(1-.1*(n-25));
end

% -------------------------------------

VDS = input('Voltage across the channel -->')

for n=1:NN
VD(n) = -(n-1)*VDS/(NN-1);
%VD(n) = -.5*(n-N_center)*VDS/(NN-1);
V(n) = V(n) + VD(n);
```

```
end

Vmin = min(V);

subplot(3,2,1)
plot(XX,V,'k')
title('Current')
axis( [ 0 10 Vmin .6 ])
grid on
TT = text(3,.4,sprintf('V_D_S = %5.2f V',VDS));
set(TT,'fontsize',12)
set(gca,'fontsize',12)
xlabel(' x (nm)')
ylabel('V (eV)')

% -- Construct the Hamiltonian --

H = zeros(NN,NN);

H(1,1) = 2*chi0+V(1);
H(1,2) = -1*chi0;

for n=2:NN-1
H(n,n-1)= -1*chi0;
H(n,n)  = 2*chi0+ V(n);
H(n,n+1)= -1*chi0;
end

H(NN,NN-1) = -1*chi0;
H(NN,NN)   = 2*chi0+V(NN);

% -- Calculate the Fermi functions at the contacts ---

Emax = 1;
Emin = 0.;
NE = 250;
EE = zeros(1,NE);
del_E = (Emax-Emin)/NE;
EE = (0:del_E:del_E*(NE-1));

fermi1 =zeros(1,NE);
fermi2 =zeros(1,NE);
TM =zeros(1,NE);
over =zeros(1,NE);
for m=1:NE
fermi1(m) = 1/(1 + exp( (EE(m) - (mu + V(1))) /kT ));
fermi2(m) = 1/(1 + exp( (EE(m) - (mu + V(NN))) /kT ));
end
```

```
subplot(3,2,2)
plot(EE,fermi1,'k')
hold on
plot(EE,fermi2,'k--')
hold off
set(gca,'fontsize',12)
ylabel('Fermi')
xlabel('E (eV)')
legend('f_1','f_2')
TT = text(.1,.6,'m_1','FontName','Symbol');
set(TT,'fontsize',12)
TT = text(.15,.61,sprintf(' = %4.3f eV',mu+V(1)));
set(TT,'fontsize',12)
TT = text(.1,.3,'m_2','FontName','Symbol');
set(TT,'fontsize',12)
TT = text(.15,.36,sprintf(' = %4.3f eV',mu+V(NN)));
set(TT,'fontsize',12)
%TT = text(.1,.1,sprintf('kT = %4.3f eV',kT));
%set(TT,'fontsize',12)
grid on

% --- Calculate the Transmission function and the current

sigma1 = zeros(NN,NN);
sigma2 = zeros(NN,NN);
gamma1 = zeros(NN,NN);
gamma2 = zeros(NN,NN);
sig1 = 0.;
sig2 = 0.;
eta = 0;
n = zeros(NN,1);
I = 0.;
for m=1:NE
k = sqrt(2*melec*(EE(m)-V(1))*eV2J)/hbar;
sig1 = exp(i*k*del_x);
k = sqrt(2*melec*(EE(m)-V(NN))*eV2J)/hbar;
sig2 = exp(i*k*del_x);
 sigma1(1,1) = -chi0*sig1;
 sigma2(NN,NN) = -chi0*sig2;
 gamma1 = i*(sigma1-sigma1');
 gamma2 = i*(sigma2-sigma2');
G = inv( (EE(m) + i*eta)*eye(NN) - H - sigma1 - sigma2);
TM(m) = real(trace(gamma1*G*gamma2*G'));
I = I + (eV2J*del_E)*(ecoul/(2*pi*hbar))*TM(m)*(fermi1(m)
- fermi2(m));
%I = I + G0*del_E*TM(m)*(fermi1(m) - fermi2(m));
over(m) = TM(m)*(fermi1(m) - fermi2(m));
end
I
```

```
subplot(3,2,5)
plot(EE,TM,'k')
grid on
axis( [ 0 1 0 1.2 ])
set(gca,'fontsize',12)
ylabel('TM')
xlabel('E (eV)')
%TT = text(.4,.3,sprintf('V_D_S = %5.2f V',VDS))'
%set(TT,'fontsize',12)
%TT = text(.4,.1,sprintf('I = %5.3f uA',1e6*I))'
%set(TT,'fontsize',12)

subplot(3,2,4)
plot(EE,fermi1-fermi2,'k--')
hold on
%plot(EE,over,'k')
bar(EE,over)
hold off
%legend('f_1 - f_2','T*(f_1-f_2)')
legend('f_1 - f_2')
%TT = text(.5,.75,sprintf('F_t_o_t = %5.3f
',G0*sum(fermi1-fermi2)));
%set(TT,'fontsize',12)
TT = text(.5,.4,sprintf(' I = %5.2f uA',1e6*I));
set(TT,'fontsize',12)
TT = text(.5,.2,sprintf(' G = %5.2f uS',1e6*(I/VDS)));
set(TT,'fontsize',12)
ylabel('(f_1 - f_2) x TM')
set(gca,'fontsize',12)
xlabel('E (eV)')
axis([ 0 1 0 1 ])
grid on


% Cal_IV.m.  Current vs. voltage plot.

clear all

NN = 50;
hbar = 1.054e-34;
m0 = 9.1e-31;
melec = 0.25*m0;
elec = 1.6e-19;
eV2J = 1.6e-19;
J2eV = 1./eV2J;

del_x = 2.e-10;          % The cell size
DX = del_x*1e9;
```

```
XX = (DX:DX:NN*DX);

% Energies are eV
chi0 = J2eV*hbar^2/(2*melec*del_x^2)

G0= elec^2/(2*pi*hbar);    % Constant to calculate I

kT = 0.025;

% --- Channel potential

V = zeros(1,NN);

% Tunneling barrier
for n=22:27
  V(n) = 0.4;
end

%Resonant barrier
for n=14:19
%  V(n) = 0.4;
end
for n=31:36
%  V(n) = 0.4;
end

% Triangle potential
for n=15:25
%V(n) = 0.05*(n-15);
end
for n=25:35
%V(n) = 0.5*(1-.1*(n-25));
end

subplot(3,2,2)
plot(XX,V)
axis( [ 0 10 0 .5 ])
xlabel('nm')
ylabel('V (eV)')
set(gca,'fontsize',12)

% -------------------------------------

Emax = 1;
Emin = 0.;
NE = 250;
EE = zeros(1,NE);
del_E = (Emax-Emin)/NE;
EE = (0:del_E:del_E*(NE-1));
```

```
mu = input('mu (left side) -->')

% -- Calculate the Fermi function at the left contact ---
fermi1 =zeros(1,NE);
for m=1:NE
fermi1(m) = 1/(1 + exp( (EE(m) - mu ) /kT ));
end

Npoints = 21;
Iout = zeros(1,Npoints);
Vin  = zeros(1,Npoints);

% ----------- This is the main loop -------------

for nvolt=1:Npoints

VDS = 0.1*(nvolt-1)+ 0.0001;
Vin(nvolt) = VDS;

% --- Change in V for VDS ------
for n=1:NN
VD(n) = -(n-1)*VDS/(NN-1);
V(n) = V(n) + VD(n);
end

Vmin = min(V);

% -- Construct the Hamiltonian --

H = zeros(NN,NN);

H(1,1) = 2*chi0+V(1);
H(1,2) = -1*chi0;

for n=2:NN-1
H(n,n-1)= -1*chi0;
H(n,n)  = 2*chi0+ V(n);
H(n,n+1)= -1*chi0;
end

H(NN,NN-1) = -1*chi0;
H(NN,NN)   = 2*chi0+V(NN);

% -- Calculate the Fermi function at the right contact ---

fermi2 =zeros(1,NE);
TM =zeros(1,NE);
over =zeros(1,NE);
```

```
for m=1:NE
fermi2(m) = 1/(1 + exp( (EE(m) - (mu + V(NN))) /kT ));
end

% --- Calculate the Transmission function and the current

sigma1 = zeros(NN,NN);
sigma2 = zeros(NN,NN);
gamma1 = zeros(NN,NN);
gamma2 = zeros(NN,NN);
sig1 = 0.;
sig2 = 0.;
eta = 0;
n = zeros(NN,1);
I = 0.;
for m=1:NE
k = sqrt(2*melec*(EE(m)-V(1))*eV2J)/hbar;
sig1 = exp(i*k*del_x);
k = sqrt(2*melec*(EE(m)-V(NN))*eV2J)/hbar;
sig2 = exp(i*k*del_x);
 sigma1(1,1) = -chi0*sig1;
 sigma2(NN,NN) = -chi0*sig2;
 gamma1 = i*(sigma1-sigma1');
 gamma2 = i*(sigma2-sigma2');
G = inv( (EE(m) + i*eta)*eye(NN) - H - sigma1 - sigma2);
TM(m) = real(trace(gamma1*G*gamma2*G'));
%I = I + (eV2J*del_E)*(elec/(2*pi*hbar))*TM(m)*(fermi1(m)
- fermi2(m));
I = I + G0*del_E*TM(m)*(fermi1(m) - fermi2(m));
over(m) = TM(m)*(fermi1(m) - fermi2(m));
end
Iout(nvolt) = I;

end

% ----------- End of the main loop --------------------

Imax = 1e6*max(Iout);
subplot(3,2,1)
plot(Vin,1e6*Iout,'k')
axis( [ 0 1 0 1.2*Imax ])
xlabel('V_D_S (eV)')
ylabel('I (uA)')
set(gca,'fontsize',12)
title('Cal-IV')
TT = text(.2,.7*Imax,'m_1','FontName','Symbol');
set(TT,'fontsize',12)
TT = text(.27,.72*Imax,sprintf(' = %4.2f eV',mu));
set(TT,'fontsize',12)
```

```
%TT = text(.2,.4*Imax,'Ballistic channel');
%TT = text(.2,.4*Imax,'Tunneling barrier');
TT = text(.2,.4*Imax,'Resonant barrier');
set(TT,'fontsize',12)
grid on


% Broadening.m Calculate the thermal broadening function

clear all

hbar = 1.06e-34;
m0 = 9.1e-31;
melec = 0.25*m0;
elec = 1.6e-19;
eV2J = 1.6e-19;
J2eV = 1./eV2J;

mu = 0.25;
kT = 0.01;

% --- Calculate the Fermi-Dirac and the broadening functions ------

Emax = .4;
Emin = 0;
NE = 250;
EE = zeros(1,NE);
del_E = (Emax-Emin)/NE;
EE = (0:del_E:del_E*(NE-1));

fermi =zeros(1,NE);
derf =zeros(1,NE);
fermi(1) = 1;
for m=2:NE
fermi(m) = 1/(1 + exp((EE(m) - mu)/kT));
derf(m) = -(fermi(m) - fermi(m-1) )/del_E;
end

subplot(3,2,1)
plot(EE,fermi,'k')
title('Broadening')

TT = text(.05,.7,'m ','FontName','Symbol');
set(TT,'fontsize',12)
TT = text(.07,.7,sprintf(' = %5.3f eV',mu))
set(TT,'fontsize',12)
TT = text(.05,.3,sprintf('kT = %5.3f eV',kT))
set(TT,'fontsize',12)
ylabel('f_T (E)')
```

```
set(gca,'fontsize',12)
grid on

dmax = max(derf);
subplot(3,2,3)
plot(EE,derf,'k')
axis( [ 0 .4 0 dmax])
%saveas(gcf,'rho.bmp')
ylabel('F_T(E)')
xlabel('E (eV)')
ylabel('F_T (E)')
set(gca,'fontsize',12)
grid on
```

D.10   CHAPTER 10

```
%find_var.m Estimate an eigenfunction
%           through the variational method.

clear all

NN = 100;
hbar = 1.054e-34;
melec = 9.11e-31;
ecoul = 1.6e-19;
eV2J = 1.6e-19;
J2eV = 1./eV2J;

del_x = .1e-9;
DX = del_x*1e9;
XX = (DX:DX:NN*DX);
mid = DX*NN/2;

% Energy is in J.
chi0 = hbar^2/(2*melec*del_x^2);

% ---- Potential -----
V = zeros(1,NN);

% V-shapted potential
Eref = .05*eV2J;
for n=1:NN
V(n) = (Eref/(NN/2))*abs(n-NN/2);
end

% Harmonic oscillator
Eref = 0.02*eV2J;
for n=1:NN
```

```
%V(n) = 0.5*melec*(Eref/hbar)^2*del_x^2*(n-NN/2)^2;
end

subplot(3,2,1)
plot(XX,J2eV*V,'k');
set(gca,'fontsize',12)
ylabel('V (eV)')
xlabel('nm')
title('Find-var')
Vmax = max(J2eV*V);
grid on
axis( [ 0  DX*NN 0 .06 ])

% ------- Create the Hamiltonian matrix ---

H = zeros(NN,NN);

H(1,1)  = 2*chi0+V(1);
H(1,2)  = -1*chi0;
for n=2:NN-1
H(n,n-1)= -1*chi0;
H(n,n)   = 2*chi0+ V(n);
H(n,n+1)= -1*chi0;
end
H(NN,NN-1) = -1*chi0;
H(NN,NN)   = 2*chi0+V(NN);
% These two lines add the PBC
%H(1,NN)  = -1*chi0
%H(NN,1)  = -1*chi0

% -----------------------------------

% This calculates the eigenfunction and
%  eigenenergies
[phi,D] = eig(H);

% Write the eigevalues in meV.
for m = 1:NN
E(m) = 1e3*J2eV*D(m,m);
end

% ----- Plot an eigenfunction ----------
subplot(3,2,2)
plot(XX,phi(:,2),'k')
TT = text( 3.5,.075,sprintf('%5.2f meV',E(2)));
set(TT,'fontsize',12);
TT = ylabel('f_2','FontName','Symbol','fontsize',12);
axis( [ 0 DX*NN -.25 .25 ])
xlabel('nm')
```

```
grid on
set(gca,'fontsize',12)

%  --- Guess at the eigenfunction  -------

prl = zeros(1,NN);

nc = NN/2;
sigma = 1.;
while sigma > 0
sigma = input('Sigma -->')
LL = input('LL(nm)  -->')
LL = LL/DX;
for n=2:NN-1
%prl(n) = exp(-.5*((n-nc)/sigma)^2);        % Ground state
prl(n) = exp(-.5*((n-nc)/sigma)^2)*sin(2*pi*(n-nc)/LL);  % 2nd
state
end
ptot = prl*prl';
prl = prl/sqrt(ptot);
prl*prl';

subplot(3,2,5)
plot(XX,prl,'k')
TT = text(4,.2,'s','FontName','Symbol')
set(TT,'fontsize',12);
TT = text( 4.5,.2,sprintf(' = %4.2f ',sigma));
set(TT,'fontsize',12);
%TT = text( 4.,.0,sprintf('L = %4.2f ',LL*0.1));
%set(TT,'fontsize',12);

Evar = J2eV*prl*H*prl'
TT = text( 2.5,-.2,sprintf('E_v_a_r = %5.2f meV',1e3*Evar));
set(TT,'fontsize',12);
axis( [ 0 DX*NN -0.25 .25 ])
xlabel('nm')
grid on
set(gca,'fontsize',12)
title('Find-var')

end


% Se2d_pert.m. Adds a perturbation gradually using a Hanning
window.

clear all

NN = 50;                % Number of points in the problem space.
```

```
MM = 50;
hbar = 1.054e-34;      % Plank's constant
melec = 9.1e-31;       % Mass of an electron
eV2J = 1.6e-19;        % Energy conversion factors
J2eV = 1/eV2J;

del_x = 2e-10;         % The cells size
dt = .05e-15;            % Time steps
ra = (0.5*hbar/melec)*(dt/del_x^2) % ra must be < .1
DX = del_x*1e9;          % Cell size in nm.
XX = (DX:DX:DX*NN);    % Length in nm for plotting
YY = (DX:DX:DX*MM);    % Length in nm for plotting

NC = NN/2;               % Starting position of the pulse
MC = NN/2;

% --- Hanning Window

han = zeros(1,3e4);
del_T = dt*1d15;
Time = (0:del_T:del_T*(3e4-1));
for n=1:10000
han(n) = 0.5*(1-cos(2*pi*n/9999))  ;
%han(n) = exp(-0.5*((n-5000)/1500)^2);
end

% --- Add the purturbing potential

V = zeros(NN,MM);
Vp = input('Vpert (eV)  -->');

for n=2:NC
for m=2:MC
V(n,m) = Vp;
end
end

subplot(2,2,3)
mesh(YY,XX,1e3*V)
view(30,30)
axis( [ 0 DX*MM 0 DX*NN 0. 1.2e3*Vp ])
xlabel('Y (nm)')
ylabel('X (nm)')
zlabel('H_p (meV)')
title('Se2d-pert')
set(gca,'fontsize',12)

subplot(3,3,1)
plot(Time,han,'k')
```

```
axis( [ 0 500 0 1 ])
xlabel('fs')
ylabel('Han(t)')
set(gca,'fontsize',12)

%---- Input -----

prl = zeros(NN,MM);
pim = zeros(NN,MM);
ptot = 0.;
for n=2:NN-1
for m=2:MM-1
%prl(n,m) = sin(pi*n/NN)*sin(2*pi*m/MM);

prl(n,m) =   sin(pi*n/NN)*sin(2*pi*m/MM) ...
           - sin(2*pi*n/NN)*sin(1*pi*m/MM);  % The Good state
   pim(n,m) = 0.;
   ptot = ptot + pim(n,m)^2 + prl(n,m)^2;
end
end
ptot

ptot1 = 0.;
for n=1:NN
for m=1:MM
prl(n,m) = prl(n,m)/sqrt(ptot);
pim(n,m) = pim(n,m)/sqrt(ptot);
   ptot1 = ptot1 + prl(n,m)^2 + pim(n,m)^2;
end
end
ptot1

saveas(gcf,'pert.png')      % This saves the picture to a file

T = 0;
n_step = 1;
while n_step > 0
n_step = input('How many time steps -->');

% -----------This is the core FDTD program -------------
for iT=1:n_step

T = T + 1;

for m=2:MM-1
for n=2:NN-1
prl(n,m) = prl(n,m) - ra*(-4*pim(n,m) + pim(n,m-1)+
pim(n,m+1)  ...
                                    + pim(n-1,m) +
pim(n+1,m) )  ...
```

```
                   + han(T)*(dt/hbar)*eV2J*V(n,m)*pim(n,m);
end
end

for m=2:MM-1
for n=2:NN-1
pim(n,m) = pim(n,m) + ra*(-4*prl(n,m) + prl(n,m-1) +
prl(n,m+1)  ...
                                     + prl(n-1,m) +
prl(n+1,m) )...
           - han(T)*(dt/hbar)*eV2J*V(n,m)*prl(n,m);
end
end

end
T
han(T)
% -------------------------

% Check normalization
ptot = 0.;
for m=1:MM
for n=1:NN
   psi(n,m) = prl(n,m) + i*pim(n,m);
   ptot = ptot + prl(n,m)^2 + pim(n,m)^2;
end
end
ptot

% Calculate the expected values

ke = 0. + j*0.;
pe = 0.;
for m=2:MM-1
for n=2:NN-1
lap_p = psi(n+1,m) - 4*psi(n,m) + psi(n-1,m)  ...
       + psi(n,m-1) + psi(n,m+1);
ke = ke + lap_p*psi(n,m)';
pe = pe + han(T)*V(n,m)*psi(n,m)*psi(n,m)';
end
end
KE = -J2eV*((hbar/del_x)^2/(2*melec))*real(ke);

subplot(2,2,2)
mesh(YY,XX,prl)
view(30,30)
axis( [ 0 DX*MM 0 DX*NN -.03 .03 ])
% TT = text(0,0,.05,sprintf('%4.0f fs',T*dt*1e15));
% set(TT,'fontsize',12)
```

```
%TT = text(0,0,.07,sprintf('KE = %5.2f meV   PE = %5.2f
meV',1e3*KE,1e3*pe));
%set(TT,'fontsize',12)
TT = zlabel('Y')
set(TT,'FontName','symbol')
set(TT,'FontName','symbol')
set(gca,'fontsize',12)
xlabel('Y (nm)')
ylabel('X (nm)')
title('Se2d-pert')

subplot(3,3,9)
contour(YY,XX,prl)
 TT = text(1.5,8,sprintf('%4.0f fs',T*dt*1e15));
 set(TT,'fontsize',12)
xlabel('Y (nm)')
ylabel('X (nm)')
set(gca,'fontsize',11)

end


% Se10_1.m Simulate the oscillator between states caused
%          by the introduction of an electric field.

clear all

NN = 200;              % Number of points in the problem
space.
hbar = 1.054e-34;      % Plank's constant
melec = 9.1e-31;       % Mass of an electron
eV2J = 1.6e-19;        % Energy conversion factors
J2eV = 1/eV2J;

del_x = .05e-9;        % The cells size
dt = .5e-17;           % Time steps
ra = (0.5*hbar/melec)*(dt/del_x^2) % ra must be < .1
DX = del_x*1e9;        % Cell size in nm.
XX = (DX:DX:DX*NN);    % Length in nm for plotting
DT = dt*1e15;          % Time steps in fs
tt = (0:DT:DT*(1e6-1));

chi0 = hbar^2/(2*melec*del_x^2);

c1time = zeros(1,1e6);
c2time = zeros(1,1e6);

V = zeros(1,NN);
```

```
%  ----- Hamiltonian matrix ---------------

H = zeros(NN,NN);
H(1,1)  = 2*chi0+V(1);
H(1,2)  = -1*chi0;
for n=2:NN-1
H(n,n-1)= -1*chi0;
H(n,n)   =  2*chi0+ V(n);
H(n,n+1)= -1*chi0;
end
H(NN,NN-1)  = -1*chi0;
H(NN,NN)    =  2*chi0+V(NN);
% Switch to PBC
%H(NN,1)  = -1*chi0;
%H(1,NN)  = -1*chi0;

%  ----- Determine the eigenfunctions

[phi,D] = eig(H);

neig = input('Initialize in state # -->')

%  ------------------------------------------------

%  ------------------------------------------------

% Initialized a sine wave in a gaussian envelope
prl = zeros(1,NN);
pim = zeros(1,NN);
ptot = 0.;
for n=2:NN-1
prl(n) = phi(n,neig);
ptot = ptot +  prl(n)^2 + pim(n)^2;
end
pnorm = sqrt(ptot);

% Normalize and check
ptot = 0.;
for n=1:NN
prl(n) = prl(n)/pnorm;
pim(n) = pim(n)/pnorm;
ptot = ptot +  prl(n)^2 + pim(n)^2;
end
ptot

%------ Perturbation potential -----
% E field
Vpert = 0.005;
for n=1:NN
```

```
V(n) = eV2J*(Vpert/(NN/2))*(NN/2-n);
end

subplot(3,2,5)
axis( [ 0 10 -10 10 ])
plot(XX,1e3*J2eV*V)
xlabel('nm')
ylabel('V_p_e_r_t (meV)')
set(gca,'fontsize',12)
grid on

% ----- Calculate Hab ---------
Hab = 0.;
for n=1:NN
Hab = Hab + J2eV*phi(n,1)*phi(n,2)*V(n)
end

% ----- Calculate the oscillation time between states --
E_dif = D(2,2) - D(1,1);
Hmag = 2*(Hab/ (J2eV*E_dif))^2
Tperiod = (hbar*2*pi/E_dif)*1e15;

T = 0;
n_step = 1;
while n_step > 0
T
n_step = input('How many time steps -->');

% --------- This is the main FDTD loop --------------
for m=1:n_step
T = T + 1;

%Vtime(T) = exp(-0.5*((T-Tcent)/
spread)^2)*cos(omega_p*dt*(T-Tcent));

for n=2:NN-1
prl(n) = prl(n) - ra*(pim(n-1) -2*pim(n) + pim(n+1)) ...
        + (dt/hbar)*V(n)*pim(n);
end

for n=2:NN-1
pim(n) = pim(n) + ra*(prl(n-1) -2*prl(n) + prl(n+1)) ...
        - (dt/hbar)*V(n)*prl(n);
end

psi = prl + i*pim;

C1T = psi*phi(:,1);
C2T = psi*phi(:,2);
```

```
c1time(T) = C1T*C1T';
c2time(T) = C2T*C2T';

end
% -----------------------

% Check normalization
ptot1 = 0.;
for n=1:NN
ptot1 = ptot1 + prl(n)^2 + pim(n)^2;
end
ptot1

% Calculate the expected values
PE = 0.;
for n=1:NN
psi(n) = prl(n) + i*pim(n);
PE = PE + psi(n)*psi(n)'*V(n);
end
PE = PE*J2eV;

ke = 0. + j*0.;
for n=2:NN-1
lap_p = psi(n+1) - 2*psi(n) + psi(n-1);
ke = ke + lap_p*psi(n)';
end
KE = -J2eV*((hbar/del_x)^2/(2*melec))*real(ke);

subplot(3,2,1)
plot(XX,prl,'k')
hold on
plot(XX,pim,'k-.')
hold off
axis( [ 0 DX*NN -.25 .25 ])
TT = text(1,.2,sprintf('%6.0f  fs',T*dt*1e15));
set(TT,'fontsize',12)
TT = text(1,-.2,sprintf('KE = %6.2f meV',1e3*KE));
set(TT,'fontsize',12)
TT = text(1,-.1,sprintf('PE = %6.2f meV',1e3*PE));
set(TT,'fontsize',12)
E_tot = KE + PE;
% TT = text(7,.15,sprintf('E_t_o_t = %6.1f meV',1e3*(KE+PE)));
% set(TT,'fontsize',12)
set(gca,'fontsize',12)
xlabel('nm')
TT = ylabel('y');
set(TT,'FontName','Symbol')
title('Se10-1')
```

```
angle = zeros(1,40);
for m=1:40
xeig(m) = m;
Crl = 0.;
Cim = 0.;
for n=1:NN
Crl = Crl + prl(n)*phi(n,m);
Cim = Cim + pim(n)*phi(n,m);
end
P_eig(m) = Crl^2 + Cim^2;
   if abs(P_eig(m)) > .1
      angle(m) = atan2(Cim,Crl);
   end
end

subplot(2,2,4)
bar(xeig,P_eig)
axis( [ 1 10 0 1.1 ])
TT = text(5,.5,sprintf('|P(%2d)| = %8.3f ',2,P_eig(2)));
set(TT,'fontsize',12)
%TT = text(5,.85,['% Eigenfunctions']);
%set(TT,'fontsize',10)
set(gca,'fontsize',12)
grid on
xlabel('Eigenfunction #')

%subplot(3,4,4)
%bar(xeig,angle)
%axis( [ 0 5 -pi pi ])
%TT = text(2,2,['Angle Eigenfunctions']);
%set(TT,'fontsize',10)
%title('Eigen Phase')
%set(gca,'fontsize',10)
%xlabel('Eigenfunction #')

subplot(2,2,2)
plot(tt,abs(c1time),'k')
hold on
plot(tt,abs(c2time),'k--')
hold off
axis( [ 0 DT*T 0 1 ])
grid on
TT = text(.05*DT*T,.65,sprintf('H_P_2_1 = %6.3f
meV',1e3*Hab));
set(TT,'fontsize',12)
%TT = text(.05*DT*T,.45,sprintf('Hmag = %6.3f ',Hmag));
%set(TT,'fontsize',12)
TT = text(.05*DT*T,.25,sprintf('T_0 = %5.0f fs',Tperiod));
set(TT,'fontsize',12)
```

```
legend('|c_1|^2','|c_2|^2')
ylabel('State fraction')
xlabel('T (fs)')
%saveas(gcf,'se.bmp')
set(gca,'fontsize',12)

end


% Se10_2.m    Simulate the oscillation between states caused
%             by a sinusoidal electric field.

clear all

NN = 200;            % Number of points in the problem
space.
hbar = 1.054e-34;    % Plank's constant
melec = 9.1e-31;     % Mass of an electron
eV2J = 1.6e-19;      % Energy conversion factors
J2eV = 1/eV2J;

del_x = .05e-9;      % The cells size
dt = .5e-17;         % Time steps
ra = (0.5*hbar/melec)*(dt/del_x^2) % ra must be < .1
DX = del_x*1e9;      % Cell size in nm.
XX = (DX:DX:DX*NN);  % Length in nm for plotting
DT = dt*1e15;        % Time steps in fs
tt = (0:DT:DT*(1e6-1));
Vtime = zeros(1,1e6);

chi0 = hbar^2/(2*melec*del_x^2);

c1time = zeros(1,1e6);
c2time = zeros(1,1e6);

V = zeros(1,NN);

% ----- Specify the Hamiltonian --------------------

H = zeros(NN,NN);
H(1,1) = 2*chi0+V(1);
H(1,2) = -1*chi0;
for n=2:NN-1
H(n,n-1)= -1*chi0;
H(n,n)  = 2*chi0+ V(n);
H(n,n+1)= -1*chi0;
end
H(NN,NN-1) = -1*chi0;
H(NN,NN)   = 2*chi0+V(NN);
```

```
% Switch to PBC
%H(NN,1) = -1*chi0;
%H(1,NN) = -1*chi0;

% ------ Determine the eigenfunctions

[phi,D] = eig(H);

% ----------------------------------------------

% Initialize
prl = zeros(1,NN);
pim = zeros(1,NN);
ptot = 0.;
for n=2:NN-1
prl(n) = phi(n,2);
ptot = ptot + prl(n)^2 + pim(n)^2;
end
pnorm = sqrt(ptot);

% Normalize and check
ptot = 0.;
for n=1:NN
prl(n) = prl(n)/pnorm;
pim(n) = pim(n)/pnorm;
ptot = ptot + prl(n)^2 + pim(n)^2;
end
ptot

%------ Perturbation potential -----
% E field
%Vpert = 0.005;  % Value used by the static pert.
 Vpert = 0.0015;
for n=1:NN
V(n) = eV2J*(Vpert/(NN/2))*(NN/2-n);
end

subplot(4,2,7)
plot(XX,J2eV*V)
axis( [ 0 10 -Vpert Vpert ])
ylabel('V_p_e_r_t (eV)')
xlabel('nm')
set(gca,'fontsize',12)
grid on

Hab = 0.;
for n=1:NN
Hab = Hab + J2eV*phi(n,1)*phi(n,2)*V(n);  % Hab integral
end
```

```
E_dif = D(2,2) - D(1,1);
omega_ab = E_dif/hbar
TT = text(1,.5*Vpert,sprintf('w_a_b = %6.1f
THz',1e-12*omega_ab));
set(TT,'fontsize',12)

f_in = input('f_in (THz)-->');
omega_in = 2*pi*1e12*f_in;
TT = text(1,-.5*Vpert,sprintf('w_i_n = %6.1f
THz',1e-12*omega_in));
set(TT,'fontsize',12)
Tperiod = (2*pi/abs(omega_ab - omega_in))*1e15;   % For
emission

Hmag = (Hab/ (J2eV*hbar*(omega_in - omega_ab)) )^2;


% These are values for the sinusoid with a Gaussian smoothed
input.
%Tcent  = 80000;
%spread = 20000;

% These are values for the Gaussian envelop
Tperiod
%spread = (1/dt)*Tperiod*1e-15*(2*pi)
spread = 80000
Tcent = 2.5*spread

T = 0;
n_step = 1;
while n_step > 0
T
n_step = input('How many time steps  -->');

% ------ This is the main FDTD loop------------------
for m=1:n_step
T = T + 1;

% Sinusoid with a Gaussian smoothed input
 Vtime(T) = cos(omega_in*dt*(T-Tcent));
if T < Tcent
 Vtime(T) =
exp(-0.5*((T-Tcent)/spread)^2)*cos(omega_in*dt*(T-Tcent));
end

% Sinusoid in an exponential envelope
%Vtime(T) =
exp(-0.5*((T-Tcent)/spread)^2)*cos(omega_in*dt*(T-Tcent));
```

```
for n=2:NN-1
prl(n) = prl(n) - ra*(pim(n-1) -2*pim(n) + pim(n+1)) ...
         + Vtime(T)*(dt/hbar)*V(n)*pim(n);
end

for n=2:NN-1
pim(n) = pim(n) + ra*(prl(n-1) -2*prl(n) + prl(n+1)) ...
         - Vtime(T)*(dt/hbar)*V(n)*prl(n);
end

psi = prl + i*pim;

C1T = psi*phi(:,1);
C2T = psi*phi(:,2);
c1time(T) = C1T*C1T';
c2time(T) = C2T*C2T';

end
% ------------------------

% Check normalization
ptot1 = 0.;
for n=1:NN
ptot1 = ptot1 + prl(n)^2 + pim(n)^2;
end
ptot1

% Calculate the expected values
PE = 0.;
for n=1:NN
psi(n) = prl(n) + i*pim(n);
PE = PE + psi(n)*psi(n)'*V(n);
end
PE = PE*J2eV;

ke = 0. + j*0.;
for n=2:NN-1
lap_p = psi(n+1) - 2*psi(n) + psi(n-1);
ke = ke + lap_p*psi(n)';
end
KE = -J2eV*((hbar/del_x)^2/(2*melec))*real(ke);

subplot(3,2,1)
plot(XX,prl,'k')
hold on
plot(XX,pim,'k-.')
hold off
axis( [ 0 DX*NN -.25 .25 ])
```

```
TT = text(1,.2,sprintf('%6.0f  fs',T*dt*1e15));
set(TT,'fontsize',12)
TT = text(1,-.2,sprintf('KE = %6.2f meV',1e3*KE));
set(TT,'fontsize',12)
TT = text(1,-.1,sprintf('PE = %6.2f meV',1e3*PE));
set(TT,'fontsize',12)
E_tot = KE + PE;
% TT = text(7,.15,sprintf('E_t_o_t = %6.1f meV',1e3*(KE+PE)));
% set(TT,'fontsize',12)
set(gca,'fontsize',12)
xlabel('nm')
title('Se10-2')
TT = ylabel('y');
set(TT,'FontName','Symbol')

angle = zeros(1,40);
for m=1:40
xeig(m) = m;
Crl = 0.;
Cim = 0.;
for n=1:NN
Crl = Crl + prl(n)*phi(n,m);
Cim = Cim + pim(n)*phi(n,m);
end
P_eig(m) = Crl^2 + Cim^2;
   if abs(P_eig(m)) > .1
      angle(m) = atan2(Cim,Crl);
   end
end

subplot(3,2,3)
bar(abs(P_eig))
axis( [ 1 5 0 .1 ])
grid on

subplot(3,2,6)
plot(tt,Vpert*(10^3)*Vtime,'k')
axis( [ 0 DT*T -2 3 ])
TT = text(.045*DT*T,2.,sprintf('f_i_n = %5.2f THz',(1e-12/
(2*pi))*omega_in));
set(TT,'fontsize',12)
grid on
xlabel('fs')
%ylabel('V_t_i_m_e (x 10^3)')
ylabel('H_p (meV)')
set(gca,'fontsize',12)
title('Se10-2')

subplot(2,2,2)
```

```
plot(tt,abs(c1time),'k')
hold on
plot(tt,abs(c2time),'k--')
hold off
amax = 1.
axis( [ 0 DT*T 0 amax ])
grid on
TT = text(.05*DT*T,.65*amax,sprintf('H_P_2_1 = %6.3f
meV',1e3*Hab));
set(TT,'fontsize',12)
TT = text(.045*DT*T,.45*amax,sprintf('f_1_2 = %5.2f
THz',(1e-12/(2*pi))*omega_ab));
set(TT,'fontsize',12)
TT = text(.05*DT*T,.25*amax,sprintf('T_0 = %5.0f
fs',Tperiod));
set(TT,'fontsize',12)
legend('|c_1|^2','|c_2|^2')
set(gca,'fontsize',12)
%ylabel('Probability')
xlabel('T (fs)')

end



% Se10_3.m Simulation of a HO potential with a gaussian
pulse
%          oscillating potential to illustrate Fermi's Golden
Rule.

clear all

NN = 400;              % Number of points in the problem
space.
hbar = 1.054e-34;      % Plank's constant
melec = 9.1e-31;       % Mass of an electron
eV2J = 1.6e-19;        % Energy conversion factors
J2eV = 1/eV2J;

del_x = .1e-9;         % The cells size
dt = 1e-17;            % Time steps
ra = (0.5*hbar/melec)*(dt/del_x^2) % ra must be < .1
DX = del_x*1e9;        % Cell size in nm.
XX = (DX:DX:DX*NN);    % Length in nm for plotting
DT = dt*1e15;          % Time steps in fs
tt = (0:DT:DT*(1e6-1));
Vtime = zeros(1,1e6);

chi0 = hbar^2/(2*melec*del_x^2);
```

```
% V potential
for n=1:NN
%U(n)  = eV2J*(0.001)*abs(NN/2-n);
end

Eref = 0.005;
% Harmonic Oscillator
Eref = Eref*eV2J;
omega0 = Eref/hbar;
f0 = omega0/(2*pi)
T0 = 1/f0
for n=1:NN
U(n)  = 0.5*melec*omega0^2*del_x^2*(n-NN/2)^2;
end

% ---- Specify the Hamiltonian --------

H = zeros(NN,NN);
H(1,1)  = 2*chi0+U(1);
H(1,2)  = -1*chi0;
for n=2:NN-1
H(n,n-1)= -1*chi0;
H(n,n)   = 2*chi0+ U(n);
H(n,n+1)= -1*chi0;
end
H(NN,NN-1) = -1*chi0;
H(NN,NN)    = 2*chi0+U(NN);
% Switch to PBC
%H(NN,1) = -1*chi0;
%H(1,NN) = -1*chi0;

% Determine the eigenfunctions

[phi,D] = eig(H);
for n=1:NN
energy(n) = J2eV*D(n,n);
end

subplot(3,2,2)
plot(energy,'ko')
axis( [ 1 20 0 .15 ])
grid on
xlabel('eigen #')
ylabel('E (eV)')
set(gca,'fontsize',12)


% ------------------------------------------------
```

```
% Initialize in an eigenstate
nc = 0;
prl = zeros(1,NN);
pim = zeros(1,NN);
ptot = 0.;
for n=2:NN-1
prl(n) = phi(n,10);
ptot = ptot + prl(n)^2 + pim(n)^2;
end
pnorm = sqrt(ptot);

% Normalize and check
ptot = 0.;
for n=1:NN
prl(n) = prl(n)/pnorm;
pim(n) = pim(n)/pnorm;
ptot = ptot + prl(n)^2 + pim(n)^2;
end
ptot

%------ Perturbation potential -----
% E field
%Vpert = 0.005;  % Value used by the static pert.
 Vpert = 0.0005;
for n=1:NN
V(n) = eV2J*(Vpert/(NN/2))*(NN/2-n);
end

subplot(3,2,1)
plot(XX,J2eV*U,'k')
%axis( [ 0 40 -Vpert Vpert ])
ylabel('V_H_O (eV)')
xlabel('nm')
set(gca,'fontsize',12)
grid on

%subplot(3,2,3)
%plot(XX,J2eV*V,'--k')
%axis( [ 0 40 -Vpert Vpert ])
%ylabel('V_p_e_r_t (eV)')
%xlabel('nm')
%set(gca,'fontsize',12)
%grid on

Hab = 0.;
for n=1:NN
Hab = Hab + J2eV*phi(n,1)*phi(n,2)*V(n);  % Hab integral
end
```

```
E_dif = D(5,5) - D(4,4);
omega_ab = E_dif/hbar
%TT = text(1,.5*Vpert,sprintf('w_a_b = %6.1f
THz',1e-12*omega_ab));
%set(TT,'fontsize',12)
Hmag = (Hab/ (J2eV*E_dif))^2
f_dif = omega_ab/(2*pi)

f_in = input('f_in (THz)-->');
omega_in = 2*pi*1e12*f_in;
%TT = text(1,-.5*Vpert,sprintf('w_i_n = %6.1f
THz',1e-12*omega_in));
%set(TT,'fontsize',12)
Tperiod = (2*pi/abs(omega_ab - omega_in))*1e15;   % For
emission

spread = 60000;
Tcent = 2.5*spread

saveas(gcf,'V.png')

c1time = zeros(1,1e6);

figure

T = 0;
n_step = 1;
while n_step > 0
T
n_step = input('How many time steps -->');

% ---------- This is the main FDTD loop --------------
for m=1:n_step
T = T + 1;

%Vtime(T) = cos(omega_in*dt*(T-Tcent));
%if T < Tcent
Vtime(T) =
exp(-0.5*((T-Tcent)/spread)^2)*cos(omega_in*dt*(T-Tcent));
%end

for n=2:NN-1
prl(n) = prl(n) - ra*(pim(n-1) -2*pim(n) + pim(n+1)) ...
         + (dt/hbar)*U(n)*pim(n) ...
         + Vtime(T)*(dt/hbar)*V(n)*pim(n);
end

for n=2:NN-1
```

```
pim(n) = pim(n) + ra*(prl(n-1) -2*prl(n) + prl(n+1)) ...
         - (dt/hbar)*U(n)*prl(n) ...
         - Vtime(T)*(dt/hbar)*V(n)*prl(n);
end

psi = prl + i*pim;

C1T = psi*phi(:,10);
cltime(T) = C1T*C1T';

end
% ------------------------

% Check normalization
ptot1 = 0.;
for n=1:NN
ptot1 = ptot1 + prl(n)^2 + pim(n)^2;
end
ptot1

% Calculate the expected values
PE = 0.;
for n=1:NN
psi(n) = prl(n) + i*pim(n);
PE = PE + psi(n)*psi(n)'*(U(n) + V(n));
end
PE = PE*J2eV;

ke = 0. + j*0.;
for n=2:NN-1
lap_p = psi(n+1) - 2*psi(n) + psi(n-1);
ke = ke + lap_p*psi(n)';
end
KE = -J2eV*((hbar/del_x)^2/(2*melec))*real(ke);

subplot(3,2,1)
plot(XX,prl,'k')
hold on
plot(XX,pim,'k-.')
hold off
axis( [ 0 DX*NN -.25 .25 ])
TT = text(1,.2,sprintf('%6.0f  fs',T*dt*1e15));
set(TT,'fontsize',12)
TT = text(1,-.2,sprintf('KE = %6.1f meV',1e3*KE));
set(TT,'fontsize',12)
TT = text(1,-.1,sprintf('PE = %6.1f meV',1e3*PE));
set(TT,'fontsize',12)
E_tot = KE + PE;
% TT = text(7,.15,sprintf('E_t_o_t = %6.1f meV',1e3*(KE+PE)));
```

```
% set(TT,'fontsize',12)
set(gca,'fontsize',12)
xlabel('nm')
title('Se10-3')
TT = ylabel('y')
set(TT,'FontName','Symbol')

angle = zeros(1,40);
for m=1:40
xeig(m) = m;
Crl = 0.;
Cim = 0.;
for n=1:NN
Crl = Crl + prl(n)*phi(n,m);
Cim = Cim + pim(n)*phi(n,m);
end
P_eig(m) = Crl^2 + Cim^2;
   if abs(P_eig(m)) > .1
      angle(m) = atan2(Cim,Crl);
   end
end

subplot(3,2,6)
Vmax = 10^3*Vpert;
plot(tt,10^3*Vpert*Vtime,'k')
axis( [ 0 DT*T -1.*Vmax 2*Vmax ])
TT = text(.045*DT*T,Vmax,sprintf('f_i_n = %5.2f THz',(1e-12/
(2*pi))*omega_in));
set(TT,'fontsize',12)
grid on
xlabel('fs')
%ylabel('V_t_i_m_e ')
ylabel('H_p (meV) ')
set(gca,'fontsize',12)
title('Se10-3')

subplot(3,2,2)
plot(tt,abs(c1time),'k')
hold on
%plot(tt,abs(c2time),'k--')
hold off
amax = 1.
axis( [ 0 DT*T 0 amax ])
grid on
%TT = text(.05*DT*T,.65*amax,sprintf('H_P_2_1 = %6.3f
meV',1e3*Hab));
%set(TT,'fontsize',12)
%TT = text(.045*DT*T,.45*amax,sprintf('f_1_2 = %5.2f
THz',(1e-12/(2*pi))*omega_ab));
```

```
%set(TT,'fontsize',12)
%TT = text(.05*DT*T,.25*amax,sprintf('T_0 = %5.0f
fs',Tperiod));
%set(TT,'fontsize',12)
legend('|c_1_0|^2',3)
set(gca,'fontsize',12)
%ylabel('Probability')
xlabel('T (fs)')

% ---- The eigenfunction decomposition ------
angle = zeros(1,NN);
P_eig = zeros(1,NN);
for m=1:NN
xeig(m) = m;
Crl = 0.;
Cim = 0.;
for n=1:NN
Crl = Crl + prl(n)*phi(n,m);
Cim = Cim + pim(n)*phi(n,m);
end
P_eig(m) = Crl +i*Cim;
end

subplot(3,2,5)
bar(xeig,abs(P_eig))
axis( [ 1 20 0 1.1 ])
grid on
xlabel('eigen #')
ylabel('%')
set(gca,'fontsize',12)

sum(P_eig*P_eig');   % This must sum to one

end
```

D.11   CHAPTER 11

```
% Gr_HO.m.   Find the eigenergies and eigenfunction of
%            a harmonic oscillator.

clear all

NN = 100
N_center = NN/2;
hbar = 1.054e-34;
m0 = 9.11e-31;
ecoul = 1.6e-19;
eV2J = 1.6e-19;
```

```
J2eV = 1./eV2J;

del_x = 4.e-10;
DX = del_x*1e9;
XX = (DX:DX:NN*DX);
mid = DX*N_center;

% Energies are eV
t0 = J2eV*hbar^2/(2*m0*del_x^2)

% ------ Specify the HO potential

V = zeros(1,NN);

Eref = .01; % Ground state energy in eV
k = m0*(eV2J*Eref/hbar)^2;
for n=1:NN
V(n) = J2eV*.5*k*((N_center-n)*del_x)^2;
end

subplot(4,2,1)
plot(XX,V,'k');
title('Gr-HO')
TT = text( 10, .1, ['E_r_e_f = 0.01 eV' ]);
set(TT,'fontsize',12)
ylabel('V_H_O(eV)')
xlabel('nm')
Vmax = max(V);
axis( [ DX DX*NN 0 Vmax ])
set(gca,'fontsize',12)

% ---- Create the Hamiltonian ----------------

H = zeros(NN,NN);

H(1,1) = 2*t0+V(1);
H(1,2) = -1*t0;

for n=2:NN-1
H(n,n-1)= -1*t0;
H(n,n)  = 2*t0+ V(n);
H(n,n+1)= -1*t0;
end

H(NN,NN-1) = -1*t0;
H(NN,NN)   = 2*t0+V(NN);

% ----- Find the eigenenergies and eigenvalues---
```

```
[phi,D] = eig(H);

for n=1:NN
   E(n) = D(n,n);
end

LL = length(E);
L = (0:1:LL-1);
subplot(4,2,2)
plot(L,E,'ko')
axis( [ 0 9 0 0.1 ])
title('Eigenenergies')
ylabel('E (meV)')
xlabel('Eigenvalue #')
set(gca,'fontsize',12)

%-------- Plot the eigenfunctions ---------

amax = .5;
subplot(4,2,3)
plot(XX,-phi(:,1),'k')
TT = ylabel('f_0','FontName','Symbol','fontsize',12);
TT = text( 1.2*mid, .2, sprintf('%5.3f eV',D(1,1)));
set(TT,'fontsize',12)
axis( [ DX DX*NN -amax amax ])
set(gca,'fontsize',12)

subplot(4,2,5)
plot(XX,phi(:,2),'k')
axis( [ DX DX*NN -amax amax ])
TT = ylabel('f_1','FontName','Symbol','fontsize',12);
TT = text( 1.2*mid, .2, sprintf('%5.3f eV',D(2,2)));
set(TT,'fontsize',12)
set(gca,'fontsize',12)

subplot(4,2,7)
plot(XX,phi(:,3),'k')
axis( [ DX DX*NN -amax amax ])
TT = ylabel('f_2','FontName','Symbol','fontsize',12);
TT = text( 1.2*mid, .2, sprintf('%5.3f eV',D(3,3)));
set(TT,'fontsize',12)
xlabel('nm')
set(gca,'fontsize',12)

subplot(4,2,4)
plot(XX,phi(:,4),'k')
axis( [ DX DX*NN -amax amax ])
TT = ylabel('f_3','FontName','Symbol','fontsize',12);
TT = text( 1.2*mid, .2, sprintf('%5.3f eV',D(4,4)));
```

```
set(TT,'fontsize',12)
set(gca,'fontsize',12)

subplot(4,2,6)
plot(XX,phi(:,5),'k')
axis( [ DX DX*NN -amax amax ])
TT = ylabel('f_4','FontName','Symbol','fontsize',12);
TT = text( 1.2*mid, .2, sprintf('%5.3f eV',D(5,5)));
set(TT,'fontsize',12)
set(gca,'fontsize',12)

subplot(4,2,8)
plot(XX,phi(:,6),'k')
axis( [ DX DX*NN -amax amax ])
TT = ylabel('f_5','FontName','Symbol','fontsize',12);
TT = text( 1.2*mid, .2, sprintf('%5.3f eV',D(6,6)));
set(TT,'fontsize',12)
xlabel('nm')
set(gca,'fontsize',12)




% Sell_well.m.   Simulation of the superposition of two
states.

clear all

NN = 100;             % Number of points in the problem
space.
hbar = 1.054e-34;     % Plank's constant
h_nobar = 2*pi*hbar
melec = 9.1e-31;      % Mass of an electron
eV2J = 1.6e-19;       % Energy conversion factors
J2eV = 1/eV2J;

del_x = .1e-9;        % The cells size
dt = 8e-17;
ra = (0.5*hbar/melec)*(dt/del_x^2)  % ra must be < .1
DX = del_x*1e9;       % Cell size in nm.
XX = (DX:DX:DX*NN);   % Length in nm for plotting

% Energies are in J.
chi0 = hbar^2/(2*melec*del_x^2);

V = zeros(1,NN);

subplot(3,2,1)
plot(XX,J2eV*V,'k');
set(gca,'fontsize',12)
```

```
ylabel('V (eV)')
xlabel('nm')
Umax = max(J2eV*V);
title('Se2-coher')
%axis( [ 0 DX*NN 0 Umax ])

% --- Create the Hamiltonian ------

H = zeros(NN,NN);
H(1,1) = 2*chi0+V(1);
H(1,2) = -1*chi0;
for n=2:NN-1
H(n,n-1)= -1*chi0;
H(n,n)  = 2*chi0+ V(n);
H(n,n+1)= -1*chi0;
end
H(NN,NN-1) = -1*chi0;
H(NN,NN)   = 2*chi0+V(NN);
% Switch to PBC
%H(NN,1) = -1*chi0;
%H(1,NN) = -1*chi0;

% ---- Find the eigenfunctions and eigenvalues ---

[phi,D] = eig(H);

%Plot the eigenfunctions
subplot(3,2,3)
m=1;
plot(XX,-phi(:,m),'k')
TT = ylabel('f_1','FontName','Symbol','fontsize',12);
TT = text(5,0.05,sprintf('%7.4f eV',J2eV*D(m,m)));
set(TT,'fontsize',12)
set(gca,'fontsize',12)
title('Se11-coher')

m=2;
subplot(3,2,4)
plot(XX,phi(:,m),'k')
TT = ylabel('f_2','FontName','Symbol','fontsize',12);
TT = text(5,.03,sprintf('%7.4f eV',J2eV*D(m,m)));
set(TT,'fontsize',12)
set(gca,'fontsize',12)

m=3;
subplot(3,2,5)
plot(XX,phi(:,m),'k')
TT = ylabel('f_2','FontName','Symbol','fontsize',12);
TT = text(5,.03,sprintf('%7.4f eV',J2eV*D(m,m)));
```

```
set(TT,'fontsize',12)
set(gca,'fontsize',12)

% ----------------------------------------------
T12 = 2*pi*hbar/(D(2,2) - D(1,1))
T23 = 2*pi*hbar/(D(3,3) - D(2,2))
T13 = 2*pi*hbar/(D(3,3) - D(1,1))

% Initialize in a superposition of the 1st two eigenstates
prl = zeros(1,NN);
pim = zeros(1,NN);
ptot = 0.;
for n=2:NN-1
   prl(n) = -phi(n,1) - phi(n,2);
   ptot = ptot + prl(n)^2 + pim(n)^2;
end
pnorm = sqrt(ptot);

% Normalize and check
ptot = 0.;
for n=1:NN
prl(n) = prl(n)/pnorm;
pim(n) = pim(n)/pnorm;
ptot = ptot + prl(n)^2 + pim(n)^2;
end
ptot

T = 0;
n_step = 1;
while n_step > 0
T
n_step = input('How many time steps -->');

% ---------This is the main FDTD loop---------------
for m=1:n_step
T = T + 1;

for n=2:NN-1
prl(n) = prl(n) - ra*(pim(n-1) -2*pim(n) + pim(n+1)) ...
       + (dt/hbar)*V(n)*pim(n);
end

for n=2:NN-1
pim(n) = pim(n) + ra*(prl(n-1) -2*prl(n) + prl(n+1)) ...
       - (dt/hbar)*V(n)*prl(n);
end

end
% -----------------------
```

```
% Check normalization
ptot1 = 0.;
for n=1:NN
ptot1 = ptot1 +  prl(n)^2 + pim(n)^2;
end
ptot1

% Calculate the expected values
PE = 0.;
for n=1:NN
psi(n) = prl(n) + i*pim(n);
PE = PE + psi(n)*psi(n)'*V(n);
end
PE = PE*J2eV;

ke = 0. + j*0.;
for n=2:NN-1
lap_p = psi(n+1) - 2*psi(n) + psi(n-1);
ke = ke + lap_p*psi(n)';
end
KE = -J2eV*((hbar/del_x)^2/(2*melec))*real(ke);

Trev = 2*pi*hbar/(D(2,2) - D(1,1));


subplot(1,1,1)      % This creates a new window

subplot(3,2,1)
plot(XX,prl,'-.k')
hold on
plot(XX,pim,'--k')
plot(XX,abs(psi),'k')
hold off
TT = text(1,.15,sprintf('%6.3f  ps',T*dt*1e12));
set(TT,'fontsize',12)
E_tot = KE + PE;
TT = text(6,.15,sprintf('E_t_o_t = %5.2f meV',1e3*(KE+PE)));
TT = text(3,-.15,sprintf('T_r_e_v = %6.4f ps',1e12*Trev));
axis( [ 0 10 -.2 .3 ])
set(TT,'fontsize',12)
set(gca,'fontsize',12)
xlabel('nm')
title('Se11-well')

% Calculate the revival time


Edif = J2eV*(D(2,2) - D(1,1))
```

```
T12 = 4.124e-15/Edif
T12/dt

% The eigenfunction decomposition
angle = zeros(1,NN);
P_eig = zeros(1,NN);
for m=1:NN
xeig(m) = m;
Crl = 0.;
Cim = 0.;
for n=1:NN
Crl = Crl + prl(n)*phi(n,m);
Cim = Cim + pim(n)*phi(n,m);
end
P_eig(m) = Crl + i*Cim;          % The complex value of cn
end

subplot(3,4,3)
bar(xeig,abs(P_eig))
axis( [ 1 10 0 1.1 ])
%TT = text(5,.85,['% Eigenfunctions']);
%set(TT,'fontsize',10)
set(gca,'fontsize',12)
title('Amplitude')
xlabel('Eigen #')

subplot(3,4,4)
%bar(xeig,angle)
bar(xeig,imag(log(P_eig)))
axis( [ 1 10 -pi pi ])
%TT = text(2,2,['Angle Eigenfunctions']);
%set(TT,'fontsize',12)
set(gca,'fontsize',12)
title('Phase')

end


% Sell_coher.m.  Simulate a particle in a HO as either the
%                superperposition of two states, or as a
coherent state.

clear all

NN = 200;           % Number of points in the problem
space.
hbar = 1.054e-34;   % Plank's constant
melec = 9.1e-31;    % Mass of an electron
eV2J = 1.6e-19;     % Energy conversion factors
```

```
J2eV = 1/eV2J;

del_x = .2e-9;         % The cells size
dt = 8e-17;
ra = (0.5*hbar/melec)*(dt/del_x^2) % ra must be < .1
DX = del_x*1e9;        % Cell size in nm.
XX = (DX:DX:DX*NN);    % Length in nm for plotting

chi0 = hbar^2/(2*melec*del_x^2);

V = zeros(1,NN);
% V shaped potential
for n=1:NN
%V(n) = eV2J*(.0005)*(abs(NN/2-n));
end
% E field
for n=1:NN
%V(n) = eV2J*(.002)*(NN/2-n);
end

% ----- Harmonic Oscillator potential
Eref = 0.01;
Eref = Eref * eV2J;
omega0 = (Eref / hbar)
f0 = omega0/(2*pi);
T0 = 1/f0
for n=1:NN
V(n) = 0.5*melec*(Eref/hbar)^2*del_x^2*(n-NN/2)^2;
end

subplot(3,2,1)
plot(XX,J2eV*V,'k');
set(gca,'fontsize',12)
ylabel('V (eV)')
xlabel('nm')
TT = text(10,0.22,sprintf('E_r_e_f = 0.01 eV'));
set(TT,'fontsize',12)
Umax = max(J2eV*V);
title('Se11-coher')
%axis( [ 0 DX*NN 0 Umax ])

% -- Create the Hamiltonian ------

H = zeros(NN,NN);
H(1,1) = 2*chi0+V(1);
H(1,2) = -1*chi0;
for n=2:NN-1
H(n,n-1)= -1*chi0;
H(n,n)  = 2*chi0+ V(n);
```

```
H(n,n+1)= -1*chi0;
end
H(NN,NN-1) = -1*chi0;
H(NN,NN)   = 2*chi0+V(NN);
% Switch to PBC
%H(NN,1) = -1*chi0;
%H(1,NN) = -1*chi0;

% Find the eigenstates and eigenenergies of H

[phi,D] = eig(H);

%Plot the eigenfunctions
subplot(3,2,3)
m=1;
plot(XX,phi(:,m),'k')
TT = ylabel('f_0','FontName','Symbol','fontsize',12)
TT = text(5,0.05,sprintf('%7.4f eV',J2eV*D(m,m)));
set(TT,'fontsize',12)
set(gca,'fontsize',12)
title('Se11-coher')
m=2;
subplot(3,2,4)
plot(XX,phi(:,m),'k')
TT = ylabel('f_1','FontName','Symbol','fontsize',12)
TT = text(5,.03,sprintf('%7.4f eV',J2eV*D(m,m)));
set(TT,'fontsize',12)
set(gca,'fontsize',12)
m=3;
subplot(3,2,5)
plot(XX,phi(:,m),'k')
TT = ylabel('f_2','FontName','Symbol','fontsize',12)
TT = text(5,.03,sprintf('%7.4f eV',J2eV*D(m,m)));
set(TT,'fontsize',12)
set(gca,'fontsize',12)
xlabel('nm')
m=4
subplot(3,2,6)
plot(XX,phi(:,m),'k')
TT = ylabel('f_3','FontName','Symbol','fontsize',12)
TT = text(5,.03,sprintf('%7.4f eV',J2eV*D(m,m)));
set(TT,'fontsize',12)
set(gca,'fontsize',12)
xlabel('nm')

% ------------Initialize a particle-------------------

prl = zeros(1,NN);
pim = zeros(1,NN);
```

```
ptot = 0.;

% Initialize as the superposition of two states
for n=2:NN-1
%    prl(n) = phi(n,1) + phi(n,2) - phi(n,3) - phi(n,4);
ptot = ptot + prl(n)^2 + pim(n)^2;
end
pnorm = sqrt(ptot);

% Initialize a coherent state
%  NOTE: The following few lines initialize the coherent
%   state according to the formulas in the book. However,
%   sometimes MATLAB will flip the amplitude of one or more
of
%   the eigenstates. Therefore, sometimes its better to set
%   a gaussian pulse "by hand," i.e., one that is the same
%   shape as the coherent state.

Nph = 10;            % The mean of the Poisson distribution
cn = zeros(1,NN);
for m = 1:50
cn(m) = sqrt( (Nph^m)*exp(-Nph)/prod(1:m) );
for n=1:NN
 prl(n) = prl(n) + cn(m)*phi(n,m);
end
end
 ptot = prl*prl';
 pnorm = sqrt(ptot);

% Normalize and check
ptot = 0.;
for n=1:NN
prl(n) = prl(n)/pnorm;
pim(n) = pim(n)/pnorm;
ptot = ptot + prl(n)^2 + pim(n)^2;
end
ptot

T = 0;
n_step = 1;
while n_step > 0
T
n_step = input('How many time steps -->');

% ---------This is the main FDTD loop---------------
for m=1:n_step
T = T + 1;

for n=2:NN-1
```

```
prl(n) = prl(n) - ra*(pim(n-1) -2*pim(n) + pim(n+1)) ...
        + (dt/hbar)*V(n)*pim(n);
end

for n=2:NN-1
pim(n) = pim(n) + ra*(prl(n-1) -2*prl(n) + prl(n+1)) ...
        - (dt/hbar)*V(n)*prl(n);
end

end
% -----------------------

% Check normalization
ptot1 = 0.;
for n=1:NN
ptot1 = ptot1 + prl(n)^2 + pim(n)^2;
end
ptot1

% Calculate the expected values
PE = 0.;
for n=1:NN
psi(n) = prl(n) + i*pim(n);
PE = PE + psi(n)*psi(n)'*V(n);
end
PE = PE*J2eV;

ke = 0. + j*0.;
for n=2:NN-1
lap_p = psi(n+1) - 2*psi(n) + psi(n-1);
ke = ke + lap_p*psi(n)';
end
KE = -J2eV*((hbar/del_x)^2/(2*melec))*real(ke);

subplot(1,1,1)      % This creates a new window

subplot(3,2,1)
plot(XX,abs(psi),'k')
hold on
plot(XX,prl,'-.k')
plot(XX,pim,'--k')
plot(XX,J2eV*V,'k')
hold off
%axis( [ 0 DX*NN -.25 .30 ])
TT = text(1,.2,sprintf('%6.3f  ps',T*dt*1e12));
set(TT,'fontsize',12)
TT = text(1,-.2,sprintf('KE = %4.1f meV',1e3*KE));
set(TT,'fontsize',12)
TT = text(22,-.2,sprintf('PE = %4.1f meV',1e3*PE));
```

```
set(TT,'fontsize',12)
E_tot = KE + PE;
TT = text(22,.2,sprintf('E_t_o_t = %4.1f meV',1e3*(KE+PE)));
set(TT,'fontsize',12)
set(gca,'fontsize',12)
%xlabel('nm')
%TT = ylabel('f_1','FontName','Symbol','fontsize',12)
%TT = ylabel('y','FontName','Symbol','fontsize',12)
xlabel('nm')
title('Se11-coher')

% ----The eigenfunction decomposition -------
angle = zeros(1,100);
P_eig = zeros(1,100);
for m=1:100
xeig(m) = m-1;
Crl = 0.;
Cim = 0.;
for n=1:NN
if m < 5
Crl = Crl + prl(n)*phi(n,m);
Cim = Cim + pim(n)*phi(n,m);
end
end
P_eig(m) = Crl + i*Cim;         % The complex value of cn
end

subplot(3,4,3)
bar(xeig,abs(P_eig))
axis( [ 0 3 0 1.1 ])
set(gca,'XTick',[0 1 2 3 ])
%TT = text(5,.85,['% Eigenfunctions']);
%set(TT,'fontsize',10)
set(gca,'fontsize',12)
title('Amplitude')
xlabel('Eigen #')

subplot(3,4,4)
%bar(xeig,angle)
bar(xeig,imag(log(P_eig)))
axis( [ 0 3 -pi pi ])
set(gca,'XTick',[0 1 2 3 ])
%TT = text(2,2,['Angle Eigenfunctions']);
%set(TT,'fontsize',12)
set(gca,'fontsize',12)
title('Phase (radians)')
xlabel('Eigen #')

Trev = 2*pi*hbar/(D(2,2) - D(1,1))
```

```
Trev/dt

T0
T0/dt

end


% Se2d_ho.m.  The two-dimensional harmonic oscillator

clear all

NN = 50;             % Number of points in the problem space.
hbar = 1.054e-34;    % Plank's constant
melec = 9.1e-31;     % Mass of an electron
eV2J = 1.6e-19;      % Energy conversion factors
J2eV = 1/eV2J;

del_x = 2e-10;       % The cells size
dt = .05e-15;        % Time steps
ra = (0.5*hbar/melec)*(dt/del_x^2)  % ra must be < .1
DX = del_x*1e9;      % Cell size in nm.
XX = (DX:DX:DX*NN);  % Length in nm for plotting

NC = NN/2;           % Starting position of the pulse
MC = NN/2;

chi0 = hbar^2/(2*melec*del_x^2);

% --- Specify the HO  potential ------
V=zeros(NN,NN);

E0 = input('E0 -->');    % This is the reference energy of
the HO
E0 = E0*eV2J;
%omega0 = 2*E0/hbar;
omega0 = E0/hbar;
k0 = melec*omega0^2;

for n=1:NN
for m=1:NN
   V(n,m) = 0.5*k0*( (NC-n)^2 + (MC-m)^2 )*del_x^2;
end
end

subplot(2,2,1)
mesh(XX,XX,J2eV*V)
axis( [ 0 DX*NN 0 DX*NN 0 1 ])
xlabel('nm')
```

```
ylabel('nm')
zlabel('V_H_O (eV)')
title('Se2d-ho')
set(gca,'fontsize',12)

subplot(2,2,2)
mesh(XX,XX,(dt/hbar)*V)

%---- Generate the Hermite polynomials -----

herm = zeros(NN,NN);

nord = input('nord -->');      % The eigenvalue numbers
lord = input('lord -->');
nord
lord

psi_eigen = zeros(NN,NN);

%  --- Generate the eigenstates ----

ptot = 0.;
for n=1:NN
for m=1:NN
   dist = sqrt( (NC-n)^2 + (MC-m)^2 )*del_x;
   xi = sqrt(melec*omega0/hbar);
if nord == 0
   if lord == 0
      psi_eigen(n,m) = xi*exp(-(xi*dist)^2/2);
% n=0, l=0
   elseif abs(lord) == 1
      psi_eigen(n,m) = xi*(xi*dist)*exp(-(xi*dist)^2/2);
% n=1
   end
end
if nord == 1
  psi_eigen(n,m) = xi*(((xi*dist)^2)-1.)*exp(-(xi*dist)^2/2);
% n=1
end
   ptot = ptot + psi_eigen(n,m)^2;
end
end

% Normalize, and add the theta variation
ptot1 = 0.;
for n=1:NN
for m=1:NN
psi_eigen(n,m) = psi_eigen(n,m)/sqrt(ptot);
   ptot1 = ptot1 + psi_eigen(n,m)^2;
```

```
xdist = (NC - n);
ydist = (MC - m);
theta = lord*atan2(-ydist,xdist);
prl(n,m) = psi_eigen(n,m)*cos(theta);
pim(n,m) = psi_eigen(n,m)*sin(theta);

end
end
ptot1

% Normalize and check
ptot = 0.;
pnorm = 1;
for n=1:NN
for m=1:NN
   prl(m,n) = prl(m,n)/pnorm;
   pim(m,n) = pim(m,n)/pnorm;
   ptot = ptot + prl(m,n)^2 + pim(m,n)^2;
end
end
ptot

subplot(2,2,3)
mesh(prl)

T = 0;
n_step = 1;
while n_step > 0
n_step = input('How many time steps -->');

% -----------This is the core FDTD program -------------
for iT=1:n_step

T = T + 1;

for m=2:NN-1
for n=2:NN-1
prl(m,n) = prl(m,n) - ra*(-4*pim(m,n) + pim(m,n-1)+
pim(m,n+1)   ...
            + pim(m-1,n) + pim(m+1,n) ) ...
            + (dt/hbar)*V(m,n)*pim(m,n);
end
end

for m=2:NN-1
for n=2:NN-1
pim(m,n) = pim(m,n) + ra*(-4*prl(m,n) + prl(m,n-1) +
prl(m,n+1)   ...
            +                prl(m-1,n) + prl(m+1,n) )...
```

```
         - (dt/hbar)*V(m,n)*prl(m,n);
end
end

end
% ----------------------

% Check normalization
ptot = 0.;
for m=1:NN
for n=1:NN
   psi(m,n) = prl(m,n) + i*pim(m,n);
   ptot = ptot + prl(m,n)^2 + pim(m,n)^2;
end
end
ptot

% Calculate the expected values

PE = 0.;
for m=1:NN
for n=1:NN
   PE = PE + psi(m,n)*psi(m,n)'*V(m,n);
end
end
PE = PE*J2eV;

ke = 0. + j*0.;
for m=2:NN-1
for n=2:NN-1
lap_p = psi(m,n+1) - 4*psi(m,n) + psi(m,n-1) ...
      + psi(m-1,n) + psi(m+1,n);
ke = ke + lap_p*psi(m,n)';
end
end
KE = -J2eV*((-hbar/del_x)^2/(2*melec))*real(ke);

Etot = PE + KE
Trev = 4.135/Etot      % Time to make one revival

Tsteps = Trev*1e-15/dt

subplot(2,2,2)
contour(XX,XX,prl)


subplot(2,2,3)
mesh(XX,XX,prl)
axis( [ 0 DX*NN 0 DX*NN -.1 .1 ])
 TT = text(0,10,.1,sprintf('%7.0f fs',T*dt*1e15));
```

```
set(TT,'fontsize',12)
TT = text(4,0,.16,sprintf('KE = %3.0f meV',1e3*KE));
set(TT,'fontsize',12)
TT = text(4,0,.12,sprintf('PE = %3.0f meV',1e3*PE));
set(TT,'fontsize',12)
%TT = text(25,.15,sprintf('E_t_o_t = %5.3f eV',PE+KE))
 set(TT,'fontsize',12)
xlabel('nm')
ylabel('nm')
zlabel('Psi_r_e_a_l')
view(-30,20)
title('Se2d-ho')
set(gca,'fontsize',12)
grid on


subplot(2,2,4)
%contour(XX,XX,pim)
mesh(XX,XX,pim)
axis( [ 0 DX*NN 0 DX*NN -.1 .1 ])
view(-30,20)
set(gca,'fontsize',12)
xlabel('nm')
ylabel('nm')
zlabel('Psi_i_m_a_g')

end
```

D.12   CHAPTER 12

```
% Se12_1.m   % This program finds the eigen energies in a
%             one-dimensional well.

clear all

NN = 50;              % Number of points in the problem
space.
hbar = 1.054e-34;     % Plank's constant
melec = 9.1e-31;      % Mass of an electron
eV2J = 1.6e-19;       % Energy conversion factors
J2eV = 1/eV2J;

del_x = .2e-9;        % The cells size
dt = 1e-16;           % Time steps
ra = (0.5*hbar/melec)*(dt/del_x^2)  % ra must be < .1
DX = del_x*1e9;       % Cell size in nm.
XX = (DX:DX:DX*NN);   % Length in nm for plotting
```

```
% --- Parameters for the FFT ----
Ntime = 2^16;          % Number of points in the FFT buffer
Ptime = zeros(1,Ntime);
Pwin =zeros(1,Ntime);
PF = zeros(1,Ntime);
FF = zeros(1,Ntime);
del_F = 1./(Ntime*dt);
del_E = J2eV*2*pi*hbar*del_F;
FF = (0:del_E:del_E*(Ntime-1));
delT = 1e12*dt;
PS = zeros(1,Ntime);
PS = (delT:delT:delT*Ntime);

% ----- Specify the Potential

% Slanted Potential
V = zeros(1,NN);
for n=1:NN
%V(n) = eV2J*(.005);
end
for n=20:40
%V(n) = eV2J*(.0005)*(abs(30-n));
end

% Plot V
subplot(3,2,2)
plot(XX,J2eV*V);
set(gca,'fontsize',12)
ylabel('V (eV)')
xlabel('nm')
Umax = max(J2eV*V);
Umax = 1
title('Se12-1')
axis( [ 0 DX*NN 0 1.1*Umax ])

% ----------------------------------------------

% Initialized the test function
sigma = 3.;              % Pulse width
nc = 12 ;              % Time domain data monitored at this point
prl = zeros(1,NN);
pim = zeros(1,NN);
ptot = 0.;
for n=2:NN-1
% prl(n) = exp(-1.*((nc-n)/sigma)^2) ;    % First test
function
   prl(n) = exp(-1.*((12.5-n)/sigma)^2) ...    % 2nd test
function
         - exp(-1.*((37.5-n)/sigma)^2);
```

```
ptot = ptot + prl(n)^2 + pim(n)^2;
end
pnorm = sqrt(ptot);

% Normalize and check
ptot = 0.;
for n=1:NN
prl(n) = prl(n)/pnorm;
pim(n) = pim(n)/pnorm;
ptot = ptot + prl(n)^2 + pim(n)^2;
end
ptot

% Plot psi
subplot(3,2,1)
plot(XX,prl,'k')
hold on
%plot(XX,pim,'-.r')
plot(XX,J2eV*V,'--k')
hold off
axis( [ 0 DX*NN -.4 .4 ])
set(gca,'fontsize',12)
xlabel('nm')
title('Se12-1')

saveas(gcf,'se.png')

T = 0;
n_step = 1;
while n_step > 0
T
n_step = input('How many time steps -->');

% ----------This is the main FDTD loop-------------
for m=1:n_step
T = T + 1;

for n=2:NN-1
prl(n) = prl(n) - ra*(pim(n-1) -2*pim(n) + pim(n+1)) ...
        + (dt/hbar)*V(n)*pim(n);
end

for n=2:NN-1
pim(n) = pim(n) + ra*(prl(n-1) -2*prl(n) + prl(n+1)) ...
        - (dt/hbar)*V(n)*prl(n);
end

Ptime(T) = prl(nc) - i*pim(nc);
```

```
end
% ----------------------

% Check normalization
ptot1 = 0.;
for n=1:NN
ptot1 = ptot1 + prl(n)^2 + pim(n)^2;
end
ptot1

% Plot psi
subplot(3,2,1)
plot(XX,prl,'k')
hold on
plot(XX,pim,'--k')
%plot(XX,J2eV*V,'--k')
hold off
axis( [ 0 DX*NN -.25 .5 ])
TT = text(7,.3,sprintf('%6.1f  ps',T*dt*1e12));
set(TT,'fontsize',12)
set(gca,'fontsize',12)
xlabel('nm')
title('Se12-1')

% FFT with windowing

% ---- Create the Hanning window for T points---
win = zeros(1,T);
for n=1:T
%win(n) = 1;
win(n) = 0.5*(1-cos(2*pi*n/T));
end

% ---- Window the saved time-domain data -------
for m=1:T
Pwin(m) = win(m)*Ptime(m);
end

% Plot the time-domain psi at one point
subplot(3,2,3)
plot(PS,real(Pwin),'k')
axis( [ 0 1e12*dt*T -.3 .3 ])
xlabel('ps')
ylabel('Windowed Psi')
set(gca,'fontsize',12)

% ---- Take the FFT of the windowed time-domain data--
% ---- to display the eigenenrgies ------------------
```

```
PF = (1/sqrt(Ntime))*abs(fft(Pwin));
Pmax = max(PF);

subplot(3,2,4)
plot(1e3*FF,PF,'k')
axis( [ 0 40 0 1.1*Pmax ])
set(gca,'XTick',[ 15 30])
%set(gca,'XTick',[ 3.75 10 20 30])
xlabel('E (meV)')
ylabel('| fft (Psi)|')
grid on
set(gca,'fontsize',12)


end


% Se12_2.m Construct the eigenfunction corresponding to the
%          eigenenergy found with Se12_1.m.

clear all

NN = 50;                % Number of points in the problem
space.
hbar = 1.054e-34;       % Plank's constant
melec = 9.1e-31;        % Mass of an electron
eV2J = 1.6e-19;         % Energy conversion factors
J2eV = 1/eV2J;

del_x = .2e-9;          % The cells size
dt = 1e-16;             % Time steps
ra = (0.5*hbar/melec)*(dt/del_x^2) % ra must be < .1
DX = del_x*1e9;         % Cell size in nm.
XX = (DX:DX:DX*NN);     % Length in nm for plotting

% ----------- Specify the potential

V = zeros(1,NN);
for n=1:NN
%V(n) = eV2J*(.005);
end
for n=20:40
%V(n) = eV2J*(.0005)*(abs(30-n));
end

subplot(3,2,2)
plot(XX,J2eV*V,'k');
set(gca,'fontsize',12)
ylabel('V (eV)')
```

```
xlabel('nm')
Umax = max(J2eV*V);
%axis( [ 0 DX*NN 0 Umax ])


% --- Initialized the test function ----

sigma = 3.;                % Pulse width
nc = NN/2 ;                % Starting position of the pulse
prl = zeros(1,NN);
pim = zeros(1,NN);
ptot = 0.;
for n=2:NN-1
   prl(n) = exp(-1.*((n-nc)/sigma)^2) ;
%   prl(n) = exp(-1.*((n-nc)/sigma)^2) ...
%          - exp(-1.*((n-nc-25)/sigma)^2) ;
ptot = ptot + prl(n)^2 + pim(n)^2;
end
pnorm = sqrt(ptot);

% Normalize and check
ptot = 0.;
for n=1:NN
prl(n) = prl(n)/pnorm;
pim(n) = pim(n)/pnorm;
ptot = ptot + prl(n)^2 + pim(n)^2;
end
ptot

% Plot the initial pulse
subplot(3,2,1)
plot(XX,prl,'k')
axis( [ 0 DX*NN -.5 .6 ])
set(gca,'fontsize',12)
xlabel('nm')
ylabel('Initial pulse')
title('Se12-2')

% ---- Parameters for the eigenfunction ---

Ein = input('Eigenenergy (eV) -->')
freq = Ein/(J2eV*2*pi*hbar);
omega = 2*pi*freq
arg  = omega*dt;
exp(-i*arg)
Time_period = 1/(freq*dt)

% ---- Soecify the Hanning window ----
```

```
Than = input('Length of Hanning window -->')
win = zeros(1,2^16);
for n=1:Than
win(n) = 0.5*(1-cos(2*pi*n/Than));
end

subplot(3,2,2)
plot(win,'k')
axis( [ 0 Than 0 1 ])
ylabel('Hanning')
xlabel('Time steps')

phi = zeros(1,NN);
phi0_rl = zeros(1,NN);
phi0_im = zeros(1,NN);
psi = zeros(1,NN);
phi_m = zeros(1,NN);

T = 0;
n_step = 1;
while n_step > 0
T
n_step = input('How many time steps -->');

% -------- This is the main FDTD loop ----------------

for m=1:n_step
T = T + 1;

for n=2:NN-1
prl(n) = prl(n) - ra*(pim(n-1) -2*pim(n) + pim(n+1)) ...
        + (dt/hbar)*V(n)*pim(n);
end

for n=2:NN-1
pim(n) = pim(n) + ra*(prl(n-1) -2*prl(n) + prl(n+1)) ...
        - (dt/hbar)*V(n)*prl(n);
end

for n=2:NN-1
psi(n) = prl(n) + i*pim(n);
phi(n) = phi(n) + win(T)*exp(-i*arg*T)*psi(n);
end

end
% -----------------------

psi*psi'
```

```
% Plot psi
subplot(3,2,1)
plot(XX,prl,'k')
hold on
plot(XX,pim,'--k')
%plot(XX,J2eV*V,'--k')
hold off
axis( [ 0 DX*NN -.25 .5 ])
TT = text(6,.3,sprintf('%6.2f  ps',T*dt*1e12));
set(TT,'fontsize',12)
set(gca,'fontsize',12)
xlabel('nm')
title('Se12-2')

% Normalize phi
ptot1 = phi*phi';
for n=1:NN
phi_m(n) = phi(n)/ptot1';   % The eigenfunction
angle(n) = atan2(imag(phi(n)),real(phi(n)));
end
ptot1

% --- Plot the complex reconstructed eigenfunction ----

pmax = max(abs(phi_m));
subplot(3,2,3)
plot(XX,real(phi_m),'k');
hold on
plot(XX,imag(phi_m),'k--');
hold off
set(gca,'fontsize',12)
TT = text(2.5,.5*pmax,sprintf('E_i_n = %6.2f  meV',Ein*1e3));
set(TT,'fontsize',12)
axis( [ 0 DX*NN -1.2*pmax 1.2*pmax ])
xlabel('nm')
ylabel('phi')
TT = ylabel('f','FontName','Symbol','fontsize',12)

% --- Set the angle at the source to zero. This gives an
%      eigenfunction that is all real.

ang0 = angle(25);
for n=1:NN
angle(n) = angle(n) - ang0;
phi0_rl(n) = abs(phi_m(n))*cos(angle(n));
phi0_im(n) = abs(phi_m(n))*sin(angle(n));
end
ptot = phi0_rl*phi0_rl'
phi0_rl = phi0_rl/sqrt(ptot);
```

```
pmax = max(phi0_rl)+.01;
pmin = min(phi0_rl);
subplot(3,2,4)
plot(XX,phi0_rl,'k')
hold on
plot(XX,phi0_im,'k--')
hold off
axis( [ 0 DX*NN 1.2*pmin 1.2*pmax ])
TT = text(2.5,.5*pmax,sprintf('E_i_n = %6.2f  meV',Ein*1e3));
set(TT,'fontsize',12)
grid on
set(gca,'fontsize',12)
xlabel('nm')
TT = ylabel('f_0','FontName','Symbol','fontsize',12)
phi0_rl*phi0_rl'  % Check normalization

end


% Se2d_find_E.m.

clear all

NN = 50;              % Number of points in the problem
space.
hbar = 1.054e-34;    % Plank's constant
melec = 9.1e-31;     % Mass of an electron
eV2J = 1.6e-19;      % Energy conversion factors
J2eV = 1/eV2J;

del_x = 2.e-10;      % The cells size
dt = .1e-15;         % Time steps
ra = (0.5*hbar/melec)*(dt/del_x^2) % ra must be < .1
DX = del_x*1e9;      % Cell size in nm.
XX = (DX:DX:DX*NN);  % Length in nm for plotting

% --- Parameters for the FFT

Ntime = 2^16         % Size of the FFT buffer
Ptime = zeros(1,Ntime);
Pwin = zeros(1,Ntime);
win  = zeros(1,Ntime);
PF = zeros(1,Ntime);
FF = zeros(1,Ntime);
del_F = 1./(Ntime*dt);
del_E = J2eV*2*pi*hbar*del_F;
FF = (0:del_E:del_E*(Ntime-1));
delT = 1e12*dt;
```

```
PS = zeros(1,Ntime);
PS = (delT:delT:delT*Ntime);

% --- Specify the potential ----

V=zeros(NN,NN);

% Slanted potential
for m=1:NN;
for n=10:NN
%    V(n,m) = k0*( (NC-n)^2 + .75*(MC-m)^2 )*del_x^2;
%    V(n,m) = eV2J*(0.1/NN)*n;
end
for n=1:9
%    V(n,m) = eV2J*.1;
end
end

% HO potential
k0 = 0.005;
for m=1:NN;
for n=1:NN
%    V(n,m) = k0*( (NC-n)^2 + (MC-m)^2 )*del_x^2;
end
end

subplot(2,2,1)
mesh(XX,XX,J2eV*V)
zlabel('V (eV)')
xlabel('nm')
ylabel('nm')

%---- Test function -----

NC = NN/2;              % Starting position of the pulse
MC = NN/2;

win2D = zeros(NN,NN);
sigma = 6
ptot = 0.;
for n=1:NN
for m=1:NN
% The test function is smoothed with a 2D Hanning window
win2D(m,n) = 0.25*(1. - cos(2*pi*m/NN))*(1.-cos(2*pi*n/NN));
% Single point
    dist = sqrt( (NC-n)^2 + (MC-m)^2 );
     prl(m,n) = win2D(m,n)*exp(-(dist/sigma)^2);
% Double point
  dist1 = sqrt( (MC-m)^2 + (NC+10-n)^2 );
```

```
   dist2 = sqrt( (MC-m)^2 + (NC-10-n)^2 );
%  prl(m,n) = exp(-(dist1/sigma)^2) ...
%            - exp(-(dist2/sigma)^2);
   ptot = ptot + prl(m,n)^2;
end
end
pnorm = sqrt(ptot)
msource = MC ;              % Position were the time-domain
nsource = NC-10;           %   data is monitored.

pim = zeros(NN,NN);
% Normalize and check
ptot = 0.;
for n=1:NN
for m=1:NN
   prl(m,n) = prl(m,n)/pnorm;
   pim(m,n) = pim(m,n)/pnorm;
   ptot = ptot + prl(m,n)^2 + pim(m,n)^2;
end
end
ptot

subplot(2,2,3)
mesh(XX,XX,prl)
axis( [ 0 DX*NN 0 DX*NN -.3 .3 ])

subplot(2,2,4)
mesh(XX,XX,pim)
axis( [ 0 DX*NN 0 DX*NN -.3 .3 ])

T = 0;
n_step = 1;
while n_step > 0
n_step = input('How many time steps -->');

% -----------This is the core FDTD program ------------
for iT=1:n_step

T = T + 1;

for m=2:NN-1
for n=2:NN-1
prl(m,n) = prl(m,n) - ra*(-4*pim(m,n) + pim(m,n-1)+
pim(m,n+1)  ...
        + pim(m-1,n) + pim(m+1,n) ) ...
        + (dt/hbar)*V(m,n)*pim(m,n);
end
end
```

```
for m=2:NN-1
for n=2:NN-1
pim(m,n) = pim(m,n) + ra*(-4*prl(m,n) + prl(m,n-1) +
prl(m,n+1)  ...
              +                  prl(m-1,n) + prl(m+1,n) )...
              - (dt/hbar)*V(m,n)*prl(m,n);
end
end

Ptime(T) = prl(msource,nsource) - i*pim(msource,nsource);

end
% ----------------------

% Check normalization
ptot = 0.;
for m=1:NN
for n=1:NN
   psi(m,n) = prl(m,n) + i*pim(m,n);
   ptot = ptot + prl(m,n)^2 + pim(m,n)^2;
end
end
ptot

subplot(2,2,1)
mesh(XX,XX,prl)
axis( [ 0 DX*NN 0 DX*NN -.1 .1 ])
  TT = text(0,9,.08,sprintf('%7.2f ps',T*dt*1e12));
xlabel('nm')
ylabel('nm')
view(-30,20)
zlabel('Real(Psi)')
title('Se2d-find-E')

subplot(2,2,2)
mesh(XX,XX,pim)
axis( [ 0 DX*NN 0 DX*NN -.2 .2 ])
view(-30,20)
set(gca,'fontsize',12)
xlabel('nm')
ylabel('nm')
zlabel('Imag(Psi)')

% Plot the time domain data and FFT.

% --- Create the Hanning window for the time-domain data

win = zeros(1,T);
for n=1:T
```

```
win(n) = 0.5*(1-cos(2*pi*n/T));
end

win = zeros(1,T);
for n=1:T
win(n) = 0.5*(1-cos(2*pi*n/T));
end

for m=1:T
Pwin(m) = win(m)*Ptime(m);
end

subplot(3,2,5)
plot(PS,real(Pwin),'k')
axis( [ 0 1e12*dt*T -.05 .05 ])
grid on
xlabel('Time (ps)')
ylabel('Real Psi')

% --- Take the FFT of the windowed time-domain data ---

PF = (1/sqrt(Ntime))*abs(fft(Pwin));
Pmax = max(PF)

subplot(3,2,6)
plot(1e3*FF,PF,'k')
axis( [ 0 50 0 1.1*Pmax ])
%axis( [ 0 50 0 1.1*Pmax ])
%set(gca,'XTick',[ 0 10 20])
grid on
xlabel('E (meV)')
ylabel('FT(Psi)')

T
end



% Se2d_find_phi1.m.  Construct the eigenfunction

clear all

NN = 50;            % Number of points in the problem space.
hbar = 1.054e-34;   % Plank's constant
melec = 9.1e-31;    % Mass of an electron
eV2J = 1.6e-19;     % Energy conversion factors
J2eV = 1/eV2J;

del_x = 2.e-10;     % The cells size
dt = .1e-15;        % Time steps
```

```
ra = (0.5*hbar/melec)*(dt/del_x^2) % ra must be < .1
DX = del_x*1e9;       % Cell size in nm.
XX = (DX:DX:DX*NN);   % Length in nm for plotting

NC = NN/2;            % Starting position of the pulse
MC = NN/2;

%Add the potential
V=zeros(NN,NN);

% Slanted potential
for m=1:NN
for n=10:NN
   V(n,m) = eV2J*(0.1/NN)*n;
end
for n=1:9
   V(n,m) = eV2J*0.1;
end
end

subplot(2,2,1)
mesh(XX,XX,J2eV*V)

%---- Test function -----

win = zeros(1,2^16);
sigma = 6;
ptot = 0.;
for n=1:NN
for m=1:NN
% Single point
   dist = sqrt( (NC-n)^2 + (MC-m)^2 );
   prl(m,n) = exp(-(dist/sigma)^2);
% Double point
  dist1 = sqrt( (MC-m)^2 + (NC-10-n)^2 );
  dist2 = sqrt( (MC-m)^2 + (NC+10-n)^2 );
%  prl(m,n) = exp(-(dist1/sigma)^2) ...
%           - exp(-(dist2/sigma)^2);
  ptot = ptot + prl(m,n)^2;
end
end
pnorm = sqrt(ptot)

pim = zeros(NN,NN);
% Normalize and check
ptot = 0.;
for n=1:NN
for m=1:NN
  prl(m,n) = prl(m,n)/pnorm;
```

```
    pim(m,n) = pim(m,n)/pnorm;
    ptot = ptot + prl(m,n)^2 + pim(m,n)^2;
end
end
ptot

subplot(2,2,2)
mesh(XX,XX,prl)
axis( [ 0 DX*NN 0 DX*NN -.3 .3 ])

Ein = input('Eigenenergy (eV) -->')
freq = Ein/(J2eV*2*pi*hbar)
omega = 2*pi*freq
arg = omega*dt
T_period = 1/(freq*dt)

Tspan = input('Window length-->')
for n=1:Tspan
win(n) = 0.5*(1 - cos(2*pi*n/Tspan));
end

psi = zeros(NN,NN);
phi = zeros(NN,NN);
phi_m = zeros(NN,NN);
angle = zeros(NN,NN);
phi0_rl = zeros(NN,NN);

T = 0;
n_step = 1;
while n_step > 0
n_step = input('How many time steps -->');

% -----------This is the core FDTD program -------------
for iT=1:n_step

T = T + 1;

for m=2:NN-1
for n=2:NN-1
prl(m,n) = prl(m,n) - ra*(-4*pim(m,n) + pim(m,n-1)+
pim(m,n+1)  ...
          + pim(m-1,n) + pim(m+1,n) ) ...
          + (dt/hbar)*V(m,n)*pim(m,n);
end
end

for m=2:NN-1
for n=2:NN-1
```

```
pim(m,n) = pim(m,n) + ra*(-4*prl(m,n) + prl(m,n-1)  +
prl(m,n+1)   ...
            +                   prl(m-1,n)  + prl(m+1,n)  )  ...
            - (dt/hbar)*V(m,n)*prl(m,n);
end
end

for m=2:NN-1
for n=2:NN-1
psi(m,n)  = prl(m,n) + i*pim(m,n);
phi(m,n)  = phi(m,n) + win(T)*exp(-i*arg*T)*psi(m,n);
end
end

end
% ------------------------

% Check normalization
ptot = 0.;
ptot_phi = 0.;
for m=1:NN
for n=1:NN
   ptot = ptot +  prl(m,n)^2 + pim(m,n)^2;
   ptot_phi = ptot_phi + phi(m,n)*phi(m,n)';
end
end
ptot

phi_m = zeros(NN,NN);
for m=2:NN-1
for n=2:NN-1
   phi_m(m,n)  = phi(m,n)/ptot_phi;
   angle(m,n)  = atan2(imag(phi(m,n)),real(phi(m,n)));
end
end

ang0 = angle(MC,NC)        % The angle at the source point
ptot0 = 0;
for m=1:NN
for n=1:NN
angle(m,n) = angle(m,n)  - ang0;
phi0_rl(m,n) = abs(phi_m(m,n))*cos(angle(m,n));
ptot0 = ptot0 + phi0_rl(m,n)^2;
end
end
phi0_rl = phi0_rl/sqrt(ptot0);

% One last check
ptot0 = 0;
```

```
for m=1:NN
for n=1:NN
ptot0 = ptot0 + phi0_rl(m,n)^2;
end
end

subplot(2,2,1)
mesh(XX,XX,prl)
axis( [ 0 DX*NN 0 DX*NN -.2 .2 ])
  TT = text(0,9,.15,sprintf('%7.2f ps',T*dt*1e12));
 set(TT,'fontsize',12)
set(gca,'fontsize',12)
xlabel('nm')
ylabel('nm')
view(-30,20)
zlabel('Prl')
%grid on
title('Se2d-find-phi')

subplot(2,2,2)
mesh(XX,XX,pim)
axis( [ 0 DX*NN 0 DX*NN -.2 .2 ])
view(-30,20)
set(gca,'fontsize',12)
xlabel('nm')
ylabel('nm')
zlabel('Pim')

subplot(2,2,4)
mesh(XX,XX,phi0_rl)
set(gca,'fontsize',12)
title('Se2d-find-phi')
axis( [ 0 10 0 10 -.05 .05 ])
xlabel('nm')
ylabel('nm')
TT = zlabel('f_o','FontName','Symbol','fontsize',12)

ptot0
T

file
end


% Se2d_read.m. % Read the eigenfunction constructed by
se2d_find_phi.m

NN = 50;              % Number of points in the problem
space.
```

```
hbar = 1.054e-34;      % Plank's constant
h_nobar = 2*pi*hbar;
melec = 9.1e-31;       % Mass of an electron
eV2J = 1.6e-19;        % Energy conversion factors
J2eV = 1/eV2J;

del_x = 2.e-10;        % The cells size
dt = .1e-15;           % Time steps
ra = (0.5*hbar/melec)*(dt/del_x^2)  % ra must be < .1
DX = del_x*1e9;        % Cell size in nm.
XX = (DX:DX:DX*NN);    % Length in nm for plotting

eV2THz = (eV2J*2*pi/hbar)

% --- Specify the potential -----

V=zeros(NN,NN);

for m=1:NN
for n=10:NN
%    V(n,m) = eV2J*(0.1/NN)*n;
end
for n=1:9
%    V(n,m) = eV2J*0.1;
end
end

subplot(2,2,3)
mesh(XX,XX,J2eV*V)

%---- Read in the eigenfunction -----

prl = zeros(NN,NN);
pim = zeros(NN,NN);

prl = phi0_rl;    % This is the output of Se2d_find_phi.

ptot = 0.;
for n=1:NN
for m=1:NN
   ptot = ptot + prl(m,n)^2 + pim(m,n)^2;
end
end
pnorm = sqrt(ptot);

% Normalize and check
ptot = 0.;
for n=1:NN
for m=1:NN
```

```
  prl(m,n) = prl(m,n)/pnorm;
  pim(m,n) = pim(m,n)/pnorm;
  ptot = ptot + prl(m,n)^2 + pim(m,n)^2;
end
end
ptot

subplot(2,2,1)
mesh(XX,XX,prl)
axis( [ 0 DX*NN 0 DX*NN -.3 .3 ])

subplot(2,2,2)
mesh(XX,XX,pim)
axis( [ 0 DX*NN 0 DX*NN -.3 .3 ])


T = 0;
n_step = 1;
while n_step > 0
n_step = input('How many time steps -->');

% -----------This is the core FDTD program -------------
for iT=1:n_step

T = T + 1;

for m=2:NN-1
for n=2:NN-1
prl(m,n) = prl(m,n) - ra*(-4*pim(m,n) + pim(m,n-1)+
pim(m,n+1)   ...
           + pim(m-1,n) + pim(m+1,n) ) ...
           + (dt/hbar)*V(m,n)*pim(m,n);
end
end

for m=2:NN-1
for n=2:NN-1
pim(m,n) = pim(m,n) + ra*(-4*prl(m,n) + prl(m,n-1) +
prl(m,n+1)   ...
           +                 prl(m-1,n) + prl(m+1,n) )...
           - (dt/hbar)*V(m,n)*prl(m,n);
end
end

end
% -----------------------


% Check normalization
```

```
ptot = 0.;
for m=1:NN
for n=1:NN
   psi(m,n) = prl(m,n) + i*pim(m,n);
   ptot = ptot + prl(m,n)^2 + pim(m,n)^2;
end
end
ptot

% Calculate the expected values

PE = 0.;
for m=1:NN
for n=1:NN
   PE = PE + psi(m,n)*psi(m,n)'*V(m,n);
end
end
PE = PE*J2eV;
ke = 0. + j*0.;
for m=2:NN-1
for n=2:NN-1
lap_p = psi(m,n+1) - 4*psi(m,n) + psi(m,n-1) ...
      + psi(m-1,n) + psi(m+1,n);
ke = ke + lap_p*psi(m,n)';
end
end
KE = -J2eV*((hbar/del_x)^2/(2*melec))*real(ke);
Trev = h_nobar/(dt*eV2J*(KE+PE))              % Revival time

pmax = 0.05;
subplot(2,2,1)
mesh(XX,XX,prl)
axis( [ 0 DX*NN 0 DX*NN -pmax pmax ])
  TT = text(0,9,pmax,sprintf('%5.2f ps',T*dt*1e12));
 set(TT,'fontsize',12)
  TT = text(5,0,1.6*pmax,sprintf('KE = %4.2f meV',1e3*KE));
 set(TT,'fontsize',12)
% TT = text(5,0,1.2*pmax,sprintf('PE = %4.1f meV',1e3*PE));
% set(TT,'fontsize',12)
xlabel('nm')
ylabel('nm')
view(-30,20)
zlabel('Real(Psi)')
%grid on
title('Se2d-read')

subplot(2,2,2)
mesh(XX,XX,pim)
% TT = text(0,0,1.7*pmax,sprintf('E_t_o_t = %5.3f eV',PE+KE))
```

```
%  set(TT,'fontsize',12)
view(-30,20)
zlabel('Imag(Psi)')
xlabel('nm')
ylabel('nm')
axis( [ 0 DX*NN 0 DX*NN -pmax pmax ])
T

end
```

D.13   APPENDIX B

```
%Ft_demo.m  FFT

N = 256;

TT = zeros(N,1);
TT = (0:1:N-1);

% ---- Create the input  ----
y = zeros(N,1);
freq = 6.4/N
freq = .1;
%freq = 1/2;
sigma = 10;
nc = N/2;
%for n=1:N
for n=1:64
y(n) = cos(2*pi*freq*(n-1));
%y(n) = exp(-0.5*((n-nc)/sigma)^2)*cos(2*pi*freq*(n-nc));
end

subplot(3,2,1)
plot(y,'k')
axis( [ 1 N -1 1 ])
set(gca,'XTick',[ 1  32  64])
xlabel('cell #')
ylabel('Time domain')
title('Ft-demo')
%TT = text(6,.5,['s = 10'],'FontName','Symbol')
%set(TT,'fontsize',12)
set(gca,'fontsize',12)

% --- Take the FFT ----

Y = (1/N)*fft(y);

subplot(3,2,3)
```

```
plot(abs(Y),'k')
axis( [ 1 N 0 .2 ])
set(gca,'XTick',[ 25 231 ])
xlabel('cell #')
ylabel('Freq. domain')
set(gca,'fontsize',12)


%Ft_win.m Domonstrate the effects of windowing

N =256;

% --- Specify a rectangular or Hanning window --

h = zeros(N,1);
for n=1:64
h(n) = .5*(1.-cos(2*pi*(n)/64));
%h(n) = 1.;
end

subplot(3,2,2)
plot(h,'k')
axis( [ 1 N -.1 2 ])
set(gca,'XTick',[ 1 64  256 ])
ylabel('Time domain')
set(gca,'Fontsize',12)
title('Ft-win')

subplot(3,2,4)
plot((1/N)*abs(fft(h)),'k')
axis( [ 1 30 -.01 .3 ])
ylabel('Freq. domain')
set(gca,'XTick',[ 1 10 20 30 ])
set(gca,'Fontsize',12)

% ---- Create the windowed input ---
y = zeros(N,1);
sigma = 1000;
nc = N/2;
freq = 1/10;
for n=1:64
y(n) = h(n)*exp(-.5*((n-nc)/sigma)^2)*cos(2*pi*freq*(n-nc));
%y(n) = h(n);
end

Y = zeros(NN,1);

% ---- Put the time-domaim data in a longer buffer ---
for n=1:N
```

```
Y(n) = y(n);
end

subplot(3,2,1)
plot(Y,'k')
axis( [ 1 N -1 1 ])
set(gca,'XTick',[ 1 64  256 ])
xlabel('cell #')
ylabel('Time domain')
set(gca,'fontsize',12)
title('Ft-win')

% --- Take the FFT ----

Y = (1/N)*fft(Y);

subplot(3,2,3)
plot(abs(Y),'k')
axis( [ 1 256 0 .1 ])
%set(gca,'XTick',[ 0 10 20 30 ] )
set(gca,'XTick',[ 26 230 ] )
xlabel('cell #')
ylabel('Freq. domain')
set(gca,'fontsize',12)


%Ft_psi.m   FFT of the state variable from the space domain
to
%           the wavelength domain.

N = 100;

y = zeros(N,1);
Y = zeros(N,1);
L = zeros(N,1);

L0 = 1/N;
L = (0:L0:(N-1)*L0);

% ---- Create the complex space-domain pulse ----
sigma = 15;
nc = N/2;
lambda = 20;
ptot = 0.;
for n=1:N
prl(n) = exp(-.5*((n-nc)/sigma)^2)*cos(2*pi*(n-nc)/lambda);
pim(n) = exp(-.5*((n-nc)/sigma)^2)*sin(2*pi*(n-nc)/lambda);
ptot = ptot + prl(n)^2 + pim(n)^2;
end
```

```
ptot
pnorm = sqrt(ptot);

% --- Write the normalized complex waveform ----
for n=1:N
Y(n) = (1./pnorm)*( prl(n) + i*pim(n));
end

subplot(3,2,1)
plot(real(Y),'k')
hold on
plot(imag(Y),'--k')
hold off
TT = text(10,.2,['lambda = ',num2str(lambda)])
set(TT,'Fontsize',12)
axis( [ 1 N -.3 .3 ])
%set(gca,'XTick',[ 32 64 ])
set(gca,'Fontsize',12)
title('Ft-psi')
xlabel('x')

% ---- Take the FFT -----------------------
Y = (1/sqrt(N))*fft(Y);
ptot = 0.;
for n=1:N
ptot = ptot + (real(Y(n)))^2 + (imag(Y(n)))^2;
end
ptot

subplot(3,2,3)
plot(L,abs(Y),'k')
axis( [ 0 1 0 1 ])
set(gca,'XTick',[.2 .5 .8] )
set(gca,'Fontsize',12)
xlabel(['1/l'],'FontName','Symbol')
```

INDEX

Note: Italicized page locators indicate illustrations; tables are noted with *t*.

Quantum Mechanics for Electrical Engineers, First Edition. Dennis M. Sullivan.
© 2012 The Institute of Electrical and Electronics Engineers, Inc.
Published 2012 by John Wiley & Sons, Inc.

IEEE PRESS SERIES ON MICROELECTRONIC SYSTEMS

The focus of the series is on all aspects of solid-state circuits and systems including the design, testing, and application of circuits and subsystems, as well as closely related topics in device technology and circuit theory. The series also focuses on scientific, technical and industrial applications, in addition to other activities that contribute to the moving the area of microelectronics forward.

R. Jacob Baker, *Series Editor*

1. *Nonvolatile Semiconductor Memory Technology: A Comprehensive Guide to Understanding and Using NVSM Devices*
   William D. Brown, Joe Brewer
2. *High-Performance System Design: Circuits and Logic*
   Vojin G Oklobdzija
3. *Low-Voltage/Low-Power Integrated Circuits and Systems: Low-Voltage Mixed-Signal Circuits*
   Sanchez-Sinencio
4. *Advanced Electronic Packaging, Second Edition*
   Richard K. Ulrich and William D. Brown
5. *DRAM Circuit Design: Fundamental and High-Speed Topics*
   Brent Keith, R. Jacob Baker, Brian Johnson, Feng Lin
6. CMOS: Mixed-Signal Circuit Design, Second Edition
   R. Jacob Baker
7. *Nonvolatile Memory Technologies with Emphasis on Flash:A Comprehensive Guide to Understanding and Using NVM Devices*
   Joseph E. Brewer, Manzur Gill
8. *Reliability Wearout Mechanisms in Advanced CMOS Technologies*
   Alvin W. Strong, Ernest Y. Wu, Rolf-Peter Vollertsen, Jordi Sune, Giuseppe La Rosa, Timothy D. Sullivan, Stewart Rauch
9. *CMOS: Circuit Design, Layout, and Simulation, Third Edition*
   R. Jacob Baker
10. *Quantum Mechanics for Electrical Engineers*
    Dennis M. Sullivan